Group Theory

Group Theory

Birdtracks, Lie's, and Exceptional Groups

Predrag Cvitanović

PRINCETON UNIVERSITY PRESS
PRINCETON AND OXFORD

Published by Princeton University Press
41 William Street, Princeton, New Jersey 08540

In the United Kingdom: Princeton University Press
6 Oxford Street, Woodstock, Oxfordshire 0X20 1TW

Library of Congress Cataloging-in-Publication Data

Cvitanović, Predrag.
Group theory: Birdtracks, Lie's, and exceptional groups / Predrag Cvitanović.
p. cm.
Includes bibliographical references and index.
ISBN: 9780691118369 (alk. paper) 0691118361 (alk. paper) 1. Group theory. I. Title.
QA174.2 .C85 2008
512′.2–dc22 2008062101

British Library Cataloging-in-Publication Data is available

This book has been composed in LaTeX

The publisher would like to acknowledge the author of this volume for providing the camera-ready copy from which this book was printed.

Printed on acid-free paper. ∞

press.princeton.edu

Printed in the United States of America

10 9 8 7 6 5 4 3 2 1

dedicated to the memory of
Boris Weisfeiler and William E. Caswell

Contents

Acknowledgments

I would like to thank Tony Kennedy for coauthoring the work discussed in chapters on spinors, spinsters and negative dimensions; Henriette Elvang for coauthoring the chapter on representations of $U(n)$; David Pritchard for much help with the early versions of this manuscript; Roger Penrose for inventing birdtracks while I was struggling through grade school; Paul Lauwers for the birdtracks 'rock around the clock'; Feza Gürsey and Pierre Ramond for the first lessons on exceptional groups; Susumu Okubo for inspiring correspondence; Bob Pearson for assorted birdtrack, Young tableaux and lattice calculations; Bernard Julia for many stimulating interactions; P. Howe and L. Brink for teaching me how to count (supergravity multiplets); W. Siegel for helpful criticisms; E. Cremmer for hospitality at École Normale Superieure; M. Kontsevich for bringing to my attention the more recent work of Deligne, Cohen, and de Man; A. J. Macfarlane, H. Pfeiffer and an anonymous referee for critical reading of the manuscript; P. Cartier for wonderful history lessons; J. Landsberg for acknowledging my sufferings; R. Abdelatif, G. M. Cicuta, A. Duncan, B. Durhuus, R. Edgar, E. Eichten, P. G. O. Freund, S. Garoufalidis, T. Goldman, R. J. Gonsalves, M. Günaydin, H. Harari, I. Khavkine, N. MacKay, M. Marino, L. Michel, D. Miličić, R. L. Mkrtchyan, K. Oblivia, M. Peskin, C. Sachrajda, G. Seligman, P. Sikivie, A. Springer, A. Taylor, D. Thurston, G. Tiktopoulos, and B. W. Westbury for discussions and/or correspondence.

The appellation "birdtracks" is due to Bernice Durand, who entered my office, saw the blackboard and asked, puzzled: "What is this? Footprints left by birds scurrying along a sandy beach?"

I am grateful to Dorte Glass for typing most of the manuscript, to the Aksel Tovborg Jensens Legat, Kongelige Danske Videnskabernes Selskab, for financial support which made the transformation from hand-drawn to drafted birdtracks possible, and, most of all, to the good Anders Johansen for drawing some 5,000 birdtracks — without him this book would still be but a collection of squiggles in my filing cabinet. Carol Monsrud and Cecile Gourgues helped with typing the early version of this manuscript.

The manuscript was written in stages, in Aspen, Chewton-Mendip, Princeton, Paris, Bures-sur-Yvette, Rome, Copenhagen, Frebbenholm, Miramare, Røros, Juelsminde, Göteborg – Copenhagen train, Cathay Pacific (Hong Kong / Paris) and innumerable other airports and planes, Sjællands Odde, Göteborg, Miramare, Kurkela, Čijovo, assorted Starbucks, Virginia Highlands, and Kostrena. I am grateful to T. Dorrian-Smith, R. de la Torre, BDC, N.-R. Nilsson, U. Selmer, E. Høsøinen, A. Wad, families Cvitanović, and family Herlin for their kind hospitality along this long road.

I would love to thank W. E. "Bill" Caswell for a perspicacious observation, and Boris Weisfeiler for delightful discussions at the Institute for Advanced Study, but it cannot be done. In 1985, while hiking in the Andes, Boris was kidnapped by the Chilean state security, then tortured and executed by Nazis at Colonia Dignidad, Chile. Bill boarded flight AA77 on September 11, 2001, and was flown into the Pentagon by a no-less-charming group of Islamic fanatics. This book is dedicated to Boris and Bill.

Group Theory

Chapter One

Introduction

This monograph offers a derivation of all classical and exceptional semisimple Lie algebras through a classification of "primitive invariants." Using somewhat unconventional notation inspired by the Feynman diagrams of quantum field theory, the invariant tensors are represented by diagrams; severe limits on what simple groups could possibly exist are deduced by requiring that irreducible representations be of integer dimension. The method provides the full Killing-Cartan list of all possible simple Lie algebras, but fails to prove the existence of F_4, E_6, E_7 and E_8.

One simple quantum field theory question started this project; what is the group-theoretic factor for the following Quantum Chromodynamics gluon self-energy diagram

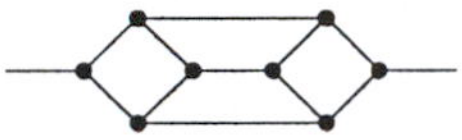

$$= ? \qquad (1.1)$$

I first computed the answer for $SU(n)$. There was a hard way of doing it, using Gell-Mann f_{ijk} and d_{ijk} coefficients. There was also an easy way, where one could doodle oneself to the answer in a few lines. This is the "birdtracks" method that will be developed here. It works nicely for $SO(n)$ and $Sp(n)$ as well. Out of curiosity, I wanted the answer for the remaining five exceptional groups. This engendered further thought, and that which I learned can be better understood as the answer to a different question. Suppose someone came into your office and asked, "On planet Z, mesons consist of quarks and antiquarks, but baryons contain three quarks in a symmetric color combination. What is the color group?" The answer is neither trivial nor without some beauty (planet Z quarks can come in 27 colors, and the color group can be E_6).

Once you know how to answer such group-theoretical questions, you can answer many others. This monograph tells you how. Like the brain, it is divided into two halves: the plodding half and the interesting half.

The plodding half describes how group-theoretic calculations are carried out for unitary, orthogonal, and symplectic groups (chapters 3–15). Except for the "negative dimensions" of chapter 13 and the "spinsters" of chapter 14, none of that is new, but the methods are helpful in carrying out daily chores, such as evaluating Quantum Chromodynamics group-theoretic weights, evaluating lattice gauge theory group integrals, computing $1/N$ corrections, evaluating spinor traces, evaluating casimirs, implementing evaluation algorithms on computers, and so on.

The interesting half, chapters 16–21, describes the "exceptional magic" (a new construction of exceptional Lie algebras), the "negative dimensions" (relations between bosonic and fermionic dimensions). Open problems, links to literature, software and other resources, and personal confessions are relegated to the epilogue, monograph's Web page birdtracks.eu. The methods used are applicable to field-theoretic model building. Regardless of their potential applications, the results are sufficiently intriguing to justify this entire undertaking. In what follows we shall forget about quarks and quantum field theory, and offer instead a somewhat unorthodox introduction to the theory of Lie algebras. If the style is not Bourbaki [29], it is not so by accident.

There are two complementary approaches to group theory. In the *canonical* approach one chooses the basis, or the Clebsch-Gordan coefficients, as simply as possible. This is the method which Killing [189] and Cartan [43] used to obtain the complete classification of semisimple Lie algebras, and which has been brought to perfection by Coxeter [67] and Dynkin [105]. There exist many excellent reviews of applications of Dynkin diagram methods to physics, such as refs. [312, 126].

In the *tensorial* approach pursued here, the bases are arbitrary, and every statement is invariant under change of basis. Tensor calculus deals directly with the invariant blocks of the theory and gives the explicit forms of the invariants, Clebsch-Gordan series, evaluation algorithms for group-theoretic weights, *etc.*

The canonical approach is often impractical for computational purposes, as a choice of basis requires a specific coordinatization of the representation space. Usually, nothing that we want to compute depends on such a coordinatization; physical predictions are pure scalar numbers ("color singlets"), with all tensorial indices summed over. However, the canonical approach can be very useful in determining chains of subgroup embeddings. We refer the reader to refs. [312, 126] for such applications. Here we shall concentrate on tensorial methods, borrowing from Cartan and Dynkin only the nomenclature for identifying irreducible representations. Extensive listings of these are given by McKay and Patera [234] and Slansky [312].

To appreciate the sense in which canonical methods are impractical, let us consider using them to evaluate the group-theoretic factor associated with diagram (1.1) for the exceptional group E_8. This would involve summations over 8 structure constants. The Cartan-Dynkin construction enables us to construct them explicitly; an E_8 structure constant has about $248^3/6$ elements, and the direct evaluation of the group-theoretic factor for diagram (1.1) is tedious even on a computer. An evaluation in terms of a canonical basis would be equally tedious for $SU(16)$; however, the tensorial approach illustrated by the example of section 2.2 yields the answer for all $SU(n)$ in a few steps.

Simplicity of such calculations is one motivation for formulating a tensorial approach to exceptional groups. The other is the desire to understand their geometrical significance. The Killing-Cartan classification is based on a mapping of Lie algebras onto a Diophantine problem on the Cartan root lattice. This yields an exhaustive classification of simple Lie algebras, but gives no insight into the associated geometries. In the 19th century, the geometries or the invariant theory were the central question, and Cartan, in his 1894 thesis, made an attempt to identify the primitive invariants. Most of the entries in his classification were the classical groups $SU(n)$, $SO(n)$, and $Sp(n)$. Of the five exceptional algebras, Cartan [44] identified G_2 as the group of octonion isomorphisms and noted already in his thesis that E_7 has a skew-symmetric quadratic and a symmetric quartic invariant. Dickson characterized E_6 as a 27-dimensional group with a cubic invariant. The fact that the orthogonal, unitary and symplectic groups were invariance groups of real, complex, and quaternion norms suggested that the exceptional groups were associated with octonions, but it took more than 50 years to establish this connection. The remaining four exceptional Lie algebras emerged as rather complicated constructions from octonions and Jordan algebras, known as the *Freudenthal-Tits construction.* A mathematician's history of this subject is given in a delightful review by Freudenthal [130]. The problem has been taken up by physicists twice, first by Jordan, von Neumann, and Wigner [173], and then in the 1970s by Gürsey and collaborators [149, 151, 152]. Jordan *et al.*'s effort was a failed attempt at formulating a new quantum mechanics that would explain the neutron, discovered in 1932. However, it gave rise to the Jordan algebras, which became a mathematics field in itself. Gürsey *et al.* took up the subject again in the hope of formulating a quantum mechanics of quark confinement; however, the main applications so far have been in building models of grand unification.

Although beautiful, the Freudenthal-Tits construction is still not practical for the evaluation of group-theoretic weights. The reason is this: the construction involves $[3 \times 3]$ octonionic matrices with octonion coefficients, and the 248-dimensional defining space of E_8 is written as a direct sum of various subspaces. This is convenient for studying subgroup embeddings [291], but awkward for group-theoretical computations.

The inspiration for the primitive invariants construction came from the axiomatic approach of Springer [314, 315] and Brown [34]: one treats the defining representation as a single vector space, and characterizes the primitive invariants by algebraic identities. This approach solves the problem of formulating efficient tensorial algorithms for evaluating group-theoretic weights, and it yields some intuition about the geometrical significance of the exceptional Lie groups. Such intuition might be of use to quark-model builders. For example, because $SU(3)$ has a cubic invariant $\epsilon^{abc} q_a q_b q_c$, Quantum Chromodynamics, based on this color group, can accommodate 3-quark baryons. Are there any other groups that could accommodate 3-quark singlets? As we shall see, G_2, F_4, and E_6 are some of the groups whose defining representations possess such invariants.

Beyond its utility as a computational technique, the primitive invariants construction of exceptional groups yields several unexpected results. First, it generates in a somewhat magical fashion a triangular array of Lie algebras, depicted in figure 1.1. This is a classification of Lie algebras different from Cartan's classification; in this new classification, all exceptional Lie groups appear in the same series (the bottom line of figure 1.1). The second unexpected result is that many groups and group representations are mutually related by interchanges of symmetrizations and antisymmetrizations and replacement of the dimension parameter n by $-n$. I call this phenomenon "negative dimensions."

For me, the greatest surprise of all is that in spite of all the magic and the strange diagrammatic notation, the resulting manuscript is in essence not very different from Wigner's [345] 1931 classic. Regardless of whether one is doing atomic, nuclear, or particle physics, all physical predictions ("spectroscopic levels") are expressed in terms of Wigner's $3n$-j coefficients, which can be evaluated by means of recursive or combinatorial algorithms.

Parenthetically, this book is *not* a book about diagrammatic methods in group

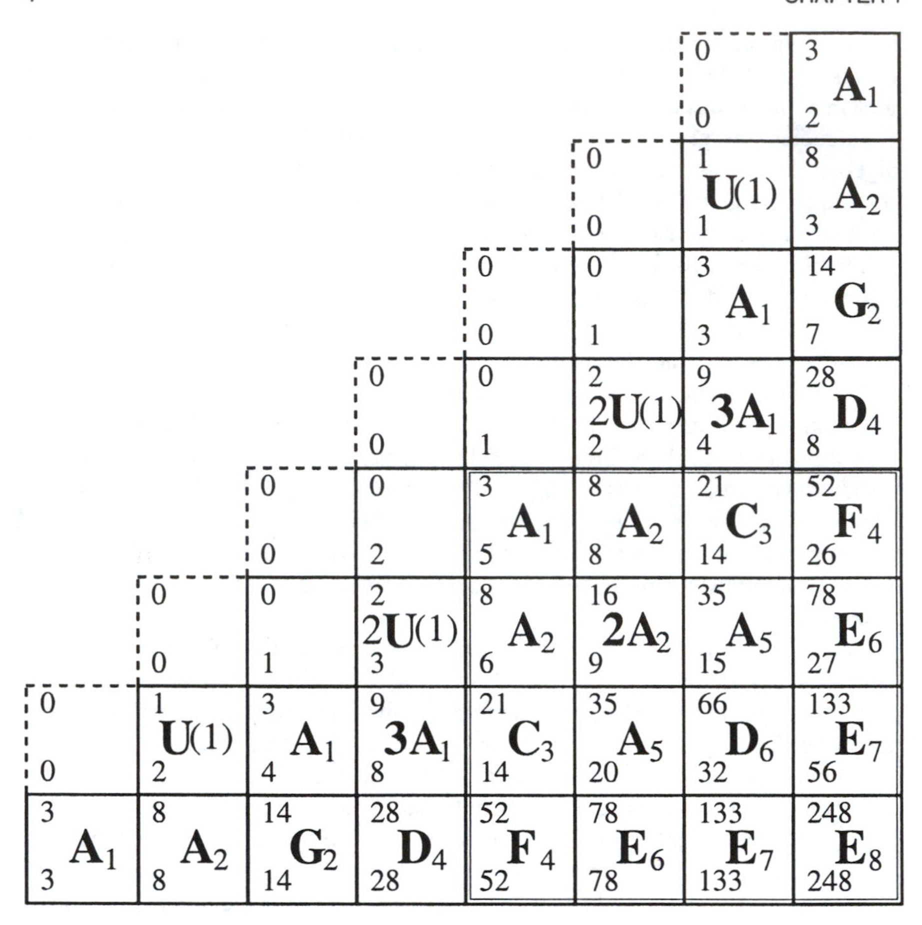

Figure 1.1 The "Magic Triangle" for Lie algebras. The "Magic Square" is framed by the double line. For a discussion, consult chapter 21.

theory. If you master a traditional notation that covers all topics in this book in a uniform way, more elegantly than birdtracks, more power to you. I would love to learn it.

Chapter Two

A preview

The theory of Lie groups presented here had mutated greatly throughout its genesis. It arose from concrete calculations motivated by physical problems; but as it was written, the generalities were collected into introductory chapters, and the applications receded later and later into the text.

As a result, the first seven chapters are largely a compilation of definitions and general results that might appear unmotivated on first reading. The reader is advised to work through the examples, section 2.2 and section 2.3 in this chapter, jump to the topic of possible interest (such as the unitary groups, chapter 9, or the E_8 family, chapter 17), and birdtrack if able or backtrack when necessary.

The goal of these notes is to provide the reader with a set of basic group-theoretic tools. They are not particularly sophisticated, and they rest on a few simple ideas. The text is long, because various notational conventions, examples, special cases, and applications have been laid out in detail, but the basic concepts can be stated in a few lines. We shall briefly state them in this chapter, together with several illustrative examples. This preview presumes that the reader has considerable prior exposure to group theory; if a concept is unfamiliar, the reader is referred to the appropriate section for a detailed discussion.

2.1 BASIC CONCEPTS

A typical quantum theory is constructed from a few building blocks, which we shall refer to as the *defining space* V. They form the defining multiplet of the theory — for example, the "quark wave functions" q_a. The group-theoretical problem consists of determining the symmetry group, *i.e.*, the group of all linear transformations

$$q'_a = G_a{}^b q_b \qquad a, b = 1, 2, \ldots, n\,,$$

which leaves invariant the predictions of the theory. The $[n \times n]$ matrices G form the *defining representation* (or "rep" for short) of the invariance group $\mathcal{G}$. The conjugate multiplet $\overline{q}$ ("antiquarks") transforms as

$$q'^a = G^a{}_b q^b\,.$$

Combinations of quarks and antiquarks transform as *tensors*, such as

$$p'_a q'_b r'^c = G_{ab}{}^c{}_{,d}{}^{ef} p_f q_e r^d\,,$$
$$G_{ab}{}^c{}_{,d}{}^{ef} = G_a{}^f G_b{}^e G_d{}^c$$

(distinction between $G_a{}^b$ and $G^a{}_b$ as well as other notational details are explained in section 3.2). Tensor reps are plagued by a proliferation of indices. These indices can either be replaced by a few collective indices:

$$\alpha = \left\{ \begin{matrix} c \\ ab \end{matrix} \right\}, \quad \beta = \left\{ \begin{matrix} ef \\ d \end{matrix} \right\},$$

$$q'_\alpha = G_\alpha{}^\beta q_\beta \,, \tag{2.1}$$

or represented diagrammatically:

a, b, c — G — f, e, d = a, b, c — f, e, d.

(Diagrammatic notation is explained in section 4.1.) Collective indices are convenient for stating general theorems; diagrammatic notation speeds up explicit calculations.

A polynomial

$$H(\overline{q}, \overline{r}, \ldots, s) = h_{ab\ldots}{}^{\ldots c} q^a r^b \ldots s_c$$

is an invariant if (and only if) for any transformation $G \in \mathcal{G}$ and for any set of vectors $q, r, s, \ldots$ (see section 3.4)

$$H(\overline{Gq}, \overline{Gr}, \ldots Gs) = H(\overline{q}, \overline{r}, \ldots, s) \,. \tag{2.2}$$

An invariance group is defined by its *primitive invariants*, *i.e.*, by a list of the elementary "singlets" of the theory. For example, the orthogonal group $O(n)$ is defined as the group of all transformations that leaves the length of a vector invariant (see chapter 10). Another example is the color $SU(3)$ of QCD that leaves invariant the mesons $(q\overline{q})$ and the baryons (qqq) (see section 15.2). A complete list of primitive invariants *defines* the invariance group via the invariance conditions (2.2); only those transformations, which respect them, are allowed.

It is not necessary to list explicitly the components of primitive invariant tensors in order to define them. For example, the $O(n)$ group is defined by the requirement that it leaves invariant a symmetric and invertible tensor $g_{ab} = g_{ba}$, $\det(g) \neq 0$. Such definition is basis independent, while a component definition $g_{11} = 1, g_{12} = 0, g_{22} = 1, \ldots$ relies on a specific basis choice. We shall define all simple Lie groups in this manner, specifying the primitive invariants only by their symmetry and by the basis-independent algebraic relations that they must satisfy.

These algebraic relations (which I shall call *primitiveness conditions*) are hard to describe without first giving some examples. In their essence they are statements of irreducibility; for example, if the primitive invariant tensors are δ^a_b, h_{abc} and h^{abc}, then $h_{abc}h^{cbe}$ must be proportional to δ^e_a, as otherwise the defining rep would be reducible. (Reducibility is discussed in section 3.5, section 3.6, and chapter 5.)

The objective of physicists' group-theoretic calculations is a description of the spectroscopy of a given theory. This entails identifying the levels (irreducible multiplets), the degeneracy of a given level (dimension of the multiplet) and the level splittings (eigenvalues of various casimirs). The basic idea that enables us to carry this program through is extremely simple: a hermitian matrix can be diagonalized. This fact has many names: Schur's lemma, Wigner-Eckart theorem, full reducibility of unitary reps, and so on (see section 3.5 and section 5.3). We exploit it by constructing invariant hermitian matrices M from the primitive invariant tensors. The

M's have collective indices (2.1) and act on tensors. Being hermitian, they can be diagonalized

$$CMC^\dagger = \begin{pmatrix} \lambda_1 & 0 & 0 & & \cdots \\ 0 & \lambda_1 & 0 & & \\ 0 & 0 & \lambda_1 & & \\ & & & \lambda_2 & \\ \vdots & & & & \ddots \end{pmatrix},$$

and their eigenvalues can be used to construct projection operators that reduce multiparticle states into direct sums of lower-dimensional reps (see section 3.5):

$$\mathbf{P}_i = \prod_{j\neq i} \frac{M - \lambda_j \mathbf{1}}{\lambda_i - \lambda_j} = C^\dagger \begin{pmatrix} \begin{array}{cc|} \ddots & \vdots \\ \cdots & 0 \\ \hline \end{array} & \cdots & 0 \\ \vdots & \boxed{\begin{array}{cccc} 1 & 0 & \cdots & 0 \\ 0 & 1 & & \\ \vdots & & \ddots & \vdots \\ 0 & & \cdots & 1 \end{array}} & \vdots \\ 0 & \cdots & \begin{array}{|cc} \hline 0 & \cdots \\ \vdots & \ddots \end{array} \end{pmatrix} C\,. \quad (2.3)$$

An explicit expression for the diagonalizing matrix C (Clebsch-Gordan coefficients or *clebsches,* section 4.2) is unnecessary — it is in fact often more of an impediment than an aid, as it obscures the combinatorial nature of group-theoretic computations (see section 4.8).

All that is needed in practice is knowledge of the characteristic equation for the invariant matrix M (see section 3.5). The characteristic equation is usually a simple consequence of the algebraic relations satisfied by the primitive invariants, and the eigenvalues λ_i are easily determined. The λ_i' s determine the projection operators $\mathbf{P}_i$, which in turn contain all relevant spectroscopic information: the rep dimension is given by tr $\mathbf{P}_i$, and the casimirs, 6-j's, crossing matrices, and recoupling coefficients (see chapter 5) are traces of various combinations of $\mathbf{P}_i$'s. All these numbers are *combinatoric*; they can often be interpreted as the number of different colorings of a graph, the number of singlets, and so on.

The invariance group is determined by considering infinitesimal transformations

$$G_a{}^b \simeq \delta^a_b + i\epsilon_i (T_i)^b_a\,.$$

The generators T_i are themselves clebsches, elements of the diagonalizing matrix C for the tensor product of the defining rep and its conjugate. They project out the adjoint rep and are constrained to satisfy the *invariance conditions* (2.2) for infinitesimal transformations (see section 4.4 and section 4.5):

$$(T_i)^{a'}_a h_{a'b\ldots}{}^{c\ldots} + (T_i)^{b'}_b h_{ab'\ldots}{}^{c\ldots} - (T_i)^c_{c'} h_{ab\ldots}{}^{c'\ldots} + \ldots = 0$$

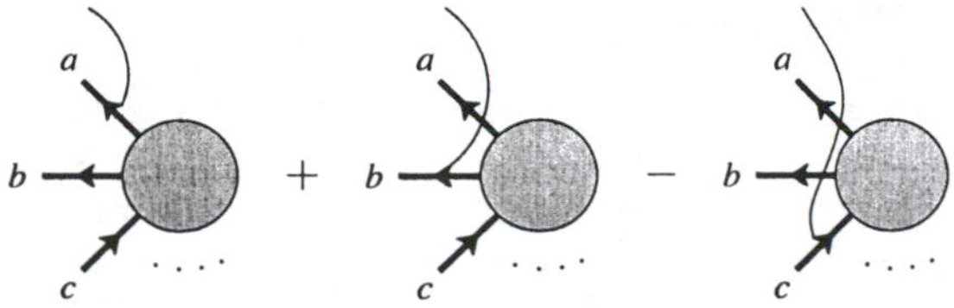

$+ \ldots = 0\,.$ (2.4)

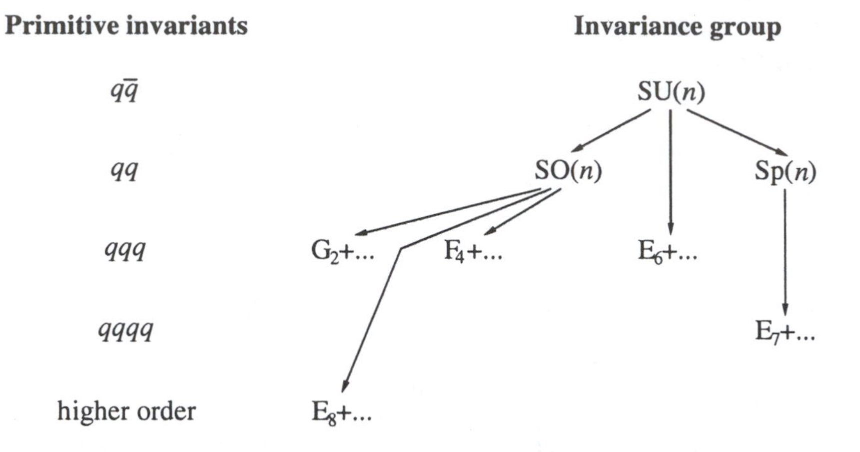

Figure 2.1 Additional primitive invariants induce chains of invariance subgroups.

As the corresponding projector operators are already known, we have an explicit construction of the symmetry group (at least infinitesimally — we will not consider discrete transformations).

If the primitive invariants are bilinear, the above procedure leads to the familiar tensor reps of classical groups. However, for trilinear or higher invariants the results are more surprising. In particular, all exceptional Lie groups emerge in a pattern of solutions which I will refer to as a *Magic Triangle.* The flow of the argument (see chapter 16) is schematically indicated in figure 2.1, with the arrows pointing to the primitive invariants that characterize a particular group. For example, E_7 primitives are a sesquilinear invariant $q\bar{q}$, a skew symmetric qp invariant, and a symmetric $qqqq$ (see chapter 20).

The strategy is to introduce the invariants one by one, and study the way in which they split up previously irreducible reps. The first invariant might be realizable in many dimensions. When the next invariant is added (section 3.6), the group of invariance transformations of the first invariant splits into two subsets; those transformations that preserve the new invariant, and those that do not. Such decompositions yield Diophantine conditions on rep dimensions. These conditions are so constraining that they limit the possibilities to a few that can be easily identified.

To summarize: in the primitive invariants approach, all simple Lie groups, classical as well as exceptional, are constructed by (see chapter 21)

1. defining a symmetry group by specifying a list of *primitive invariants*;

2. using *primitiveness* and *invariance* conditions to obtain algebraic relations between primitive invariants;

3. constructing *invariant matrices* acting on tensor product spaces;

4. constructing *projection operators* for reduced rep from characteristic equations for invariant matrices.

Once the projection operators are known, all interesting spectroscopic numbers can be evaluated.

The foregoing run through the basic concepts was inevitably obscure. Perhaps working through the next two examples will make things clearer. The first example illustrates computations with classical groups. The second example is more interesting; it is a sketch of construction of irreducible reps of E_6.

2.2 FIRST EXAMPLE: $SU(n)$

How do we describe the invariance group that preserves the norm of a complex vector? The *list of primitives* consists of a single primitive invariant,

$$m(p,q) = \delta_b^a p^b q_a = \sum_{a=1}^{n} (p_a)^* q_a \, .$$

The Kronecker δ_b^a is the only primitive invariant tensor. We can immediately write down the two *invariant matrices* on the tensor product of the defining space and its conjugate,

$$\text{identity}: \mathbf{1}_{d,b}^{a\,c} = \delta_b^a \delta_d^c =$$

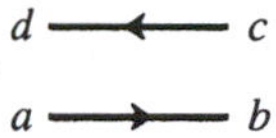

$$\text{trace}: T_{d,b}^{a\,c} = \delta_d^a \delta_b^c = \, .$$

The *characteristic equation* for T written out in the matrix, tensor, and birdtrack notations is

$$T^2 = nT$$

$$T_{d,e}^{a\,f} T_{f,b}^{e\,c} = \delta_d^a \delta_e^f \delta_f^e \delta_b^c = n\, T_{d,b}^{a\,c}$$

$$= \quad = n \quad .$$

Here we have used $\delta_e^e = n$, the dimension of the defining vector space. The roots are $\lambda_1 = 0$, $\lambda_2 = n$, and the corresponding *projection operators* are

$$SU(n) \text{ adjoint rep:} \quad \mathbf{P}_1 = \frac{T - n\mathbf{1}}{0 - n} = \mathbf{1} - \frac{1}{n}T$$

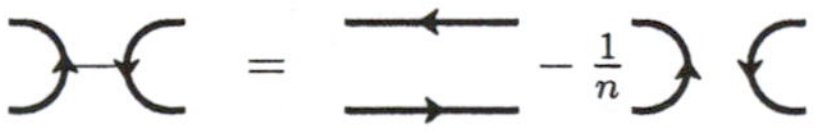

$$= \quad - \frac{1}{n} \qquad (2.5)$$

$$U(n) \text{ singlet:} \quad \mathbf{P}_2 = \frac{T - 0 \cdot \mathbf{1}}{n - 0} = \frac{1}{n}T = \frac{1}{n} \quad .$$

Now we can evaluate any number associated with the $SU(n)$ adjoint rep, such as its dimension and various casimirs.

The *dimensions* of the two reps are computed by tracing the corresponding projection operators (see section 3.5):

$$SU(n) \text{ adjoint:} \quad d_1 = \operatorname{tr} \mathbf{P}_1 = \quad = \quad - \frac{1}{n} \quad = \delta_b^b \delta_a^a - \frac{1}{n} \delta_a^b \delta_b^a$$

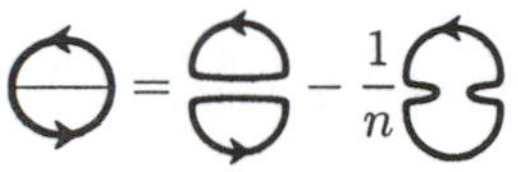

$$= n^2 - 1$$

$$\text{singlet:} \quad d_2 = \operatorname{tr} \mathbf{P}_2 = \frac{1}{n} \quad = 1 \, .$$

To evaluate *casimirs*, we need to fix the overall normalization of the generators T_i of $SU(n)$. Our convention is to take

$$\delta_{ij} = \mathrm{tr}\, T_i T_j = \quad .$$

The value of the quadratic casimir for the defining rep is computed by substituting the adjoint projection operator:

$$SU(n): \quad C_F \delta_a^b = (T_i T_i)_a^b = \quad = \quad - \frac{1}{n}$$ 
$$= \frac{n^2-1}{n} \quad = \frac{n^2-1}{n}\delta_a^b \,. \tag{2.6}$$

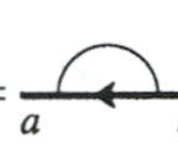

In order to evaluate the quadratic casimir for the adjoint rep, we need to replace the structure constants iC_{ijk} by their *Lie algebra* definition (see section 4.5)

$$T_i T_j - T_j T_i = i C_{ij\ell} T_\ell$$

$$- \quad = \quad .$$

Tracing with T_k, we can express C_{ijk} in terms of the defining rep traces:

$$iC_{ijk} = \mathrm{tr}(T_i T_j T_k) \; - \; \mathrm{tr}(T_j T_i T_k)$$

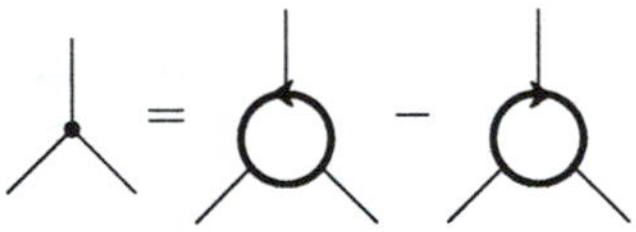

$$= \quad - \quad .$$

The adjoint quadratic casimir $C_{imn}C^{nmj}$ is now evaluated by first eliminating C_{ijk}'s in favor of the defining rep:

$$\delta_{ij} C_A = \quad = 2 \quad .$$

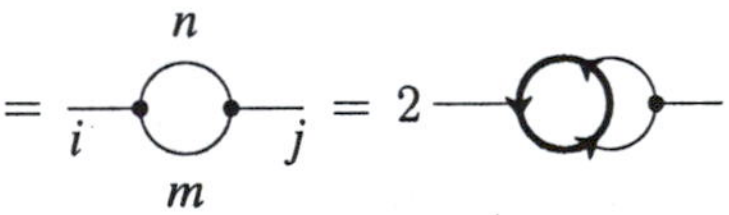

The remaining C_{ijk} can be unwound by the Lie algebra commutator:

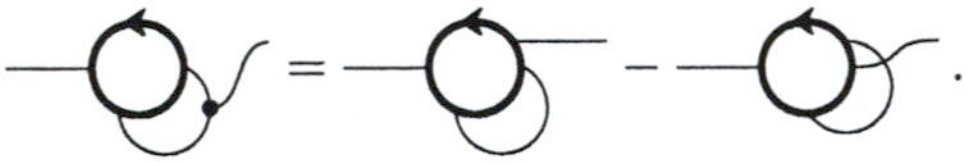

$$= \quad - \quad .$$

We have already evaluated the quadratic casimir (2.6) in the first term. The second term we evaluate by substituting the adjoint projection operator

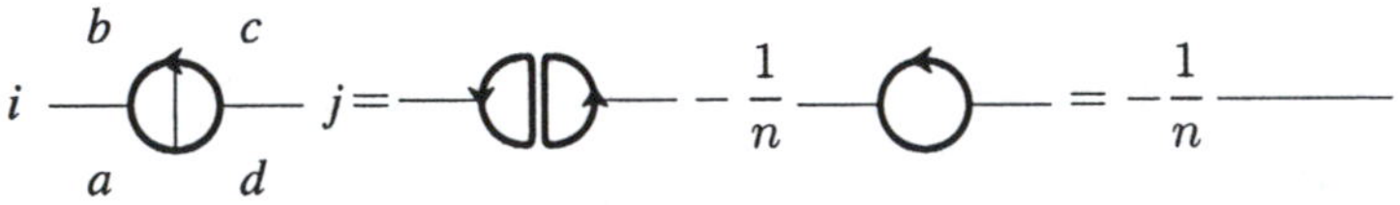

$$= \quad - \frac{1}{n} \quad = -\frac{1}{n}$$

$$\mathrm{tr}(T_i T_k T_j T_k) = (T_i)_a^b (P_1)_{d,}{}^a{}_b{}^c (T_j)_c^d = (T_i)_a^a (T_j)_c^c - \frac{1}{n}(T_i)_a^b (T_j)_b^a \,.$$

The $(T_i)_a^a (T_j)_c^c$ term vanishes by the tracelessness of T_i's. This is a consequence of the orthonormality of the two projection operators $\mathbf{P}_1$ and $\mathbf{P}_2$ in (2.5) (see (3.50)):

$$0 = \mathbf{P}_1 \mathbf{P}_2 = \quad \Rightarrow \mathrm{tr}\, T_i = \quad = 0 \,.$$

Combining the above expressions we finally obtain

$$C_A = 2\left(\frac{n^2-1}{n} + \frac{1}{n}\right) = 2n\ .$$

The problem (1.1) that started all this is evaluated the same way. First we relate the adjoint quartic casimir to the defining casimirs:

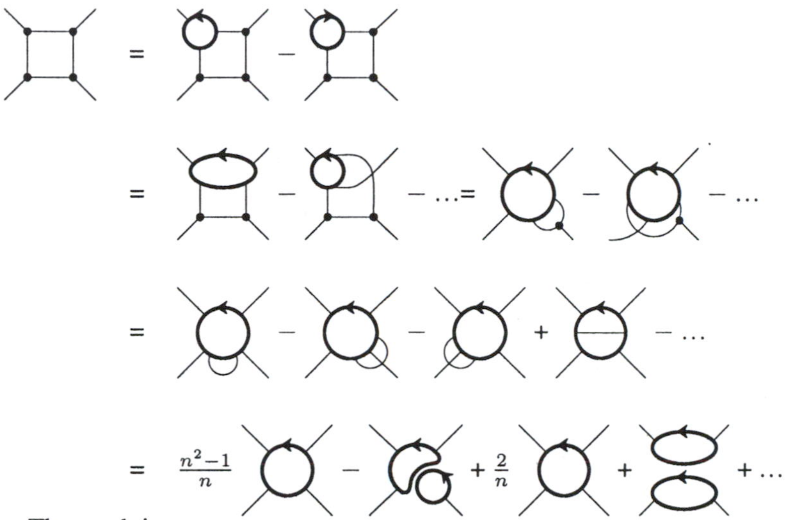

$$= \quad - \quad$$

$$= \quad - \quad - \ldots = \quad - \quad - \ldots$$

and so

$$= \quad - \quad - \quad + \quad - \ldots$$

$$= \frac{n^2-1}{n} \quad - \quad + \frac{2}{n} \quad + \quad + \ldots$$

on. The result is

$$SU(n): \quad = n\left\{ \quad + \quad \right\} + 2\left\{ \quad + \quad + \quad \right\}.$$

The diagram (1.1) is now reexpressed in terms of the defining rep casimirs:

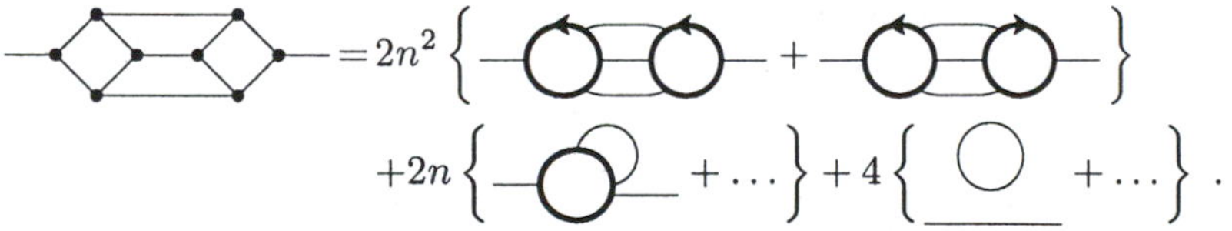

$$= 2n^2\left\{ \quad + \quad \right\}$$

$$+2n\left\{ \quad + \ldots \right\} + 4\left\{ \quad + \ldots \right\}.$$

The first two terms are evaluated by inserting the adjoint rep projection operators:

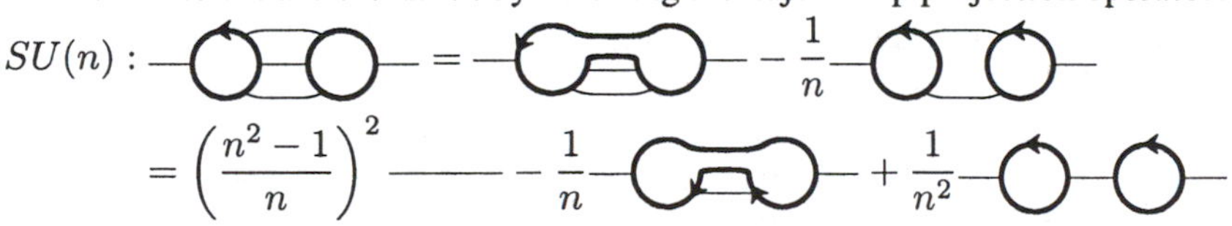

$$SU(n): \quad = \quad - \frac{1}{n} \quad$$

$$= \left(\frac{n^2-1}{n}\right)^2 \quad - \frac{1}{n} \quad + \frac{1}{n^2} \quad$$

$$= \left(n^2 - 2 + \frac{1}{n^2} - \frac{1}{n}\left(n - \frac{1}{n}\right) + \frac{1}{n^2}\right)$$

$$= \left(n^2 - 3 + \frac{3}{n^2}\right) \quad ,$$

and the remaining terms have already been evaluated. Collecting everything together, we finally obtain

$$SU(n): \quad = 2n^2(n^2+12) \quad .$$

This example was unavoidably lengthy; the main point is that the evaluation is performed by a substitution algorithm and is easily automated. Any graph, no matter how complicated, is eventually reduced to a polynomial in traces of $\delta_a^a = n$, *i.e.*, the dimension of the defining rep.

2.3 SECOND EXAMPLE: E_6 FAMILY

What invariance group preserves norms of complex vectors, as well as a symmetric cubic invariant,

$$D(p,q,r) = d^{abc} p_a q_b r_c = D(q,p,r) = D(p,r,q) \ ?$$

We analyze this case following the steps of the summary of section 2.1:

i) *Primitive invariant tensors*

$$\delta_a^b = a \longrightarrow b\,, \quad d_{abc} = \ , \quad d^{abc} = (d_{abc})^* = \ .$$

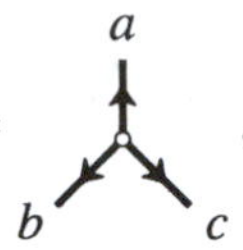

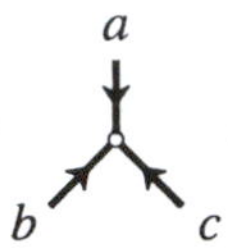

ii) *Primitiveness.* $d_{aef} d^{efb}$ must be proportional to δ_b^a, the only primitive 2-index tensor. We use this to fix the overall normalization of d_{abc}'s:

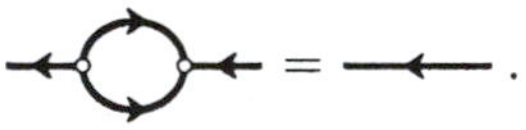

$$= \ .$$

iii) *Invariant hermitian matrices.* We shall construct here the adjoint rep projection operator on the tensor product space of the defining rep and its conjugate. All invariant matrices on this space are

$$\delta_b^a \delta_d^c = \ , \quad \delta_d^a \delta_b^c = \ , \quad d^{ace} d_{ebd} = \ .$$

They are hermitian in the sense of being invariant under complex conjugation and transposition of indices (see (3.21)). The crucial step in constructing this basis is the primitiveness assumption: 4-leg diagrams containing loops are not primitive (see section 3.3).

The adjoint rep is always contained in the decomposition of $V \otimes \bar{V} \to V \otimes \bar{V}$ into (ir)reducible reps, so the adjoint projection operator must be expressible in terms of the 4-index invariant tensors listed above:

$$(T_i)_b^a (T_i)_c^d = A(\delta_c^a \delta_b^d + B \delta_b^a \delta_c^d + C d^{ade} d_{bce})$$

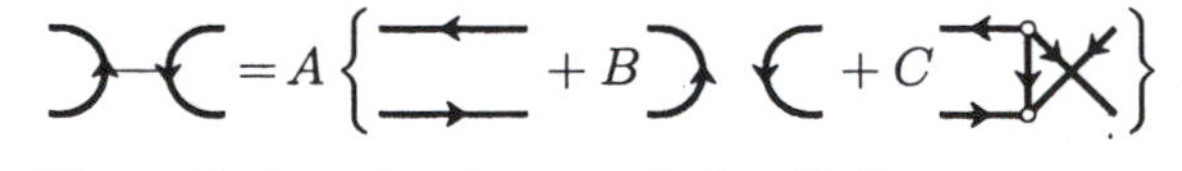

$$= A \left\{ \ + B \ + C \ \right\} .$$

iv) *Invariance.* The cubic invariant tensor satisfies (2.4)

$$+ \ + \ = 0 \, .$$

Contracting with d^{abc}, we obtain

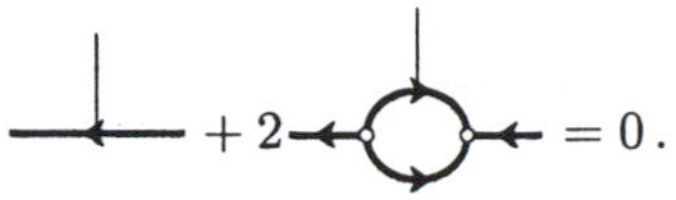

$$+ 2 \quad = 0 .$$

Contracting next with $(T_i)_a^b$, we get an invariance condition on the adjoint projection operator,

$$+ 2 \quad = 0 .$$

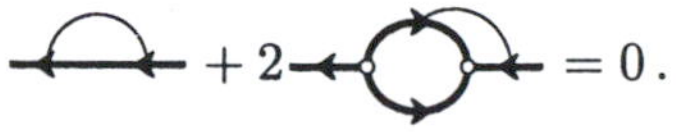

Substituting the adjoint projection operator yields the first relation between the coefficients in its expansion:

$$0 = (n + B + C) \quad + 2 \left\{ \quad + B \quad + C \quad \right\}$$

$$0 = B + C + \frac{n+2}{3} .$$

v) *The projection operators* should be orthonormal, $\mathbf{P}_\mu \mathbf{P}_\sigma = \mathbf{P}_\mu \delta_{\mu\sigma}$. The adjoint projection operator is orthogonal to (2.5), the singlet projection operator $\mathbf{P}_2$. This yields the second relation on the coefficients:

$$0 = \mathbf{P}_2 \mathbf{P}_A$$

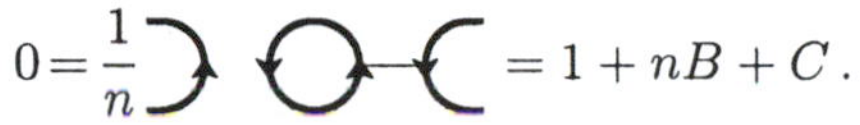

$$0 = \frac{1}{n} \quad = 1 + nB + C .$$

Finally, the overall normalization factor A is fixed by $\mathbf{P}_A \mathbf{P}_A = \mathbf{P}_A$:

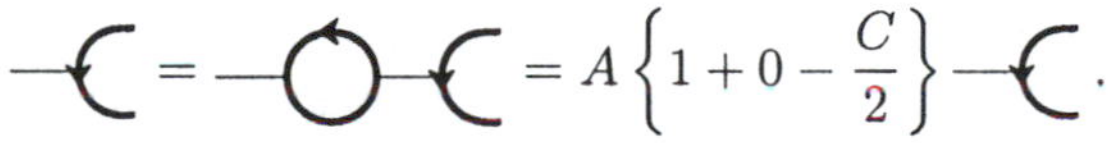

$$= \quad = A \left\{ 1 + 0 - \frac{C}{2} \right\} \quad .$$

Combining the above three relations, we obtain the adjoint projection operator for the invariance group of a symmetric cubic invariant:

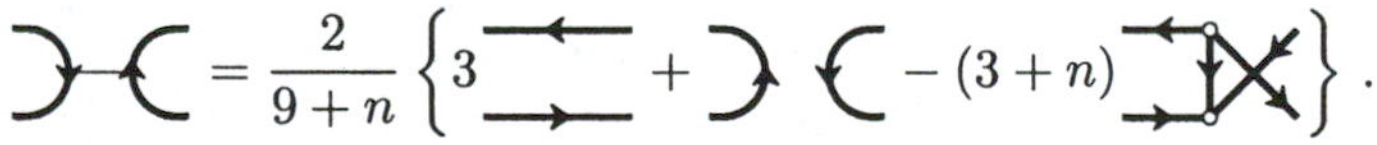

$$= \frac{2}{9+n} \left\{ 3 \quad + \quad - (3+n) \quad \right\} . \qquad (2.7)$$

The corresponding *characteristic equation*, mentioned in the point iv) of the summary of section 2.1, is given in (18.10).

The dimension of the adjoint rep is obtained by tracing the projection operator:

$$N = \delta_{ii} = \quad = \quad = nA(n + B + C) = \frac{4n(n-1)}{n+9} .$$

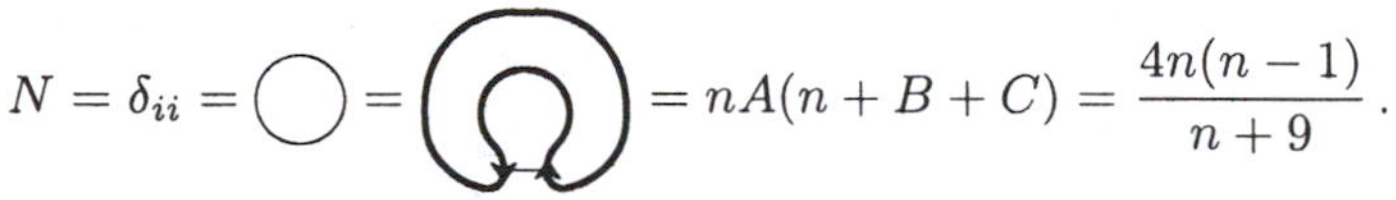

This *Diophantine condition* is satisfied by a small family of invariance groups, discussed in chapter 18. The most interesting member of this family is the exceptional Lie group E_6, with $n = 27$ and $N = 78$.

The solution to problem (1.1) requires further computation, but for exceptional Lie groups the answer, given in table 7.4, turns out to be surprisingly simple. The part of the 4-loop that cannot be simplified by Lie algebra manipulations vanishes identically for all exceptional Lie groups (see chapter 17).

Chapter Three

Invariants and reducibility

Basic group-theoretic notions are introduced: groups, invariants, tensors, the diagrammatic notation for invariant tensors.

The key results are the construction of projection operators from invariant matrices, the Clebsch-Gordan coefficients rep of projection operators (4.18), the invariance conditions (4.35) and the Lie algebra relations (4.47).

The basic idea is simple: a hermitian matrix can be diagonalized. If this matrix is an invariant matrix, it decomposes the reps of the group into direct sums of lower-dimensional reps. Most of computations to follow implement the spectral decomposition

$$\mathbf{M} = \lambda_1 \mathbf{P}_1 + \lambda_2 \mathbf{P}_2 + \cdots + \lambda_r \mathbf{P}_r \,,$$

which associates with each distinct root λ_i of invariant matrix $\mathbf{M}$ a projection operator (3.48):

$$\mathbf{P}_i = \prod_{j \neq i} \frac{\mathbf{M} - \lambda_j \mathbf{1}}{\lambda_i - \lambda_j} \,.$$

The exposition given here in sections. 3.5–3.6 is taken from refs. [73, 74]. Who wrote this down first I do not know, but I like Harter's exposition [155, 156, 157] best.

What follows is a bit dry, so we start with a motivational quote from Hermann Weyl on the "so-called first main theorem of invariant theory":

"*All invariants are expressible in terms of a finite number among them.* We cannot claim its validity for every group $\mathcal{G}$; rather, it will be our chief task to investigate for each particular group whether a finite integrity basis exists or not; the answer, to be sure, will turn out affirmative in the most important cases."

3.1 PRELIMINARIES

In this section we define basic building blocks of the theory to be developed here: groups, vector spaces, algebras, *etc.* This material is covered in any introduction to linear algebra [135, 211, 253] or group theory [324, 153]. Most of the material reviewed here is probably known to the reader and can be profitably skipped on the first reading. Nevertheless, it seems that a refresher is needed here, as an expert (more so than a novice to group theory) tends to find the first exposure to the diagrammatic rewriting of elementary properties of linear vector spaces (chapter 4) hard to digest.

3.1.1 Groups

Definition. A set of elements $g \in \mathcal{G}$ forms a group with respect to multiplication $\mathcal{G} \times \mathcal{G} \to \mathcal{G}$ if

(a) the set is *closed* with respect to multiplication; for any two elements $a, b \in \mathcal{G}$, the product $ab \in \mathcal{G}$;

(b) multiplication is *associative*

$$(ab)c = a(bc)$$

for any three elements $a, b, c \in \mathcal{G}$;

(c) there exists an *identity* element $\mathbf{e} \in \mathcal{G}$ such that

$$\mathbf{e}g = g\mathbf{e} \qquad \text{for any } g \in \mathcal{G}\,;$$

(d) for any $g \in \mathcal{G}$ there exists an *inverse* g^{-1} such that

$$g^{-1}g = gg^{-1} = \mathbf{e}\,.$$

If the group is finite, the number of elements is called the *order* of the group and denoted $|\mathcal{G}|$. If the multiplication $ab = ba$ is commutative for all $a, b \in \mathcal{G}$, the group is *abelian*.

Definition. A *subgroup* $\mathcal{H} \subset \mathcal{G}$ is a subset of $\mathcal{G}$ that forms a group under multiplication. $\mathbf{e}$ is always a subgroup; so is $\mathcal{G}$ itself.

3.1.2 Vector spaces

Definition. A set V of elements $\mathbf{x}, \mathbf{y}, \mathbf{z}, \ldots$ is called a *vector* (or *linear*) *space* over a field $\mathbb{F}$ if

(a) *vector addition* "+" is defined in V such that V is an abelian group under addition, with identity element $\mathbf{0}$;

(b) the set is *closed* with respect to *scalar multiplication* and vector addition

$$\begin{aligned} a(\mathbf{x} + \mathbf{y}) &= a\mathbf{x} + a\mathbf{y}\,, \qquad a, b \in \mathbb{F}\,, \quad \mathbf{x}, \mathbf{y} \in V \\ (a + b)\mathbf{x} &= a\mathbf{x} + b\mathbf{x} \\ a(b\mathbf{x}) &= (ab)\mathbf{x} \\ 1\,\mathbf{x} &= \mathbf{x}\,, \qquad 0\,\mathbf{x} = \mathbf{0}\,. \end{aligned}$$

Here the field $\mathbb{F}$ is either $\mathbb{R}$, the field of reals numbers, or $\mathbb{C}$, the field of complex numbers. Given a subset $V_0 \subset V$, the set of all linear combinations of elements of V_0, or the *span* of V_0, is also a vector space.

Definition. A *basis* $\{\mathbf{e}^1, \cdots, \mathbf{e}^n\}$ is any linearly independent subset of V whose span is V. n, the number of basis elements, is called the *dimension* of the vector space V.

In calculations to be undertaken a vector $\mathbf{x} \in V$ is often specified by the n-tuple $(x_1, \cdots, x_n)$ in $\mathbb{F}^n$, its coordinates $\mathbf{x} = \sum \mathbf{e}^a x_a$ in a given basis. We will rarely, if ever, actually fix an explicit basis $\{\mathbf{e}^1, \cdots, \mathbf{e}^n\}$, but thinking this way makes it often easier to manipulate tensorial objects.

Repeated index summation. Throughout this text, the repeated pairs of upper/lower indices are always summed over

$$G_a{}^b x_b \equiv \sum_{b=1}^{n} G_a{}^b x_b \,, \tag{3.1}$$

unless explicitly stated otherwise.

Let $GL(n, \mathbb{F})$ be the group of general linear transformations,

$$GL(n, \mathbb{F}) = \{G : \mathbb{F}^n \to \mathbb{F}^n \mid \det(G) \neq 0\} \,. \tag{3.2}$$

Under $GL(n, \mathbb{F})$ a basis set of V is mapped into another basis set by multiplication with a $[n \times n]$ matrix G with entries in $\mathbb{F}$,

$$\mathbf{e}'^{\,a} = \mathbf{e}^b (G^{-1})_b{}^a \,.$$

As the vector $\mathbf{x}$ is what it is, regardless of a particular choice of basis, under this transformation its coordinates must transform as

$$x'_a = G_a{}^b x_b \,.$$

Definition. We shall refer to the set of $[n \times n]$ matrices G as a *standard rep* of $GL(n, \mathbb{F})$, and the space of all n-tuples $(x_1, x_2, \ldots, x_n)^t$, $x_i \in \mathbb{F}$ on which these matrices act as the *standard representation space* V.

Under a general linear transformation $G \in GL(n, \mathbb{F})$, the row of basis vectors transforms by right multiplication as $\mathbf{e}' = \mathbf{e}\, G^{-1}$, and the column of x_a's transforms by left multiplication as $x' = Gx$. Under left multiplication the column (row transposed) of basis vectors $\mathbf{e}^t$ transforms as $\mathbf{e}'^t = G^\dagger \mathbf{e}^t$, where the *dual rep* $G^\dagger = (G^{-1})^t$ is the transpose of the inverse of G. This observation motivates introduction of a *dual* representation space $\bar{V}$, the space on which $GL(n, \mathbb{F})$ acts via the dual rep $G^\dagger$.

Definition. If V is a vector representation space, then the *dual space* $\bar{V}$ is the set of all linear forms on V over the field $\mathbb{F}$.

If $\{\mathbf{e}^1, \cdots, \mathbf{e}^n\}$ is a basis of V, then $\bar{V}$ is spanned by the *dual basis* $\{\mathbf{f}_1, \cdots, \mathbf{f}_n\}$, the set of n linear forms $\mathbf{f}_a$ such that

$$\mathbf{f}_a(\mathbf{e}^b) = \delta_a^b \,,$$

where δ_a^b is the Kronecker symbol, $\delta_a^b = 1$ if $a = b$, and zero otherwise. The components of dual representation space vectors will here be distinguished by upper indices

$$(y^1, y^2, \ldots, y^n) \,. \tag{3.3}$$

They transform under $GL(n,\mathbb{F})$ as

$$y'^a = (G^\dagger)_b{}^a y^b \,. \tag{3.4}$$

For $GL(n,\mathbb{F})$ no complex conjugation is implied by the † notation; that interpretation applies only to unitary subgroups of $GL(n,\mathbb{C})$. G can be distinguished from $G^\dagger$ by meticulously keeping track of the relative ordering of the indices,

$$G_a^b \to G_a{}^b \,, \qquad (G^\dagger)_a^b \to G^b{}_a \,. \tag{3.5}$$

3.1.3 Algebra

Definition. A set of r elements $\mathbf{t}_\alpha$ of a vector space $\mathcal{T}$ forms an algebra if, in addition to the vector addition and scalar multiplication,

(a) the set is *closed* with respect to multiplication $\mathcal{T}\cdot\mathcal{T}\to\mathcal{T}$, so that for any two elements $\mathbf{t}_\alpha, \mathbf{t}_\beta \in \mathcal{T}$, the product $\mathbf{t}_\alpha\cdot\mathbf{t}_\beta$ also belongs to $\mathcal{T}$:

$$\mathbf{t}_\alpha\cdot\mathbf{t}_\beta = \sum_{\gamma=0}^{r-1} \tau_{\alpha\beta}{}^\gamma \mathbf{t}_\gamma \,, \qquad \tau_{\alpha\beta}{}^\gamma \in \mathbb{C}\,; \tag{3.6}$$

(b) the multiplication operation is *distributive*:

$$(\mathbf{t}_\alpha+\mathbf{t}_\beta)\cdot\mathbf{t}_\gamma = \mathbf{t}_\alpha\cdot\mathbf{t}_\gamma+\mathbf{t}_\beta\cdot\mathbf{t}_\gamma$$
$$\mathbf{t}_\alpha\cdot(\mathbf{t}_\beta+\mathbf{t}_\gamma) = \mathbf{t}_\alpha\cdot\mathbf{t}_\beta+\mathbf{t}_\alpha\cdot\mathbf{t}_\gamma \,.$$

The set of numbers $\tau_{\alpha\beta}{}^\gamma$ are called the *structure constants* of the algebra. They form a matrix rep of the algebra,

$$(\mathbf{t}_\alpha)_\beta{}^\gamma \equiv \tau_{\alpha\beta}{}^\gamma \,, \tag{3.7}$$

whose dimension is the dimension of the algebra itself.

Depending on what further assumptions one makes on the multiplication, one obtains different types of algebras. For example, if the multiplication is associative

$$(\mathbf{t}_\alpha\cdot\mathbf{t}_\beta)\cdot\mathbf{t}_\gamma = \mathbf{t}_\alpha\cdot(\mathbf{t}_\beta\cdot\mathbf{t}_\gamma)\,,$$

the algebra is *associative*. Typical examples of products are the *matrix product*

$$(\mathbf{t}_\alpha\cdot\mathbf{t}_\beta)_a^c = (t_\alpha)_a^b(t_\beta)_b^c \,, \qquad \mathbf{t}_\alpha \in V\otimes\bar{V}\,, \tag{3.8}$$

and the *Lie product*

$$(\mathbf{t}_\alpha\cdot\mathbf{t}_\beta)_a^c = (t_\alpha)_a^b(t_\beta)_b^c - (t_\alpha)_c^b(t_\beta)_b^a \,, \qquad \mathbf{t}_\alpha \in V\otimes\bar{V}\,. \tag{3.9}$$

As a plethora of vector spaces, indices and dual spaces looms large in our immediate future, it pays to streamline the notation now, by singling out one vector space as "defining" and indicating the dual vector space by raised indices.

The next two sections introduce the three key notions in our construction of invariance groups: *defining rep*, section 3.2 (see also comments on page 23); *invariants*, section 3.4; and *primitiveness assumption*, page 21. Chapter 4 introduces diagrammatic notation, the computational tool essential to understanding all computations to come. As these concepts can be understood only in relation to one another, not singly, and an exposition of necessity progresses linearly, the reader is asked to be patient, in the hope that the questions that naturally arise upon first reading will be addressed in due course.

3.2 DEFINING SPACE, TENSORS, REPS

Definition. In what follows V will always denote the *defining* n-dimensional complex vector representation space, that is to say the initial, "elementary multiplet" space within which we commence our deliberations. Along with the defining vector representation space V comes the *dual* n-dimensional vector representation space $\bar{V}$. We shall denote the corresponding element of $\bar{V}$ by raising the index, as in (3.3), so the components of defining space vectors, resp. dual vectors, are distinguished by lower, resp. upper indices:

$$x = (x_1, x_2, \ldots, x_n) \,, \qquad \mathbf{x} \in V$$
$$\bar{x} = (x^1, x^2, \ldots, x^n) \,, \qquad \bar{\mathbf{x}} \in \bar{V} \,. \tag{3.10}$$

Definition. Let $\mathcal{G}$ be a group of transformations acting linearly on V, with the action of a group element $g \in \mathcal{G}$ on a vector $x \in V$ given by an $[n \times n]$ matrix G

$$x'_a = G_a{}^b x_b \qquad a, b = 1, 2, \ldots, n \,. \tag{3.11}$$

We shall refer to $G_a{}^b$ as the *defining rep* of the group $\mathcal{G}$. The action of $g \in \mathcal{G}$ on a vector $\bar{q} \in \bar{V}$ is given by the *dual rep* $[n \times n]$ matrix $G^\dagger$:

$$x'^a = x^b (G^\dagger)_b{}^a = G^a{}_b x^b \,. \tag{3.12}$$

In the applications considered here, the group $\mathcal{G}$ will almost always be assumed to be a subgroup of the *unitary group*, in which case $G^{-1} = G^\dagger$, and † indicates hermitian conjugation:

$$(G^\dagger)_a{}^b = (G_b{}^a)^* = G^b{}_a \,. \tag{3.13}$$

Definition. A *tensor* $x \in V^p \otimes \bar{V}^q$ transforms under the action of $g \in \mathcal{G}$ as

$$x'^{a_1 a_2 \ldots a_q}_{b_1 \ldots b_p} = G^{a_1 a_2 \ldots a_q}_{b_1 \ldots b_p}{}^{d_p \ldots d_1}_{\;,\, c_q \ldots c_2 c_1} \, x^{c_1 c_2 \ldots c_q}_{d_1 \ldots d_p} \,, \tag{3.14}$$

where the $V^p \otimes \bar{V}^q$ *tensor rep* of $g \in \mathcal{G}$ is defined by the group acting on all indices of x.

$$G^{a_1 a_2 \ldots a_p}_{b_1 \ldots b_q}{}^{d_q \ldots d_1}_{\;,\, c_p \ldots c_2 c_1} \equiv G^{a_1}{}_{c_1} G^{a_2}{}_{c_2} \ldots G^{a_p}{}_{c_p} G_{b_q}{}^{d_q} \ldots G_{b_2}{}^{d_2 1} G_{b_1}{}^{d_1} \,. \tag{3.15}$$

Tensors can be combined into other tensors by

(a) *addition:*

$$z^{ab \ldots c}_{d \ldots e} = \alpha x^{ab \ldots c}_{d \ldots e} + \beta y^{ab \ldots c}_{d \ldots e} \,, \qquad \alpha, \beta \in \mathbb{C} \,, \tag{3.16}$$

(b) *product:*

$$z^{abcd}_{efg} = x^{abc}_e y^d_{fg} \,, \tag{3.17}$$

(c) *contraction:* Setting an upper and a lower index equal and summing over all of its values yields a tensor $z \in V^{p-1} \otimes \bar{V}^{q-1}$ without these indices:

$$z^{bc \ldots d}_{e \ldots f} = x^{abc \ldots d}_{e \ldots af} \,, \qquad z^{ad}_e = x^{abc}_e y^d_{cb} \,. \tag{3.18}$$

A tensor $x \in V^p \otimes \bar{V}^q$ transforms linearly under the action of g, so it can be considered a vector in the $d = n^{p+q}$-dimensional vector space $\tilde{V} = V^p \otimes \bar{V}^q$. We can replace the array of its indices by one collective index:

$$x_\alpha = x^{a_1 a_2 \ldots a_q}_{b_1 \ldots b_p} \,. \tag{3.19}$$

One could be more explicit and give a table like

$$x_1 = x^{11\ldots1}_{1\ldots1}\,,\; x_2 = x^{21\ldots1}_{1\ldots1}, \ldots,\; x_d = x^{nn\ldots n}_{n\ldots n}\,, \tag{3.20}$$

but that is unnecessary, as we shall use the compact index notation only as a shorthand.

Definition. *Hermitian conjugation* is effected by complex conjugation and index transposition:

$$(h^\dagger)^{ab}_{cde} \equiv (h^{edc}_{ba})^*\,. \tag{3.21}$$

Complex conjugation interchanges upper and lower indices, as in (3.10); transposition reverses their order. A matrix is *hermitian* if its elements satisfy

$$(\mathbf{M}^\dagger)^a_b = M^a_b\,. \tag{3.22}$$

For a hermitian matrix there is no need to keep track of the relative ordering of indices, as $M_b{}^a = (\mathbf{M}^\dagger)_b{}^a = M^a{}_b$.

Definition. The tensor dual to x_α defined by (3.19) has form

$$x^\alpha = x^{b_p\ldots b_1}_{a_q\ldots a_2 a_1}\,. \tag{3.23}$$

Combined, the above definitions lead to the hermitian conjugation rule for collective indices: a collective index is raised or lowered by interchanging the upper and lower indices and reversing their order:

$$_\alpha = \left\{ \begin{matrix} a_1 a_2 \ldots a_q \\ b_1 \ldots b_p \end{matrix} \right\} \quad\leftrightarrow\quad ^\alpha = \left\{ \begin{matrix} b_p \ldots b_1 \\ a_q \ldots a_2 a_1 \end{matrix} \right\}. \tag{3.24}$$

This transposition convention will be motivated further by the diagrammatic rules of section 4.1.

The tensor rep (3.15) can be treated as a $[d \times d]$ matrix

$$G_\alpha{}^\beta = G^{a_1 a_2 \ldots a_q}{}_{b_1\ldots b_p}\,,{}^{d_p\ldots d_1}{}_{c_q\ldots c_2 c_1}\,, \tag{3.25}$$

and the tensor transformation (3.14) takes the usual matrix form

$$x'_\alpha = G_\alpha{}^\beta x_\beta\,. \tag{3.26}$$

3.3 INVARIANTS

Definition. The vector $q \in V$ is an *invariant vector* if for any transformation $g \in \mathcal{G}$

$$q = Gq\,. \tag{3.27}$$

Definition. A tensor $x \in V^p \otimes \bar{V}^q$ is an *invariant tensor* if for any $g \in G$

$$x^{a_1 a_2 \ldots a_p}_{b_1\ldots b_q} = G^{a_1}{}_{c_1} G^{a_2}{}_{c_2} \ldots G_{b_1}{}^{d_1} \ldots G_{b_q}{}^{d_q} x^{c_1 c_2 \ldots c_p}_{d_1\ldots d_q}\,. \tag{3.28}$$

We can state this more compactly by using the notation of (3.25)

$$x_\alpha = G_\alpha{}^\beta x_\beta\,. \tag{3.29}$$

Here we treat the tensor $x^{a_1 a_2 \ldots a_p}_{b_1\ldots b_q}$ as a vector in $[d{\times}d]$-dimensional space, $d = n^{p+q}$.

If a bilinear form $m(\bar{x}, y) = x^a M_a{}^b y_b$ is invariant for all $g \in \mathcal{G}$, the matrix

$$M_a{}^b = G_a{}^c G^b{}_d M_c{}^d \tag{3.30}$$

is an *invariant matrix*. Multiplying with $G_b{}^e$ and using the unitary condition (3.13), we find that the invariant matrices *commute* with all transformations $g \in \mathcal{G}$:

$$[G, \mathbf{M}] = 0 \,. \tag{3.31}$$

If we wish to treat a tensor with equal number of upper and lower indices as a matrix $\mathbf{M} : V^p \otimes \bar{V}^q \to V^p \otimes \bar{V}^q$,

$$M_\alpha{}^\beta = M^{a_1 a_2 \ldots a_q}_{b_1 \ldots b_p} {}_{, c_q \ldots c_2 c_1}^{\; d_p \ldots d_1} \,, \tag{3.32}$$

then the invariance condition (3.29) will take the commutator form (3.31). Our convention of separating the two sets of indices by a comma, and reversing the order of the indices to the right of the comma, is motivated by the diagrammatic notation introduced below (see (4.6)).

Definition. We shall refer to an invariant relation between p vectors in V and q vectors in $\bar{V}$, which can be written as a homogeneous polynomial in terms of vector components, such as

$$h(x, y, \bar{z}, \bar{r}, \bar{s}) = h^{ab}{}_{cde} x_b y_a s^e r^d z^c \,, \tag{3.33}$$

as an *invariant* in $V^q \otimes \bar{V}^p$ (repeated indices, as always, summed over). In this example, the coefficients $h^{ab}{}_{cde}$ are components of invariant tensor $h \in V^3 \otimes \bar{V}^2$, obeying the invariance condition (3.28).

Diagrammatic representation of tensors, such as

$$h^{ab}{}_{cde} = \tag{3.34}$$

a b c d e

makes it easier to distinguish different types of invariant tensors. We shall explain in great detail our conventions for drawing tensors in section 4.1; sketching a few simple examples should suffice for the time being.

The standard example of a defining vector space is our 3-dimensional Euclidean space: $V = \bar{V}$ is the space of all 3-component real vectors ($n = 3$), and examples of invariants are the length $L(x, x) = \delta_{ij} x_i x_j$ and the volume $V(x, y, z) = \epsilon_{ijk} x_i y_j z_k$. We draw the corresponding invariant tensors as

$$\delta_{ij} = i \;\text{———}\; j \,, \qquad \epsilon_{ijk} = \;. \tag{3.35}$$

i j k

Definition. A *composed* invariant tensor can be written as a product and/or contraction of invariant tensors.

Examples of composed invariant tensors are

$$\delta_{ij}\epsilon_{klm} = \,, \qquad \epsilon_{ijm}\delta_{mn}\epsilon_{nkl} = \;. \tag{3.36}$$

i; j k l m; m n; i j k l

The first example corresponds to a product of the two invariants $L(x,y)V(z,r,s)$. The second involves an index *contraction*; we can write this as $V(x,y,\frac{d}{dz})V(z,r,s)$.

In order to proceed, we need to distinguish the "primitive" invariant tensors from the infinity of composed invariants. We begin by defining a finite basis for invariant tensors in $V^p \otimes \bar{V}^q$:

Definition. A *tree invariant* can be represented diagrammatically as a product of invariant tensors involving no loops of index contractions. We shall denote by $T = \{\mathbf{t}_0, \mathbf{t}_1 \ldots \mathbf{t}_{r-1}\}$ a (maximal) set of r linearly independent tree invariants $\mathbf{t}_\alpha \in V^p \otimes \bar{V}^q$. As any linear combination of $\mathbf{t}_\alpha$ can serve as a basis, we clearly have a great deal of freedom in making informed choices for the basis tensors.

Example: Tensors (3.36) are tree invariants. The tensor

$$h_{ijkl} = \epsilon_{ims}\epsilon_{jnm}\epsilon_{krn}\epsilon_{\ell sr} =$$ 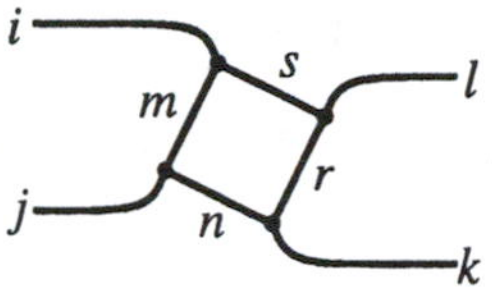 , (3.37)

with intermediate indices m, n, r, s summed over, is not a tree invariant, as it involves a loop.

Definition. An invariant tensor is called a *primitive* invariant tensor if it cannot be expressed as a linear combination of tree invariants composed from lower-rank primitive invariant tensors. Let $\mathcal{P} = \{\mathbf{p}_1, \mathbf{p}_2, \ldots \mathbf{p}_k\}$ be the set of all primitives.

For example, the Kronecker delta and the Levi-Civita tensor (3.35) are the primitive invariant tensors of our 3-dimensional space. The loop contraction (3.37) is not a primitive, because by the Levi-Civita completeness relation (6.28) it reduces to a sum of tree contractions:

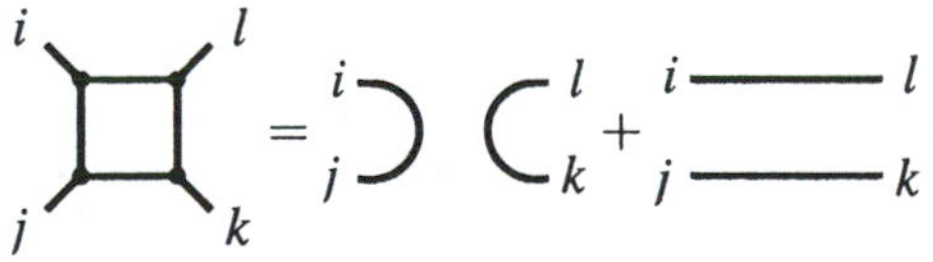 $= \delta_{ij}\delta_{kl} + \delta_{il}\delta_{jk}$, (3.38)

(The Levi-Civita tensor is discussed in section 6.3.)

Primitiveness assumption. Any invariant tensor $h \in V^p \otimes \bar{V}^q$ can be expressed as a linear sum over the tree invariants $T \subset V^q \otimes \bar{V}^p$:

$$h = \sum_{\alpha \in T} h^\alpha \mathbf{t}_\alpha \,. \tag{3.39}$$

In contradistinction to arbitrary composite invariant tensors, the number of tree invariants for a fixed number of external indices is finite. For example, given bilinear and trilinear primitives $\mathcal{P} = \{\delta_{ij}, f_{ijk}\}$, any invariant tensor $h \in V^p$ (here denoted by a blob) must be expressible as

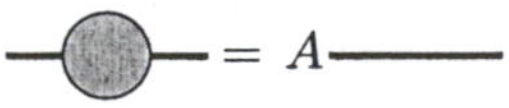 , $(p = 2)$ (3.40)

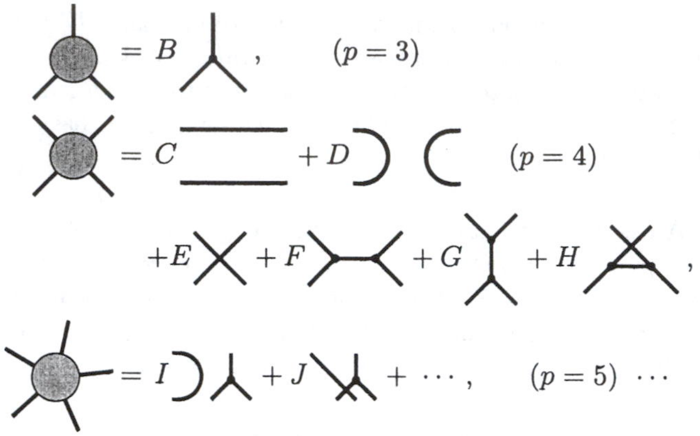

$= B$, $(p = 3)$

$= C$ $+ D$ $(p = 4)$

$+E$ $+F$ $+G$ $+H$,

$= I$ $+ J$ $+ \cdots,$ $(p = 5)$ $\cdots$ (3.41)

3.3.1 Algebra of invariants

Any invariant tensor of matrix form (3.32)

$$M_\alpha{}^\beta = M^{a_1 a_2 \ldots a_q}_{b_1 \ldots b_p}{}_{, c_q \ldots c_2 c_1}^{\;d_p \ldots d_1}$$

that maps $V^q \otimes \bar{V}^p \to V^q \otimes \bar{V}^p$ can be expanded in the basis (3.39). In this case the basis tensors $\mathbf{t}_\alpha$ are themselves matrices in $V^q \otimes \bar{V}^p \to V^q \otimes \bar{V}^p$, and the matrix product of two basis elements is also an element of $V^q \otimes \bar{V}^p \to V^q \otimes \bar{V}^p$ and can be expanded in an r element basis:

$$\mathbf{t}_\alpha \mathbf{t}_\beta = \sum_{\mathbf{t} \in T} (\tau_\alpha)_\beta{}^\gamma \mathbf{t}_\gamma \,. \tag{3.42}$$

As the number of tree invariants composed from the primitives is finite, under matrix multiplication the bases $\mathbf{t}_\alpha$ form a finite r-dimensional algebra, with the coefficients $(\tau_\alpha)_\beta{}^\gamma$ giving their multiplication table. As in (3.7), the structure constants $(\tau_\alpha)_\beta{}^\gamma$ form a $[r \times r]$-dimensional matrix rep of $\mathbf{t}_\alpha$ acting on the vector $(\mathbf{e}, \mathbf{t}_1, \mathbf{t}_2, \cdots \mathbf{t}_{r-1})$. Given a basis, we can evaluate the matrices $\mathbf{e}_\beta{}^\gamma$, $(\tau_1)_\beta{}^\gamma$, $(\tau_2)_\beta{}^\gamma, \cdots (\tau_{r-1})_\beta{}^\gamma$ and their eigenvalues. For at least one of combinations of these matrices all eigenvalues will be distinct (or we have failed to choose a good basis). The projection operator technique of section 3.5 will enable us to exploit this fact to decompose the $V^q \otimes \bar{V}^p$ space into r irreducible subspaces.

This can be said in another way; the choice of basis $\{\mathbf{e}, \mathbf{t}_1, \mathbf{t}_2 \cdots \mathbf{t}_{r-1}\}$ is arbitrary, the only requirement being that the basis elements are linearly independent. Finding a $(\tau_\alpha)_\beta{}^\gamma$ with all eigenvalues distinct is all we need to construct an orthogonal basis $\{\mathbf{P}_0, \mathbf{P}_1, \mathbf{P}_2, \cdots \mathbf{P}_{r-1}\}$, where the basis matrices $\mathbf{P}_i$ are the projection operators, to be constructed below in section 3.5. For an application of this algebra, see section 9.11.

3.4 INVARIANCE GROUPS

So far we have defined invariant tensors as the tensors invariant under transformations of a given group. Now we proceed in reverse: given a set of tensors, what is the group of transformations that leaves them invariant?

Given a full set of primitives (3.33), $\mathcal{P} = \{\mathbf{p}_1, \mathbf{p}_2, \ldots, \mathbf{p}_k\}$, meaning that *no other* primitives exist, we wish to determine all possible transformations that preserve this given set of invariant relations.

Definition. An *invariance group* $\mathcal{G}$ is the set of all linear transformations (3.28) that preserve the primitive invariant relations (and, by extension, *all* invariant relations)

$$\begin{aligned} p_1(x, \bar{y}) &= p_1(Gx, \bar{y}G^\dagger) \\ p_2(x, y, z, \ldots) &= p_2(Gx, Gy, Gz \ldots) \,, \quad \ldots . \end{aligned} \tag{3.43}$$

Unitarity (3.13) guarantees that all contractions of primitive invariant tensors, and hence all composed tensors $h \in H$, are also invariant under action of $\mathcal{G}$. As we assume unitary $\mathcal{G}$, it follows from (3.13) that the list of primitives must always include the Kronecker delta.

Example 1. If $p^a q_a$ is the only invariant of $\mathcal{G}$

$$p'^a q'_a = p^b (G^\dagger G)_b{}^c q_c = p^a q_a \,, \tag{3.44}$$

then $\mathcal{G}$ is the full *unitary group* $U(n)$ (invariance group of the complex norm $|x|^2 = x^b x_a \delta^a_b$), whose elements satisfy

$$G^\dagger G = 1 \,. \tag{3.45}$$

Example 2. If we wish the z-direction to be invariant in our 3-dimensional space, $q = (0, 0, 1)$ is an invariant vector (3.27), and the invariance group is $O(2)$, the group of all rotations in the x-y plane.

Which rep is "defining"?

1. The defining space V *need not* carry the lowest-dimensional rep of $\mathcal{G}$; it is merely the space in terms of which we chose to define the primitive invariants.

2. We shall always assume that the Kronecker delta δ^b_a is one of the primitive invariants, *i.e.*, that $\mathcal{G}$ is a unitary group whose elements satisfy (3.45). This restriction to unitary transformations is not essential, but it simplifies proofs of full reducibility. The results, however, apply as well to the finite-dimensional reps of noncompact groups, such as the Lorentz group $SO(3, 1)$.

3.5 PROJECTION OPERATORS

For $\mathbf{M}$, a hermitian matrix, there exists a diagonalizing unitary matrix C such that

$$
C\mathbf{M}C^{\dagger} = \left(\begin{array}{ccc} \begin{array}{ccc} \lambda_1 & \dots & 0 \\ & \ddots & \\ 0 & \dots & \lambda_1 \end{array} & 0 & 0 \\ 0 & \boxed{\begin{array}{cccc} \lambda_2 & 0 & \dots & 0 \\ 0 & \lambda_2 & & \\ \vdots & & \ddots & \vdots \\ 0 & & \dots & \lambda_2 \end{array}} & 0 \\ 0 & 0 & \begin{array}{cc} \lambda_3 & \dots \\ \vdots & \ddots \end{array} \end{array} \right) . \tag{3.46}
$$

Here $\lambda_i \neq \lambda_j$ are the r distinct roots of the minimal characteristic polynomial

$$
\prod_{i=1}^{r} (\mathbf{M} - \lambda_i \mathbf{1}) = 0 \tag{3.47}
$$

(the characteristic equations will be discussed in section 6.6).

In the matrix $C(\mathbf{M} - \lambda_2 \mathbf{1})C^{\dagger}$ the eigenvalues corresponding to λ_2 are replaced by zeroes:

$$
\left(\begin{array}{ccc} \begin{array}{ccc} \lambda_1 - \lambda_2 & & \\ & \lambda_1 - \lambda_2 & \\ & & \lambda_1 - \lambda_2 \end{array} & & \\ & \boxed{\begin{array}{cc} 0 & \\ & \ddots \\ & & 0 \end{array}} & \\ & & \begin{array}{ccc} \lambda_3 - \lambda_2 & & \\ & \lambda_3 - \lambda_2 & \\ & & \ddots \end{array} \end{array} \right) ,
$$

and so on, so the product over all factors $(\mathbf{M} - \lambda_2 \mathbf{1})(\mathbf{M} - \lambda_3 \mathbf{1}) \dots$, with exception of the $(\mathbf{M} - \lambda_1 \mathbf{1})$ factor, has nonzero entries only in the subspace associated with λ_1:

$$
C \prod_{j \neq 1} (\mathbf{M} - \lambda_j \mathbf{1}) C^{\dagger} = \prod_{j \neq 1} (\lambda_1 - \lambda_j) \left(\begin{array}{cc} \begin{array}{ccc} 1 & 0 & 0 \\ 0 & 1 & 0 \\ 0 & 0 & 1 \end{array} & 0 \\ 0 & \begin{array}{cccc} 0 & & & \\ & 0 & & \\ & & 0 & \\ & & & \ddots \end{array} \end{array} \right) .
$$

In this way, we can associate with each distinct root λ_i a *projection operator* $\mathbf{P}_i$,

$$
\mathbf{P}_i = \prod_{j \neq i} \frac{\mathbf{M} - \lambda_j \mathbf{1}}{\lambda_i - \lambda_j} , \tag{3.48}
$$

which acts as identity on the ith subspace, and zero elsewhere. For example, the projection operator onto the λ_1 subspace is

$$\mathbf{P}_1 = C^\dagger \left(\begin{array}{ccc|cccc} 1 & & & & & & \\ & \ddots & & & & & \\ & & 1 & & & & \\ \hline & & & 0 & & & \\ & & & & 0 & & \\ & & & & & \ddots & \\ & & & & & & 0 \end{array} \right) C \,. \tag{3.49}$$

The matrices $\mathbf{P}_i$ are *orthogonal*

$$\mathbf{P}_i \mathbf{P}_j = \delta_{ij} \mathbf{P}_j \,, \qquad (\text{no sum on } j) \,, \tag{3.50}$$

and satisfy the *completeness relation*

$$\sum_{i=1}^{r} \mathbf{P}_i = \mathbf{1} \,. \tag{3.51}$$

As $\mathrm{tr}(C\mathbf{P}_i C^\dagger) = \mathrm{tr}\,\mathbf{P}_i$, the dimension of the ith subspace is given by

$$d_i = \mathrm{tr}\,\mathbf{P}_i \,. \tag{3.52}$$

It follows from the characteristic equation (3.47) and the form of the projection operator (3.48) that λ_i is the eigenvalue of $\mathbf{M}$ on $\mathbf{P}_i$ subspace:

$$\mathbf{M}\mathbf{P}_i = \lambda_i \mathbf{P}_i \,, \qquad (\text{no sum on } i) \,. \tag{3.53}$$

Hence, any matrix polynomial $f(\mathbf{M})$ takes the scalar value $f(\lambda_i)$ on the $\mathbf{P}_i$ subspace

$$f(\mathbf{M})\mathbf{P}_i = f(\lambda_i)\mathbf{P}_i \,. \tag{3.54}$$

This, of course, is the reason why one wants to work with irreducible reps: they reduce matrices and "operators" to pure numbers.

3.6 SPECTRAL DECOMPOSITION

Suppose there exist several linearly independent invariant $[d \times d]$ hermitian matrices $\mathbf{M}_1, \mathbf{M}_2, \ldots$, and that we have used $\mathbf{M}_1$ to decompose the d-dimensional vector space $\tilde{V} = \Sigma \oplus V_i$. Can $\mathbf{M}_2, \mathbf{M}_3, \ldots$ be used to further decompose V_i? This is a standard problem of quantum mechanics (simultaneous observables), and the answer is that further decomposition is possible if, and only if, the invariant matrices commute:

$$[\mathbf{M}_1, \mathbf{M}_2] = 0 \,, \tag{3.55}$$

or, equivalently, if projection operators $\mathbf{P}_j$ constructed from $\mathbf{M}_2$ commute with projection operators $\mathbf{P}_i$ constructed from $\mathbf{M}_1$,

$$\mathbf{P}_i \mathbf{P}_j = \mathbf{P}_j \mathbf{P}_i \,. \tag{3.56}$$

Usually the simplest choices of independent invariant matrices do not commute. In that case, the projection operators $\mathbf{P}_i$ constructed from $\mathbf{M}_1$ can be used to project commuting pieces of $\mathbf{M}_2$:

$$\mathbf{M}_2^{(i)} = \mathbf{P}_i \mathbf{M}_2 \mathbf{P}_i \,, \qquad \text{(no sum on } i) \,.$$

That $\mathbf{M}_2^{(i)}$ commutes with $\mathbf{M}_1$ follows from the orthogonality of $\mathbf{P}_i$:

$$[\mathbf{M}_2^{(i)}, \mathbf{M}_1] = \sum_j \lambda_j [\mathbf{M}_2^{(i)}, \mathbf{P}_j] = 0 \,. \tag{3.57}$$

Now the characteristic equation for $\mathbf{M}_2^{(i)}$ (if nontrivial) can be used to decompose V_i subspace.

An invariant matrix $\mathbf{M}$ induces a decomposition only if its diagonalized form (3.46) has more than one distinct eigenvalue; otherwise it is proportional to the unit matrix and commutes trivially with all group elements. A rep is said to be *irreducible* if all invariant matrices that can be constructed are proportional to the unit matrix.

In particular, the primitiveness relation (3.40) is a statement that the defining rep is *assumed* irreducible.

An invariant matrix $\mathbf{M}$ commutes with group transformations $[G, \mathbf{M}] = 0$, see (3.31). Projection operators (3.48) constructed from $\mathbf{M}$ are polynomials in $\mathbf{M}$, so they also commute with all $g \in \mathcal{G}$:

$$[G, \mathbf{P}_i] = 0 \tag{3.58}$$

(remember that $\mathbf{P}_i$ are also invariant $[d \times d]$ matrices). Hence, a $[d \times d]$ matrix rep can be written as a direct sum of $[d_i \times d_i]$ matrix reps:

$$G = \mathbf{1} G \mathbf{1} = \sum_{i,j} \mathbf{P}_i G \mathbf{P}_j = \sum_i \mathbf{P}_i G \mathbf{P}_i = \sum_i G_i \,. \tag{3.59}$$

In the diagonalized rep (3.49), the matrix G has a block diagonal form:

$$CGC^\dagger = \begin{bmatrix} G_1 & 0 & 0 \\ 0 & G_2 & 0 \\ 0 & 0 & \ddots \end{bmatrix} \,, \qquad G = \sum_i C^i G_i C_i \,. \tag{3.60}$$

The rep G_i acts only on the d_i-dimensional subspace V_i consisting of vectors $\mathbf{P}_i q$, $q \in \tilde{V}$. In this way an invariant $[d \times d]$ hermitian matrix $\mathbf{M}$ with r distinct eigenvalues induces a decomposition of a d-dimensional vector space $\tilde{V}$ into a direct sum of d_i-dimensional vector subspaces V_i:

$$\tilde{V} \overset{\mathbf{M}}{\rightarrow} V_1 \oplus V_2 \oplus \ldots \oplus V_r \,. \tag{3.61}$$

For a discussion of recursive reduction, consult appendix A. The theory of class algebras [155, 156, 157] offers a more elegant and systematic way of constructing the maximal set of commuting invariant matrices $\mathbf{M}_i$ than the sketch offered in this section.

Chapter Four

Diagrammatic notation

Some aspects of the representation theory of Lie groups are the subject of this monograph. However, it is not written in the conventional tensor notation but instead in terms of an equivalent diagrammatic notation. We shall refer to this style of carrying out group-theoretic calculations as *birdtracks* (and so do reputable journals [51]). The advantage of diagrammatic notation will become self-evident, I hope. Two of the principal benefits are that it eliminates "dummy indices," and that it does not force group-theoretic expressions into the 1-dimensional tensor format (both being means whereby identical tensor expressions can be made to look totally different). In contradistinction to some of the existing literature in this manuscript I strive to keep the diagrammatic notation as simple and elegant as possible.

4.1 BIRDTRACKS

We shall often find it convenient to represent agglomerations of invariant tensors by birdtracks, a group-theoretical version of Feynman diagrams. Tensors will be represented by *vertices* and contractions by *propagators*.

Diagrammatic notation has several advantages over the tensor notation. Diagrams do not require dummy indices, so explicit labeling of such indices is unnecessary. More to the point, for a human eye it is easier to identify topologically identical diagrams than to recognize equivalence between the corresponding tensor expressions.

If readers find birdtrack notation abhorrent, they can surely derive all results of this monograph in more conventional algebraic notations. To give them a sense of how that goes, we have covered our tracks by algebra in the derivation of the E_7 family, chapter 20, where not a single birdtrack is drawn. It it is like speaking Italian without moving hands, if you are into that kind of thing.

In the birdtrack notation, the Kronecker delta is a propagator:

$$\delta_b^a = b \longleftarrow a\,. \tag{4.1}$$

For a *real* defining space there is no distinction between V and $\bar{V}$, or up and down indices, and the lines do not carry arrows.

Any invariant tensor can be drawn as a generalized vertex:

$$X_\alpha = X_{de}^{abc} = \begin{array}{c} d \leftarrow \\ e \leftarrow \\ a \rightarrow \\ b \rightarrow \\ c \rightarrow \end{array}\!\boxed{\;X\;}\,. \tag{4.2}$$

Whether the vertex is drawn as a box or a circle or a dot is a matter of taste. The orientation of propagators and vertices in the plane of the drawing is likewise irrelevant. The only rules are as follows:

1. Arrows point *away from the upper* indices and *toward the lower* indices; the line flow is "downward," from upper to lower indices:

$$h_{ab}^{cd} = \qquad . \tag{4.3}$$

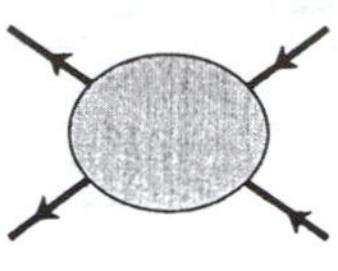

2. Diagrammatic notation must indicate which in (out) arrow corresponds to the first upper (lower) index of the tensor (unless the tensor is cyclically symmetric);

$$R^e_{abcd} = \qquad . \tag{4.4}$$

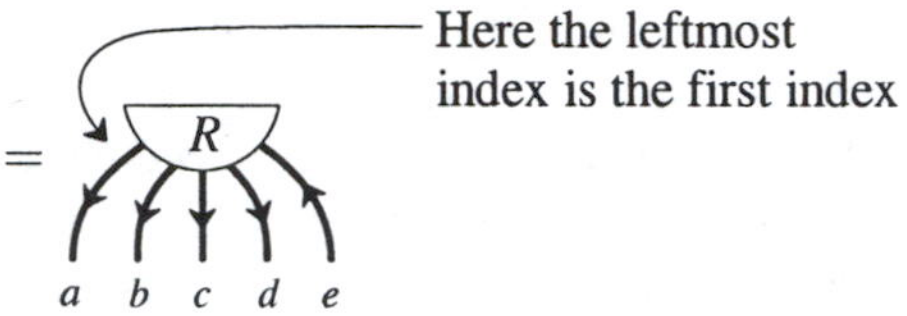

3. The indices are read in the *counterclockwise* order around the vertex:

$$X_{ad}^{bce} = \qquad . \tag{4.5}$$

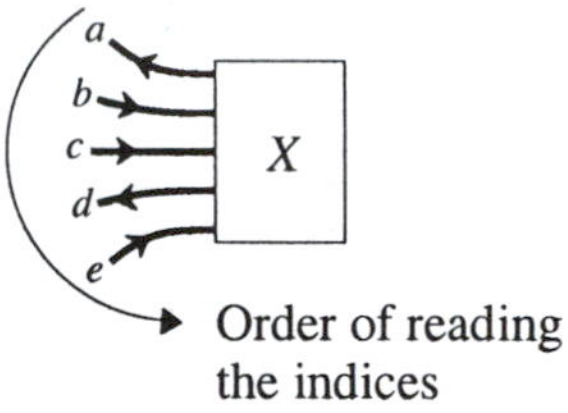

(The upper and the lower indices are read separately in the counterclockwise order; their relative ordering does not matter.)

In the examples of this section we index the external lines for the reader's convenience, but indices can always be omitted. An internal line implies a summation over corresponding indices, and for external lines the equivalent points on each diagram represent the same index in all terms of a diagrammatic equation.

Hermitian conjugation (3.21) does two things:

1. It exchanges the upper and the lower indices, *i.e.*, it reverses the directions of the arrows.

2. It reverses the order of the indices, *i.e.*, it transposes a diagram into its mirror image. For example, $X^\dagger$, the tensor conjugate to (4.5), is drawn as

$$X^\alpha = X^{ed}_{cba} = \qquad , \tag{4.6}$$

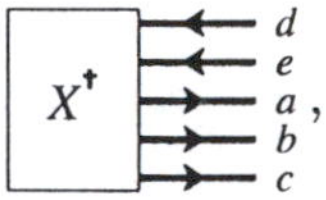

and a contraction of tensors $X^\dagger$ and Y is drawn as

$$X^\alpha Y_\alpha = X^{b_p \ldots b_1}_{a_q \ldots a_2 a_1} Y^{a_1 a_2 \ldots a_q}_{b_1 \ldots b_p} = \qquad . \tag{4.7}$$

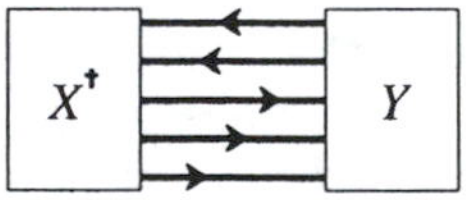

In sections. 3.1–3.2 and here we define the hermitian conjugation and (3.32) matrices $\mathbf{M} : V^p \otimes \bar{V}^q \to V^p \otimes \bar{V}^q$ in the multi-index notation

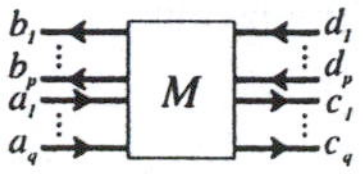

$$\text{[birdtrack diagram: box } M \text{ with lines } b_1 \ldots b_p, a_1 \ldots a_q \text{ on the left and } d_1 \ldots d_p, c_1 \ldots c_q \text{ on the right]} \tag{4.8}$$

in such a way that the matrix multiplication

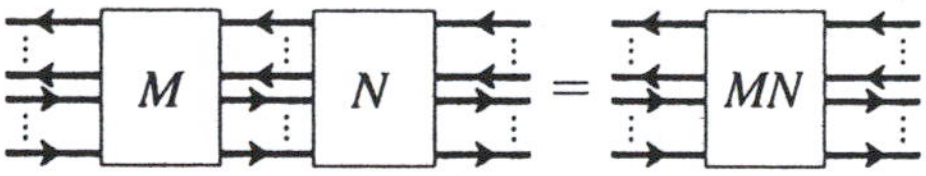

$$\text{[diagram: } M \text{ box connected to } N \text{ box]} = \text{[diagram: } MN \text{ box]} \tag{4.9}$$

and the trace of a matrix

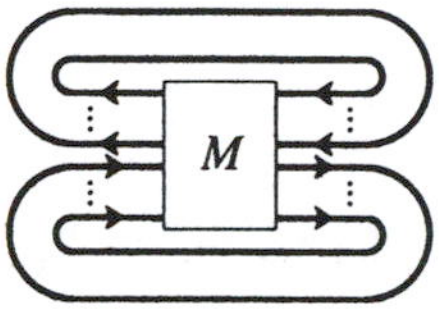

$$\text{[diagram: box } M \text{ with all outgoing lines looped back to incoming lines]} \tag{4.10}$$

can be drawn in the plane. Notation in which all internal lines are maximally crossed at each multiplication [318] is equally correct, but less pleasing to the eye.

4.2 CLEBSCH-GORDAN COEFFICIENTS

Consider the product

$$\begin{pmatrix} \begin{array}{cc|} 0 & \\ & 0 \\ \hline \end{array} & & \\ & \boxed{\begin{array}{ccc} 1 & & \\ & 1 & \\ & & 1 \end{array}} & \\ & & \begin{array}{|cccc} \hline 0 & & & \\ & 0 & & \\ & & 0 & \\ & & & \ddots \end{array} \end{pmatrix} C \tag{4.11}$$

of the two terms in the diagonal representation of a projection operator (3.49). This matrix has nonzero entries only in the d_λ rows of subspace V_λ. We collect them in a $[d_\lambda \times d]$ rectangular matrix $(C_\lambda)^\alpha_\sigma$, $\alpha = 1, 2, \ldots d$, $\sigma = 1, 2, \ldots d_\lambda$:

$$C_\lambda = \underbrace{\begin{pmatrix} (C_\lambda)^1_1 & \cdots & (C_\lambda)^d_1 \\ \vdots & & \vdots \\ & & (C_\lambda)^d_{d_\lambda} \end{pmatrix}}_{d} \Bigg\} d_\lambda \,. \tag{4.12}$$

The index α in $(C_\lambda)^\alpha_\sigma$ stands for all tensor indices associated with the $d = n^{p+q}$-dimensional tensor space $V^p \otimes \bar{V}^q$. In the birdtrack notation these indices are explicit:

$$(C_\lambda)_{\sigma,}{}^{b_p \ldots b_1}_{a_q \ldots a_2 a_1} = \text{[diagram: line } \lambda \text{ entering a triangle with outgoing lines } b_1 \ldots a_q] \,. \tag{4.13}$$

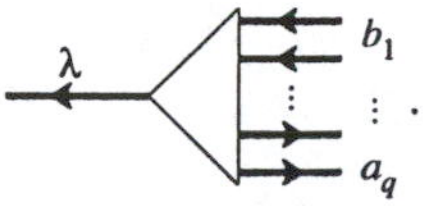

Such rectangular arrays are called *Clebsch-Gordan coefficients* (hereafter referred to as *clebsches* for short). They are explicit mappings $V \to V_\lambda$. The conjugate mapping $V_\lambda \to \bar{V}$ is provided by the product

$$C^\dagger \begin{pmatrix} \begin{matrix} 0 & \\ & 0 \end{matrix} & & \\ & \begin{matrix} 1 & & \\ & 1 & \\ & & 1 \end{matrix} & \\ & & \begin{matrix} 0 & & & \\ & 0 & & \\ & & 0 & \\ & & & \ddots \end{matrix} \end{pmatrix}, \tag{4.14}$$

which defines the $[d \times d_\lambda]$ rectangular matrix $(C^\lambda)_\alpha^\sigma, \alpha = 1, 2, \ldots d, \sigma = 1, 2, \ldots d_\lambda$:

$$C^\lambda = \underbrace{\begin{pmatrix} (C^\lambda)_1^1 & \ldots & (C^\lambda)_1^{d_\lambda} \\ \vdots & & \vdots \\ & & (C^\lambda)_d^{d_\lambda} \end{pmatrix}}_{d_\lambda} \Bigg\} d$$

$$(C^\lambda)_{b_1 \ldots b_p}^{a_1 a_2 \ldots a_q}{}^{,\,\sigma} = \tag{4.15}$$

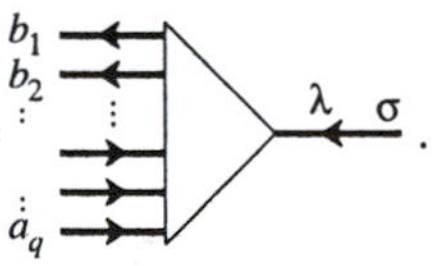

The two rectangular Clebsch-Gordan matrices C^λ and C_λ are related by hermitian conjugation.

The tensors, which we have considered in section 3.10, transform as tensor products of the defining rep (3.14). In general, tensors transform as tensor products of various reps, with indices running over the corresponding rep dimensions:

$$x_{a_1 a_2 \ldots a_p}^{a_{p+1} \ldots a_{p+q}} \qquad \text{where} \qquad \begin{aligned} a_1 &= 1, 2, \ldots, d_1 \\ a_2 &= 1, 2, \ldots, d_2 \\ &\vdots \\ a_{p+q} &= 1, 2, \ldots, d_{p+q}\,. \end{aligned} \tag{4.16}$$

The action of the transformation g on the index a_k is given by the $[d_k \times d_k]$ matrix rep G_k.

Clebsches are notoriously index overpopulated, as they require a rep label and a tensor index for each rep in the tensor product. Diagrammatic notation alleviates this index plague in either of two ways:

1. One can indicate a rep label on each line:

$$C_{a_\lambda}^{a_\mu a_\nu}{}^{,\,a_\sigma} = \tag{4.17}$$

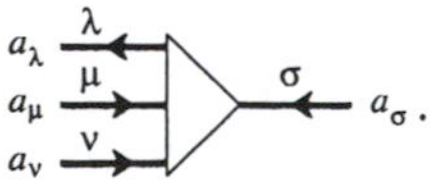

(An index, if written, is written at the end of a line; a rep label is written above the line.)

2. One can draw the propagators (Kronecker deltas) for different reps with different kinds of lines. For example, we shall usually draw the adjoint rep with a thin line.

By the definition of clebsches (3.49), the λ rep projection operator can be written out in terms of Clebsch-Gordan matrices $C^\lambda C_\lambda$:

$$C^\lambda C_\lambda = \mathbf{P}_\lambda\,, \qquad \text{(no sum on } i\text{)}$$

$$(C^\lambda)_{b_1\ldots b_q}^{a_1a_2\ldots a_p},{}^{\alpha}(C_\lambda)_{\alpha},{}_{c_p\ldots c_2c_1}^{d_q\ldots d_1} = (\mathbf{P}_\lambda)_{b_1\ldots b_q}^{a_1a_2\ldots d_p},{}_{c_p\ldots c_2c_1}^{d_q\ldots d_1} \tag{4.18}$$

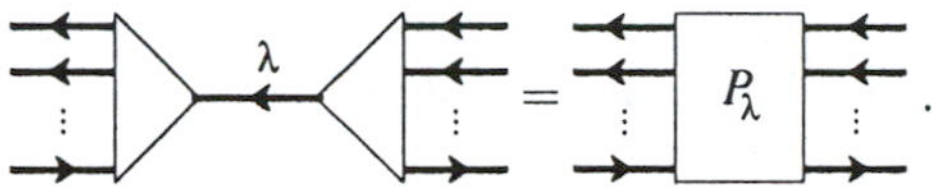

A specific choice of clebsches is quite arbitrary. All relevant properties of projection operators (orthogonality, completeness, dimensionality) are independent of the explicit form of the diagonalization transformation C. Any set of C_λ is acceptable as long as it satisfies the orthogonality and completeness conditions. From (4.11) and (4.14) it follows that C_λ are *orthonormal*:

$$C_\lambda C^\mu = \delta_\lambda^\mu \mathbf{1}\,,$$

$$(C_\lambda)_{\beta},{}_{b_1\ldots b_q}^{a_1a_2\ldots a_p}(C^\mu)_{a_p\ldots a_2a_1}^{b_q\ldots b_1},{}^{\alpha} = \delta_\beta^\alpha \delta_\lambda^\mu \tag{4.19}$$

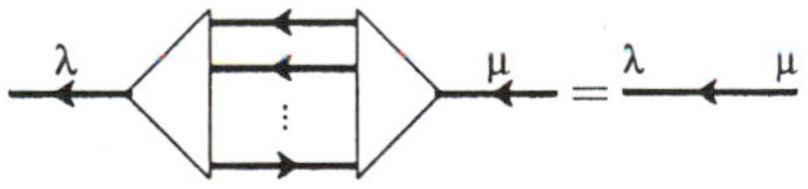

Here $\mathbf{1}$ is the $[d_\lambda \times d_\lambda]$ unit matrix, and C_λ's are multiplied as $[d_\lambda \times d]$ rectangular matrices.

The *completeness relation* (3.51)

$$\sum_\lambda C^\lambda C_\lambda = \mathbf{1}\,, \qquad ([d \times d] \text{ unit matrix})\,,$$

$$\sum_\lambda (C^\lambda)_{b_1\ldots b_q}^{a_1a_2\ldots a_p},{}^{\alpha}(C_\lambda)_{\alpha},{}_{c_p\ldots c_2c_1}^{d_q\ldots d_1} = \delta_{c_1}^{a_1}\delta_{c_2}^{a_2}\ldots\delta_{b_q}^{d_q} \tag{4.20}$$

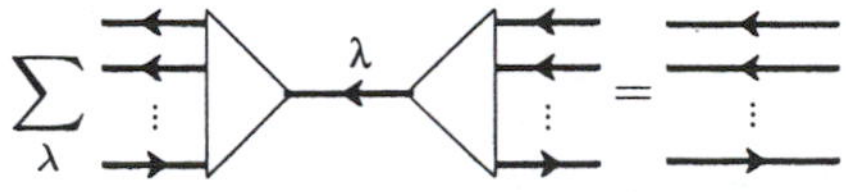

$$C^\lambda \mathbf{P}_\mu = \delta_\lambda^\mu C^\lambda\,,$$

$$\mathbf{P}_\lambda C^\mu = \delta_\lambda^\mu C^\mu\,, \qquad \text{(no sum on } \lambda, \mu)\,, \tag{4.21}$$

follows immediately from (3.50) and (4.19).

4.3 ZERO- AND ONE-DIMENSIONAL SUBSPACES

If a projection operator projects onto a zero-dimensional subspace, it must vanish identically:

$$d_\lambda = 0 \quad \Rightarrow \quad \mathbf{P}_\lambda = \quad \lambda \quad = 0\,. \tag{4.22}$$

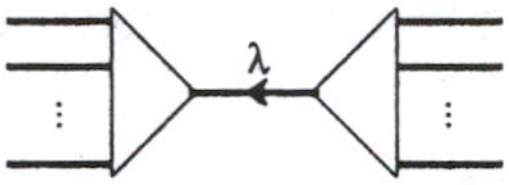

This follows from (3.49); d_λ is the number of 1's on the diagonal on the right-hand side. For $d_\lambda = 0$ the right-hand side vanishes. The general form of $\mathbf{P}_\lambda$ is

$$\mathbf{P}_\lambda = \sum_{k=1}^{r} c_k \mathbf{M}_k\,, \tag{4.23}$$

where $\mathbf{M}_k$ are the invariant matrices used in construction of the projector operators, and c_k are numerical coefficients. Vanishing of $\mathbf{P}_\lambda$ therefore implies a relation among invariant matrices $\mathbf{M}_k$.

If a projection operator projects onto a 1-dimensional subspace, its expression, in terms of the clebsches (4.18), involves no summation, so we can omit the intermediate line

$$d_\lambda = 1 \quad \Rightarrow \quad \mathbf{P}_\lambda = \quad = (C^\lambda)^{a_1 a_2 \ldots a_p}_{b_1 \ldots b_q} (C_\lambda)^{d_q \ldots d_1}_{c_p \ldots c_2 c_1}\,. \tag{4.24}$$

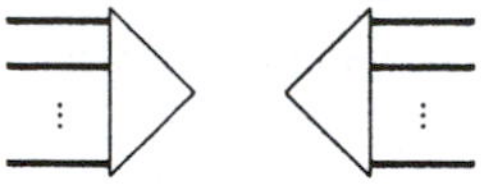

For any subgroup of $SU(n)$, the reps are unitary, with unit determinant. On the 1-dimensional spaces, the group acts trivially, $G = 1$. Hence, if $d_\lambda = 1$, the clebsch C_λ in (4.24) is an invariant tensor in $V^p \otimes \bar{V}^q$.

4.4 INFINITESIMAL TRANSFORMATIONS

A unitary transformation G infinitesimally close to unity can be written as

$$G_a{}^b = \delta_a^b + iD_a^b\,, \tag{4.25}$$

where D is a hermitian matrix with small elements, $|D_a^b| \ll 1$. The action of $g \in \mathcal{G}$ on the conjugate space is given by

$$(G^\dagger)_b{}^a = G^a{}_b = \delta_b^a - iD_b^a\,. \tag{4.26}$$

D can be parametrized by $N \leq n^2$ real parameters. N, the maximal number of independent parameters, is called the *dimension* of the group (also the dimension of the Lie algebra, or the dimension of the adjoint rep).

In this monograph we shall consider only infinitesimal transformations of form $G = 1 + iD$, $|D_b^a| \ll 1$. We do not study the entire group of invariances, but only the transformations (3.11) connected to the identity. For example, we shall not consider invariances under coordinate reflections.

The generators of infinitesimal transformations (4.25) are hermitian matrices and belong to the $D_b^a \in V \otimes \bar{V}$ space. However, not any element of $V \otimes \bar{V}$ generates an allowed transformation; indeed, one of the main objectives of group theory is to define the class of allowed transformations.

In section 3.5 we have described the decomposition of a tensor space into (ir)reducible subspaces. As a particular case, consider the decomposition of $V\otimes\bar{V}$. The corresponding projection operators satisfy the completeness relation (4.20):

$$\mathbf{1} = \frac{1}{n}T + \mathbf{P}_A + \sum_{\lambda\neq A}\mathbf{P}_\lambda$$

$$\delta^a_d\delta^c_b = \frac{1}{n}\delta^a_b\delta^c_d + (\mathbf{P}_A)^a_b,{}^c_d + \sum_{\lambda\neq A}(\mathbf{P}_\lambda)^a_b,{}^c_d$$

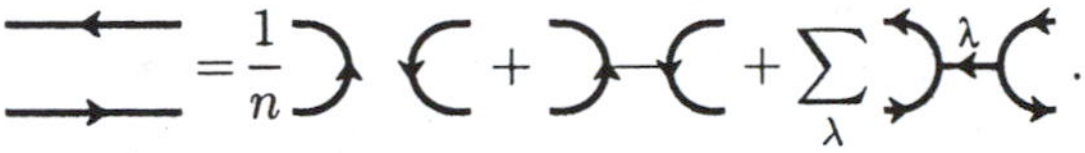

(4.27)

If δ^μ_λ is the only primitive invariant tensor, then $V\otimes\bar{V}$ decomposes into two subspaces, and there are no other irreducible reps. However, if there are further primitive invariant tensors, $V\otimes\bar{V}$ decomposes into more irreducible reps, indicated by the sum over λ. Examples will abound in what follows. The singlet projection operator T/n always figures in this expansion, as $\delta^a_b,{}^c_d$ is always one of the invariant matrices (see the example worked out in section 2.2). Furthermore, the infinitesimal generators D^a_b must belong to at least one of the irreducible subspaces of $V\otimes\bar{V}$.

This subspace is called the *adjoint* space, and its special role warrants introduction of special notation. We shall refer to this vector space by letter A, in distinction to the defining space V of (3.10). We shall denote its dimension by N, label its tensor indices by $i, j, k\ldots$, denote the corresponding Kronecker delta by a thin, straight line,

$$\delta_{ij} = i \;\text{———}\; j\,, \qquad i,j = 1,2,\ldots,N\,, \tag{4.28}$$

and the corresponding clebsches by

$$(C_A)_i,{}^a_b = \frac{1}{\sqrt{a}}(T_i)^a_b = $$ 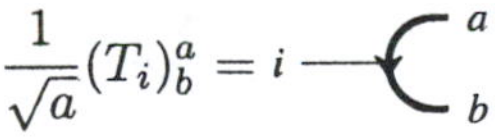 $$a,b=1,2,\ldots,n \qquad i=1,2,\ldots,N\,.$$

Matrices T_i are called the *generators* of infinitesimal transformations. Here a is an (uninteresting) overall normalization fixed by the orthogonality condition (4.19):

$$(T_i)^a_b(T_j)^b_a = \mathrm{tr}(T_iT_j) = a\,\delta_{ij}$$

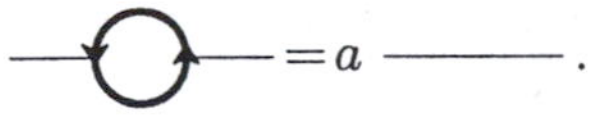

(4.29)

The scale of T_i is not set, as any overall rescaling can be absorbed into the normalization a. For our purposes it will be most convenient to use $a = 1$ as the normalization convention. Other normalizations are commonplace. For example, $SU(2)$ Pauli matrices $T_i = \frac{1}{2}\sigma_i$ and $SU(n)$ Gell-Mann [137] matrices $T_i = \frac{1}{2}\lambda_i$ are conventionally normalized by fixing $a = 1/2$:

$$\mathrm{tr}(T_iT_j) = \frac{1}{2}\delta_{ij}\,. \tag{4.30}$$

The projector relation (4.18) expresses the adjoint rep projection operators in terms of the generators:

$$(\mathbf{P}_A)_{b}^{a}{}_{,}{}^{c}_{d} = \frac{1}{a}(T_i)^a_b(T_i)^c_d = \frac{1}{a} \quad . \tag{4.31}$$

Clearly, the adjoint subspace is always included in the sum (4.27) (there must exist some allowed infinitesimal generators D^b_a, or otherwise there is no group to describe), but how do we determine the corresponding projection operator?

The adjoint projection operator is singled out by the requirement that the group transformations do not affect the invariant quantities. (Remember, the group is *defined* as the totality of all transformations that leave the invariants invariant.) For every invariant tensor q, the infinitesimal transformations $G = 1 + iD$ must satisfy the invariance condition (3.27). Parametrizing D as a projection of an arbitrary hermitian matrix $H \in V \otimes \bar{V}$ into the adjoint space, $D = \mathbf{P}_A H \in V \otimes \bar{V}$,

$$D^a_b = \frac{1}{a}(T_i)^a_b \epsilon_i \,, \qquad \epsilon_i = \frac{1}{a}\,\mathrm{tr}(T_i H)\,, \tag{4.32}$$

we obtain the *invariance condition*, which the *generators* must satisfy: they *annihilate* invariant tensors:

$$T_i q = 0\,. \tag{4.33}$$

To state the invariance condition for an arbitrary invariant tensor, we need to define the generators in the tensor reps. By substituting $G = 1 + i\epsilon \cdot T + O(\epsilon^2)$ into (3.15) and keeping only the terms linear in ϵ, we find that the generators of infinitesimal transformations for tensor reps act by touching one index at a time:

$$\begin{aligned}(T_i)^{a_1 a_2 \ldots a_p}_{b_1 \ldots b_q}{}_{,}{}^{d_q \ldots d_1}_{c_p \ldots c_2 c_1} &= (T_i)^{a_1}_{c_1} \delta^{a_2}_{c_2} \ldots \delta^{a_p}_{c_p} \delta^{d_1}_{b_1} \ldots \delta^{d_q}_{b_q} \\ &+ \delta^{a_1}_{c_1} (T_i)^{a_2}_{c_2} \ldots \delta^{a_p}_{c_p} \delta^{d_1}_{b_1} \ldots \delta^{d_q}_{b_q} + \ldots + \delta^{a_1}_{c_1} \delta^{a_2}_{c_2} \ldots (T_i)^{a_p}_{c_p} \delta^{d_1}_{b_1} \ldots \delta^{d_q}_{b_q} \\ &- \delta^{a_1}_{c_1} \delta^{a_2}_{c_2} \ldots \delta^{a_p}_{c_p} (T_i)^{d_1}_{b_1} \ldots \delta^{d_q}_{b_q} - \ldots - \delta^{a_1}_{c_1} \delta^{a_2}_{c_2} \ldots \delta^{a_p}_{c_p} \delta^{d_1}_{b_1} \ldots (T_i)^{d_q}_{b_q}\,.\end{aligned} \tag{4.34}$$

This forest of indices vanishes in the birdtrack notation, enabling us to visualize the formula for the generators of infinitesimal transformations for any tensor representation:

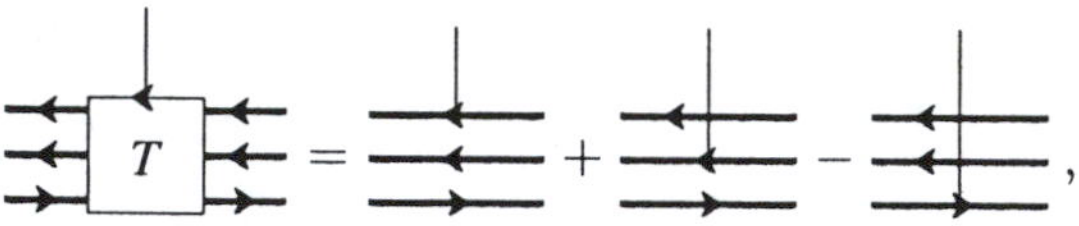

$$T \;=\; \;+\; \;-\; \,, \tag{4.35}$$

with a relative minus sign between lines flowing in opposite directions. The reader will recognize this as the Leibnitz rule.

Tensor reps of the generators decompose in the same way as the group reps (3.60):

$$T_i = \sum_\lambda C^\lambda T_i^{(\lambda)} C_\lambda$$

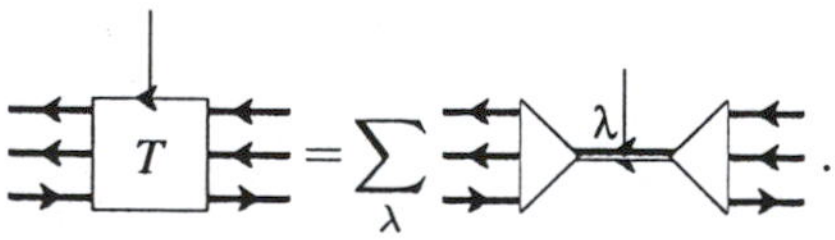

$$T = \sum_\lambda \quad \lambda \quad .$$

The invariance conditions take a particularly suggestive form in the diagrammatic notation. Equation (4.33) amounts to the insertion of a generator into all external legs of the diagram corresponding to the invariant tensor q:

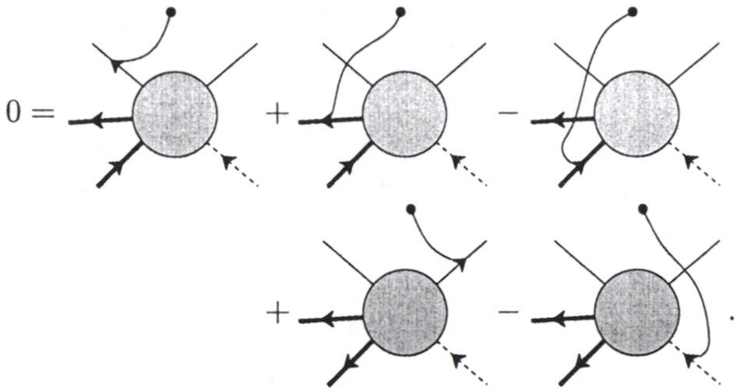

$$0 = \ldots \qquad (4.36)$$

The insertions on the lines going into the diagram carry a minus sign relative to the insertions on the outgoing lines.

Clebsches are themselves invariant tensors. Multiplying both sides of (3.60) with C_λ and using orthogonality (4.19), we obtain

$$C_\lambda G = G_\lambda C_\lambda\,, \qquad (\text{no sum on } \lambda)\,. \tag{4.37}$$

The Clebsch-Gordan matrix C_λ is a rectangular $[d_\lambda \times d]$ matrix, hence $g \in \mathcal{G}$ acts on it with a $[d_\lambda \times d_\lambda]$ rep from the left, and a $[d \times d]$ rep from the right. (3.48) is the statement of invariance for rectangular matrices, analogous to (3.30), the statement of invariance for square matrices:

$$\begin{aligned} C_\lambda &= G_\lambda^\dagger C_\lambda G\,, \\ C^\lambda &= G^\dagger C^\lambda G_\lambda\,. \end{aligned} \tag{4.38}$$

The invariance condition for the clebsches is a special case of (4.36), the invariance condition for any invariant tensor:

$$0 = -T_i^{(\lambda)} C_\lambda + C_\lambda T_i$$

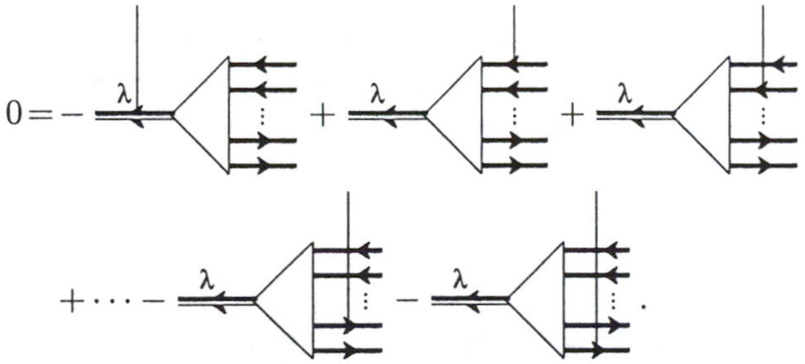

$$0 = - \ldots + \ldots + \ldots + \cdots - \ldots - \ldots\,. \tag{4.39}$$

The orthogonality condition (4.19) now yields the generators in λ rep in terms of the defining rep generators:

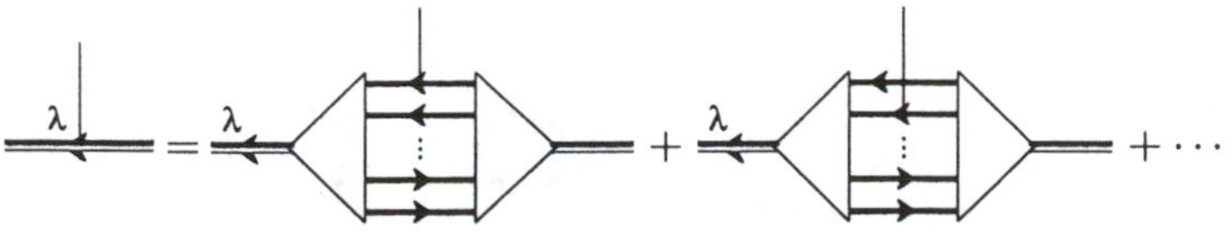

$$\ldots = \ldots + \ldots + \cdots$$

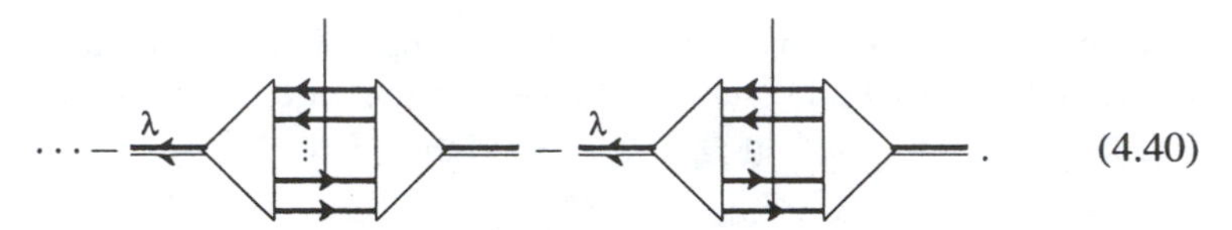

(4.40)

The reality of the adjoint rep. For hermitian generators, the adjoint rep is real, and the upper and lower indices need not be distinguished; the "propagator" needs no arrow. For nonhermitian choices of generators, the adjoint rep is complex ("gluon" lines carry arrows), but A and $\bar{A}$ are equivalent, as indices can be raised and lowered by the Cartan-Killing form,

$$g_{ij} = \mathrm{tr}(T_i^\dagger T_j)\,. \tag{4.41}$$

The Cartan canonical basis $D = \epsilon_i H_i + \epsilon_\alpha E_\alpha + \epsilon^*_\alpha E_{-\alpha}$ is an example of a nonhermitian choice. Here we shall always assume that T_i are chosen hermitian.

4.5 LIE ALGEBRA

As the simplest example of computation of the generators of infinitesimal transformations acting on spaces other than the defining space, consider the adjoint rep. Using (4.40) on the $V \otimes \bar{V} \to A$ adjoint rep clebsches (*i.e.*, generators T_i), we obtain

(4.42)

$$(T_i)_{jk} = (T_i)^c_a (T_k)^b_c (T_j)^a_b - (T_i)^c_a (T_j)^b_c (T_k)^a_b\,.$$

Our convention is always to assume that the generators T_i have been chosen hermitian. That means that ϵ_i in the expansion (4.32) is real; A is a real vector space, there is no distinction between upper and lower indices, and there is no need for arrows on the adjoint rep lines (4.28). However, the arrow on the adjoint rep generator (4.42) is necessary to define correctly the overall sign. If we interchange the two legs, the right-hand side changes sign:

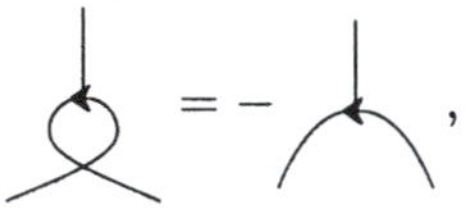

(4.43)

(the generators for real reps are always antisymmetric). This arrow has no absolute meaning; its direction is *defined* by (4.42). Actually, as the right-hand side of (4.42) is antisymmetric under interchange of any two legs, it is convenient to replace the arrow in the vertex by a more symmetric symbol, such as a dot:

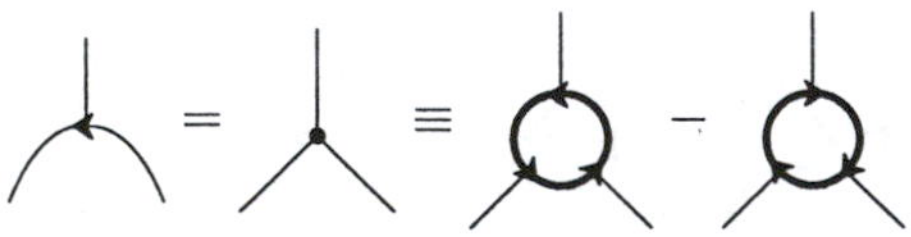

$$(T_i)_{jk} \equiv -iC_{ijk} = -\,\mathrm{tr}[T_i, T_j]T_k\,, \tag{4.44}$$

and replace the adjoint rep generators $(T_i)_{jk}$ by the fully antisymmetric structure constants iC_{ijk}. The factor i ensures their reality (in the case of hermitian generators T_i), and we keep track of the overall signs by always reading indices *counterclockwise* around a vertex:

$$-iC_{ijk} = \tag{4.45}$$

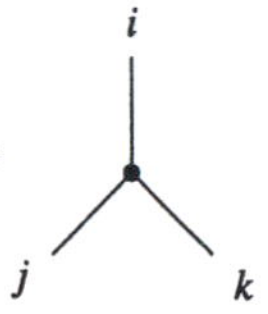

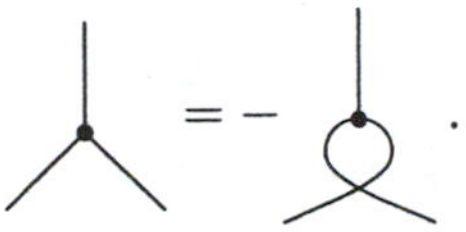

$$= - \;. \tag{4.46}$$

As all other clebsches, the generators must satisfy the invariance conditions (4.39):

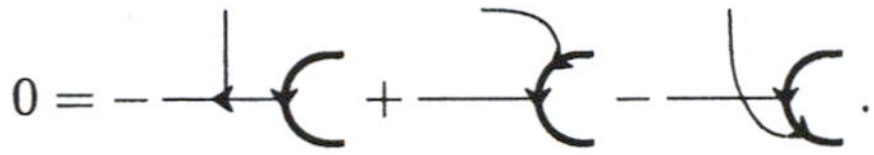

$$0 = - \quad + \quad - \quad .$$

Redrawing this a little and replacing the adjoint rep generators (4.44) by the structure constants, we find that the generators obey the *Lie algebra* commutation relation

$$T_iT_j - T_jT_i = iC_{ijk}T_k\,. \tag{4.47}$$

In other words, the Lie algebra is simply a statement that T_i, the generators of invariance transformations, are themselves invariant tensors. The invariance condition for structure constants C_{ijk} is likewise

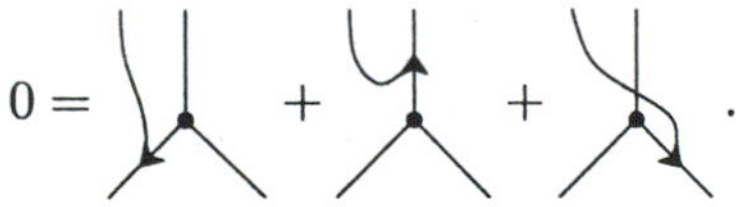

$$0 = \quad + \quad + \quad .$$

Rewriting this with the dot-vertex (4.44), we obtain

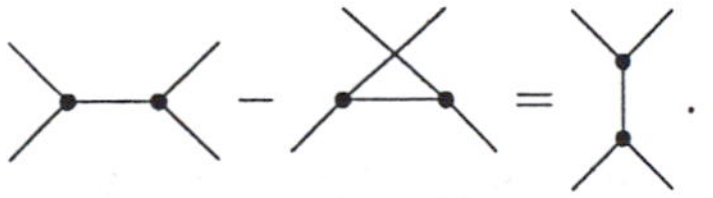

$$- \quad = \quad . \tag{4.48}$$

This is the Lie algebra commutator for the adjoint rep generators, known as the *Jacobi relation* for the structure constants

$$C_{ijm}C_{mkl} - C_{ljm}C_{mki} = C_{iml}C_{jkm}\,. \tag{4.49}$$

Hence, the Jacobi relation is also an invariance statement, this time the statement that the structure constants are invariant tensors.

Sign convention for C_{ijk}. A word of caution about using (4.47): vertex C_{ijk} is an oriented vertex. If the arrows are reversed (matrices T_i, T_j multiplied in reverse order), the right-hand side acquires an overall minus sign.

4.6 OTHER FORMS OF LIE ALGEBRA COMMUTATORS

In our calculations we shall never need explicit generators; we shall instead use the projection operators for the adjoint rep. For rep λ they have the form

$$(\mathbf{P}_A)_{b,\ \alpha}^{a,\ \beta} = \qquad a,b=1,2,\ldots,n$$

$$\alpha,\beta=1,\ldots,d_\lambda\,. \tag{4.50}$$

The invariance condition (4.36) for a projection operator is

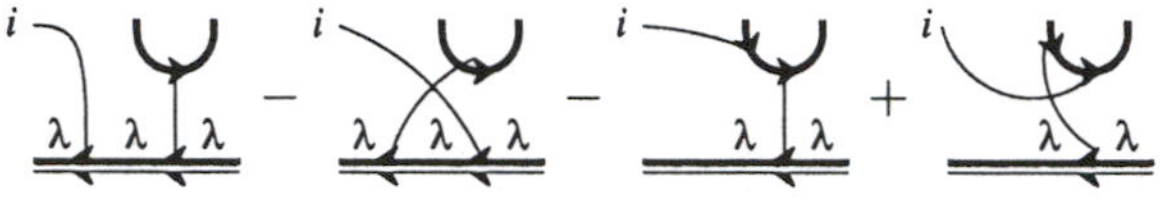

$$- \quad - \quad + \quad = 0\,. \tag{4.51}$$

Contracting with $(T_i)_b^a$ and defining $[d_\lambda \times d_\lambda]$ matrices $(T_b^a)_\alpha^\beta \equiv (\mathbf{P}_A)_{b,\ \alpha}^{a,\ \beta}$, we obtain

$$[T_b^a, T_d^c] = (\mathbf{P}_A)_{b,\ e}^{a,\ c} T_d^e - T_e^c (\mathbf{P}_A)_{b,\ d}^{a,\ e}$$

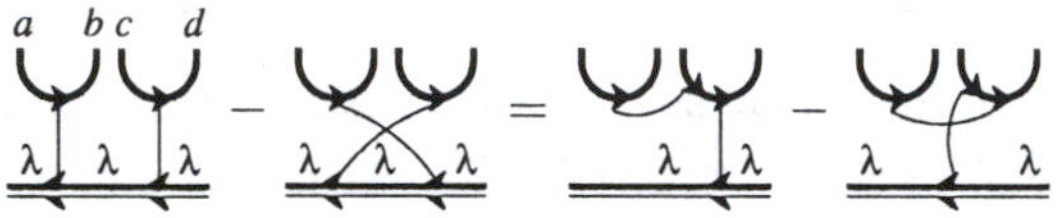

$$- \quad = \quad - \quad . \tag{4.52}$$

This is a common way of stating the Lie algebra conditions for the generators in an arbitrary rep λ. For example, for $U(n)$ the adjoint projection operator is simply a unit matrix (any hermitian matrix is a generator of unitary transformation; *cf.* chapter 9), and the right-hand side of (4.52) is given by

$$U(n), SU(n): \qquad [T_b^a, T_d^c] = \delta_b^c T_d^a - T_b^c \delta_d^a\,. \tag{4.53}$$

For the orthogonal groups the generators of rotations are antisymmetric matrices, and the adjoint projection operator antisymmetrizes generator indices:

$$SO(n): \qquad [T_{ab}, T_{cd}] = \frac{1}{2}\left\{ \begin{array}{c} g_{ac}T_{bd} - g_{ad}T_{bc} \\ -g_{bc}T_{ad} + g_{bd}T_{ac} \end{array} \right\}. \tag{4.54}$$

Apart from the normalization convention, these are the familiar Lorentz group commutation relations (we shall return to this in chapter 10).

4.7 CLASSIFICATION OF LIE ALGEBRAS BY THEIR PRIMITIVE INVARIANTS

There is a natural hierarchy to invariance groups, hinted at in sections. 2.1–3.6, that can perhaps already be grasped at this stage. Suppose we have constructed the invariance group G_1, which preserves primitives (3.39). Adding a new primitive, let us say a quartic invariant, means that we have imposed a new constraint; only those transformations of G_1 that also preserve the additional primitive constitute G_2, the invariance group of —, ⅄, ✕. Hence, $G_2 \subseteq G_1$ is a subgroup of G_1. Suppose now that you think that the primitiveness assumption is too strong, and that some quartic invariant, let us say (3.37), *cannot* be reduced to a sum of tree invariants (3.41), *i.e.*, it is of form

$$[\text{box diagram}] = [\text{X}] + (\text{rest of (3.41)}),$$

where [X] is a new primitive, not included in the original list of primitives. By the above argument only a subgroup G_3 of transformations in G_2 preserve the additional invariant, $G_3 \subseteq G_2$. If G_3 does not exist (the invariant relations are so stringent that there remain no transformations that would leave them invariant), the maximal set of primitives has been identified.

4.8 IRRELEVANCY OF CLEBSCHES

As was emphasized in section 4.2, an explicit choice of clebsches is highly arbitrary; it corresponds to a particular coordinatization of the d_λ-dimensional subspace V_λ. For computational purposes clebsches are largely irrelevant. Nothing that a physicist wants to compute depends on an explicit coordinatization. For example, in QCD the physically interesting objects are color singlets, and all color indices are summed over: one needs only an expression for the projection operators (4.31), not for the C_λ's separately.

Again, a nice example is the Lie algebra generators T_i. Explicit matrices are often constructed (Gell-Mann λ_i matrices, Cartan's canonical weights); however, in any singlet they always appear summed over the adjoint rep indices, as in (4.31). The summed combination of clebsches is just the adjoint rep projection operator, a very simple object compared with explicit T_i matrices ($\mathbf{P}_A$ is typically a combination of a few Kronecker deltas), and much simpler to use in explicit evaluations. As we shall show by many examples, all rep dimensions, casimirs, *etc.*. are computable once the projection operators for the reps involved are known. Explicit clebsches are superfluous from the computational point of view; we use them chiefly to state general theorems without recourse to any explicit realizations.

However, if one has to compute noninvariant quantities, such as subgroup embeddings, explicit clebsches might be very useful. Gell-Mann [137] invented λ_i matrices in order to embed $SU(2)$ of isospin into $SU(3)$ of the eightfold way. Cartan's canonical form for generators, summarized by Dynkin labels of a rep (table 7.6) is a very powerful tool in the study of symmetry-breaking chains [312, 126]. The same can be achieved with decomposition by invariant matrices (a nonvanishing expectation value for a direction in the defining space defines the little group of transformations in the remaining directions), but the tensorial technology in this context is underdeveloped compared to the canonical methods. And, as Stedman [317] rightly points out, if you need to check your calculations against the existing literature, keeping track of phase conventions is a necessity.

4.9 A BRIEF HISTORY OF BIRDTRACKS

> Ich wollte nicht eine abstracte Logik in Formeln darstellen, sondern einen Inhalt durch geschriebene Zeichen in genauerer und übersichtlicherer Weise zum Ausdruck bringen, als es durch Worte möglich ist.
>
> — Gottlob Frege

In this monograph, conventional subjects — symmetric group, Lie algebras (and, to a lesser extent, continuous Lie groups) — are presented in a somewhat unconventional way, in a flavor of diagrammatic notation that I refer to as "birdtracks." Similar diagrammatic notations have been invented many times before, and continue to be invented within new research areas. The earliest published example of diagrammatic notation as a language of computation, not a mere mnemonic device, appears to be F.L.G. Frege's 1879 *Begriffsschrift* [127], at its time a revolution that laid the foundation of modern logic. The idiosyncratic symbolism was not well received, ridiculed as "incorporating ideas from Japanese." Ruined by costs of typesetting, Frege died a bitter man, preoccupied by a deep hatred of the French, of Catholics, and of Jews.

According to Abdesselam and Chipalkatti [4], another precursor of diagrammatic methods was the invariant theory discrete combinatorial structures introduced by Cayley [50], Sylvester [321], and Clifford [61, 183], reintroduced in a modern, diagrammatic notation by Olver and Shakiban [264, 265].

In his 1841 fundamental paper [167] on the determinants today known as "Jacobians," Jacobi initiated the theory of irreps of the symmetric group $\mathcal{S}_k$. Schur used the $\mathcal{S}_k$ irreps to develop the representation theory of $GL(n;\mathbb{C})$ in his 1901 dissertation [306], and already by 1903 the Young tableaux [356, 338] (discussed here in chapter 9) came into use as a powerful tool for reduction of both $\mathcal{S}_k$ and $GL(n;\mathbb{C})$ representations. In quantum theory the group of choice [342] is the unitary group $U(n)$, rather than the general linear group $GL(n;\mathbb{C})$. Today this theory forms the core of the representation theory of both discrete and continuous groups, described in many excellent textbooks [238, 64, 348, 138, 26, 11, 316, 132, 133, 228]. Permutations and their compositions lend themselves naturally to diagrammatic representation developed here in chapter 6. In his extension of the $GL(n;\mathbb{C})$ Schur theory to representations of $SO(n)$, R. Brauer [31] introduced diagrammatic notation for δ_{ij} in order to represent "Brauer algebra" permutations, index contractions, and matrix multiplication diagrammatically, in the form developed here in chapter 10. His equation (39)

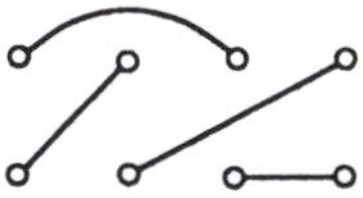

(send index 1 to 2, 2 to 4, contract ingoing $(3 \cdot 4)$, outgoing $(1 \cdot 3)$) is the earliest published proto-birdtrack I know about.

R. Penrose's papers are the first (known to me) to cast the Young projection operators into a diagrammatic form. In this monograph I use Penrose diagrammatic notation for symmetrization operators [280], Levi-Civita tensors [282], and "strand networks" [281]. For several specific, few-index tensor examples, diagrammatic Young projection operators were constructed by Canning [41], Mandula [227], and Stedman [318].

It is quite likely that since Sophus Lie's days many have doodled birdtracks in private without publishing them, partially out of a sense of gravitas and in no insignificant part because preparing these doodles for publications is even today a painful thing. I have seen unpublished 1960s course notes of J. G. Belinfante [6, 19], very much like the birdtracks drawn here in chapters 6–9, and there are surely many other such doodles lost in the mists of time. But, citing Frege [128], "the comfort of the typesetter is certainly not the *summum bonum,*" and now that the typesetter is gone, it is perhaps time to move on.

The methods used here come down to us along two distinct lineages, one that can be traced to Wigner, and the other to Feynman.

Wigner's 1930s theory, elegantly presented in his group theory monograph [345], is still the best book on what physics is to be extracted from symmetries, be it atomic, nuclear, statistical, many-body, or particle physics: all physical predictions ("spectroscopic levels") are expressed in terms of Wigner's $3n$-j coefficients, which can be evaluated by means of recursive or combinatorial algorithms. As explained here in chapter 5, decomposition (5.8) of tensor products into irreducible reps implies that any invariant number characterizing a physical system with a given symmetry corresponds to one or several "vacuum bubbles," trivalent graphs (a graph in which every vertex joins three links) with no external legs, such as those listed in table 5.1.

Since the 1930s much of the group-theoretical work on atomic and nuclear physics had focused on explicit construction of clebsches for the rotation group $SO(3) \simeq SU(2)$. The first paper recasting Wigner's theory in graphical form appears to be a 1956 paper by I. B. Levinson [213], further developed in the influental 1960 monograph by A. P. Yutsis (later A. Jucys), I. Levinson and V. Vanagas [357], published in English in 1962 (see also refs. [109, 33]). A recent contribution to this tradition is the book by G. E. Stedman [318], which covers a broad range of applications, including the methods introduced in the 1984 version of the present monograph [82]. The pedagogical work of computer graphics pioneer J. F. Blinn [25], who was inspired by Stedman's book, also deserves mention.

The main drawback of such diagrammatic notations is lack of standardization, especially in the case of clebsches. In addition, the diagrammatic notations designed for atomic and nuclear spectroscopy are complicated by various phase conventions.

R. P. Feynman went public with Feynman diagrams on my second birthday, April 1, 1948, at the Pocono Conference. The idiosyncratic symbolism (Gleick [141] describes it as "chicken-wire diagrams") was not well received by Bohr, Dirac, and Teller, leaving Feynman a despondent man [141, 307, 237]. The first Feynman diagram appeared in print in Dyson's article [106, 308] on the equivalence of (at that time) the still unpublished Feynman theory and the theories of Schwinger and Tomonaga.

If diagrammatic notation is to succeed, it need be not only precise, but also beautiful. It is in this sense that this monograph belongs to the tradition of R. P. Feynman, whose sketches of the very first "Feynman diagrams" in his fundamental 1948 Q.E.D.

paper [119, 308] are beautiful to behold. Similarly, R. Penrose's [280, 281] way of drawing symmetrizers and antisymmetrizers, adopted here in chapter 6, is imbued with a very Penrose aesthetics, and even though the print is black and white, one senses that he had drawn them in color.

In developing the "birdtrack" notation in 1975 I was inspired by Feynman diagrams and by the elegance of Penrose's binors [280]. I liked G. 't Hooft's 1974 double-line notation for $U(n)$ gluon group-theory weights [163], and have introduced analogous notation for $SU(n)$, $SO(n)$ and $Sp(n)$ in my 1976 paper [73]. In an influential paper, M. Creutz [69] has applied such notation to the evaluation of $SU(n)$ lattice gauge integrals (described here in chapter 8). The challenge was to develop diagrammatic notation for the exceptional Lie algebras, and I succeeded [73], except for E_8, which came later.

In the quantum groups literature, graphs composed of vertices (4.44) are called *trivalent*. The Jacobi relation (4.48) in diagrammatic form was first published [73] in 1976; though it seems surprising, I have not found it in the earlier literature. This set of diagrams has since been given the moniker "IHX" by D. Bar-Natan [14]. In his Ph.D. thesis Bar-Natan has also renamed the Lie algebra commutator (4.47) the "STU relation," by analogy to Mandelstam's scattering cross-channel variables (s, t, u), and the full antisymmetry of structure constants (4.46) the "AS relation."

So why call this "birdtracks" and not "Feynman diagrams"? The difference is that here diagrams are not a mnemonic device, an aid in writing down an integral that is to be evaluated by other techniques. In our applications, explicit construction of clebsches would be superfluous, and we need no phase conventions. Here "birdtracks" are everything—unlike Feynman diagrams, here all calculations are carried out in terms of birdtracks, from start to finish. Left behind are blackboards and pages of squiggles of the kind that made Bernice Durand exclaim: "What are these birdtracks!?" and thus give them the name.

Chapter Five

Recouplings

Clebsches discussed in section 4.2 project a tensor in $V^p \otimes \bar{V}^q$ onto a subspace λ. In practice one usually reduces a tensor step by step, decomposing a 2-particle state at each step. While there is some arbitrariness in the order in which these reductions are carried out, the final result is invariant and highly elegant: any group-theoretical invariant quantity can be expressed in terms of Wigner 3- and 6-j coefficients.

5.1 COUPLINGS AND RECOUPLINGS

We denote the clebsches for $\mu \otimes \nu \to \lambda$ by

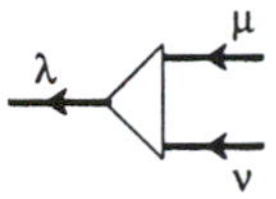

$, \quad \mathbf{P}_\lambda =$
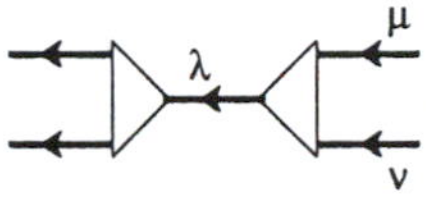

. (5.1)

Here λ, μ, ν are rep labels, and the corresponding tensor indices are suppressed. Furthermore, if μ and ν are irreducible reps, the same clebsches can be used to project $\mu \otimes \bar{\lambda} \to \bar{\nu}$

$\mathbf{P}_\nu = \dfrac{d_\nu}{d_\lambda}$
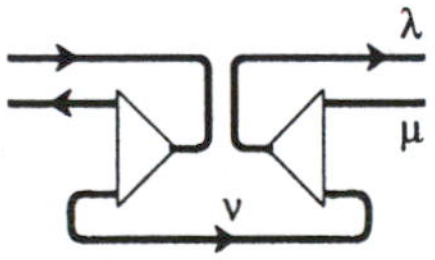

, (5.2)

and $\nu \otimes \bar{\lambda} \to \bar{\mu}$

$\mathbf{P}_\mu = \dfrac{d_\mu}{d_\lambda}$
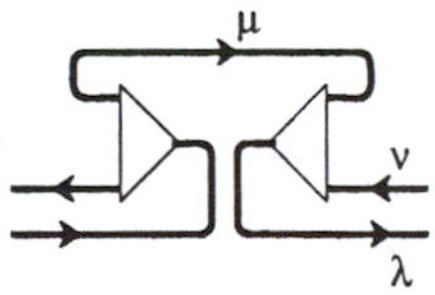

. (5.3)

Here the normalization factors come from $P^2 = P$ condition. In order to draw the projection operators in a more symmetric way, we replace clebsches by 3-vertices:

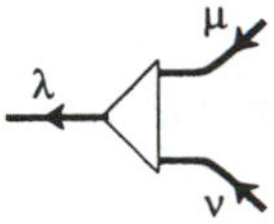

$\equiv \dfrac{1}{\sqrt{a_\lambda}}$

. (5.4)

In this definition one has to keep track of the ordering of the lines around the vertex. If in some context the birdtracks look better with two legs interchanged, one can use Yutsis's notation [357]:

$$\lambda \ \mu \ \nu \equiv \lambda \ \mu \ \nu \quad . \tag{5.5}$$

While all sensible clebsches are normalized by the orthonormality relation (4.19), in practice no two authors ever use the same normalization for 3-vertices (in other guises known as 3-j coefficients, Gell-Mann λ matrices, Cartan roots, Dirac γ matrices, *etc.*). For this reason we shall usually not fix the normalization

$$= a_\lambda \qquad , \quad a_\lambda = \frac{\quad}{d_\lambda} \,, \tag{5.6}$$

leaving the reader the option of substituting his or her favorite choice (such as $a = \frac{1}{2}$ if the 3-vertex stands for Gell-Mann $\frac{1}{2}\lambda_i$, *etc.*).

To streamline the discussion, we shall drop the arrows and most of the rep labels in the remainder of this chapter — they can always easily be reinstated.

The above three projection operators now take a more symmetric form:

$$\mathbf{P}_\lambda = \frac{1}{a_\lambda}$$

$$\mathbf{P}_\mu = \frac{1}{a_\mu}$$

$$\mathbf{P}_\nu = \frac{1}{a_\nu} \quad . \tag{5.7}$$

In terms of 3-vertices, the completeness relation (4.20) is

$$= \sum_\lambda \frac{d_\lambda}{\quad} \quad . \tag{5.8}$$

Any tensor can be decomposed by successive applications of the completeness relation:

$$= \sum_\lambda \frac{1}{a_\lambda} \quad = \sum_{\lambda,\mu} \frac{1}{a_\lambda}\frac{1}{a_\mu}$$

$$= \sum_{\lambda,\mu,\nu} \frac{1}{a_\lambda}\frac{1}{a_\mu}\frac{1}{a_\nu} \quad . \tag{5.9}$$

Hence, if we know clebsches for $\lambda \otimes \mu \to \nu$, we can also construct clebsches for $\lambda \otimes \mu \otimes \nu \otimes \ldots \to \rho$. However, there is no unique way of building up the clebsches; the above state can equally well be reduced by a different coupling scheme

$$= \sum_{\lambda,\mu,\nu} \frac{1}{a_\lambda}\frac{1}{a_\mu}\frac{1}{a_\nu} \quad . \tag{5.10}$$

Consider now a process in which a particle in the rep μ interacts with a particle in the rep ν by exchanging a particle in the rep ω:

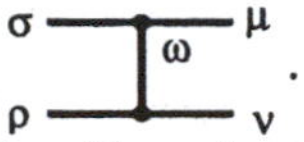

$$\qquad (5.11)$$

The final particles are in reps ρ and σ. To evaluate the contribution of this exchange to the spectroscopic levels of the μ-ν particles system, we insert the Clebsch-Gordan series (5.8) twice, and eliminate one of the sums by the orthonormality relation (5.6):

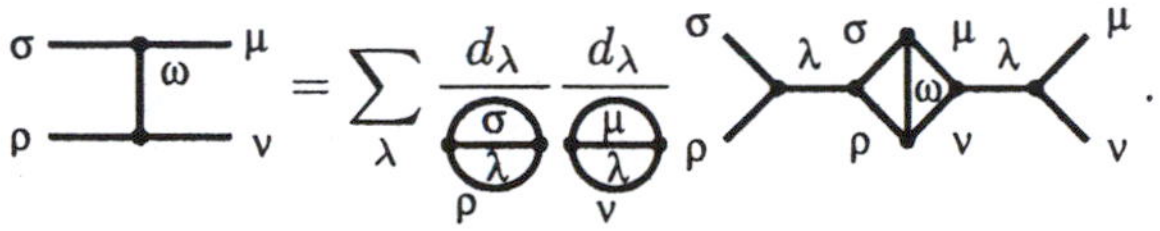

$$= \sum_\lambda \frac{d_\lambda}{\quad} \frac{d_\lambda}{\quad} \qquad (5.12)$$

By assumption λ is an irrep, so we have a recoupling relation between the exchanges in "s" and "t channels":

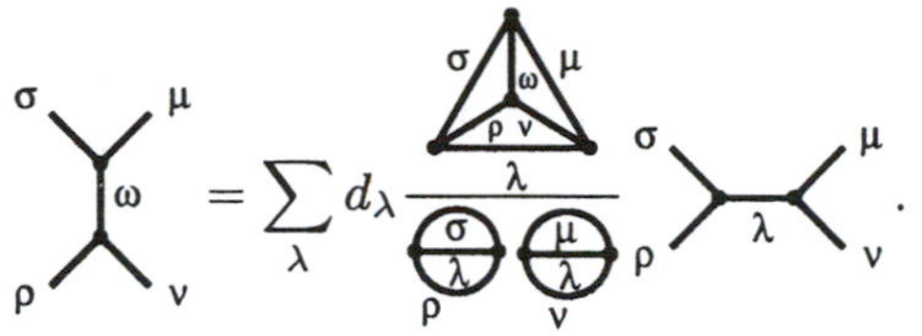

$$= \sum_\lambda d_\lambda \qquad (5.13)$$

We shall refer to as 3-j coefficients and as 6-j coefficients, and commit ourselves to no particular normalization convention.

In atomic physics it is customary to absorb into the 3-vertex and define a 3-j symbol [238, 286, 345]

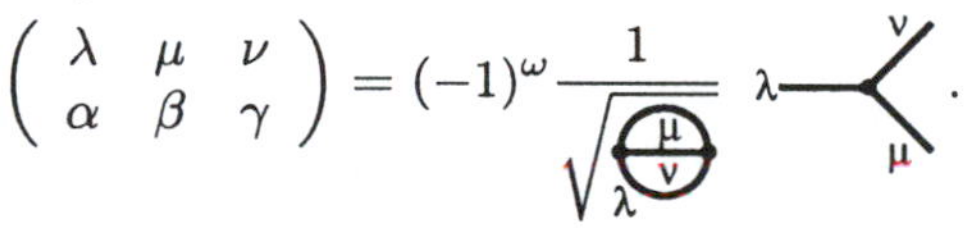

$$\begin{pmatrix} \lambda & \mu & \nu \\ \alpha & \beta & \gamma \end{pmatrix} = (-1)^\omega \frac{1}{\sqrt{\quad}} \qquad (5.14)$$

Here $\alpha = 1, 2, \ldots, d_\lambda$, *etc.*, are indices, λ, μ, ν rep labels and ω the phase convention. Fixing a phase convention is a waste of time, as the phases cancel in summed-over quantities. All the ugly square roots, one remembers from quantum mechanics, come from sticking $\sqrt{\quad}$ into 3-j symbols. Wigner [345] 6-j *symbols* are related to our 6-j *coefficients* by

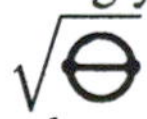

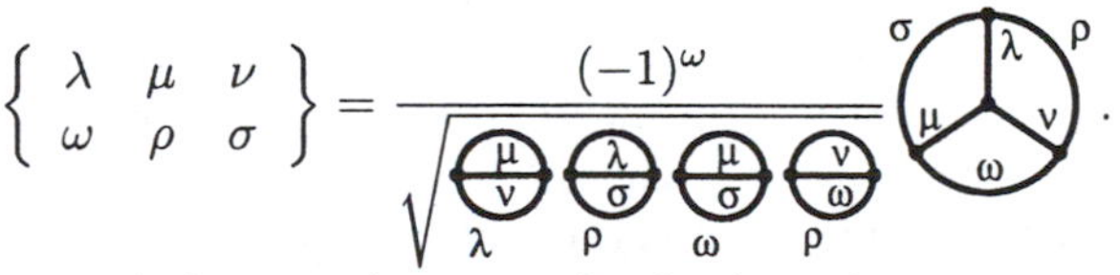

$$\begin{Bmatrix} \lambda & \mu & \nu \\ \omega & \rho & \sigma \end{Bmatrix} = \frac{(-1)^\omega}{\sqrt{\quad}} \qquad (5.15)$$

The name $3n$-j symbol comes from atomic physics, where a recoupling involves $3n$ angular momenta $j_1, j_2, \ldots, j_{3n}$ (see section 14.2).

Most of the textbook symmetries of and relations between 6-j symbols are obvious from looking at the corresponding diagrams; others follow quickly from completeness relations.

If we know the necessary 6-j's, we can compute the level splittings due to single particle exchanges. In the next section we shall show that a far stronger claim can be made: given the 3- and 6-j coefficients, we can compute *all* multiparticle matrix elements.

	Skeletons	Vertex insertions	Self-energy insertions	Total number
1-j				1
$3j$				1
6-j				2
9-j				5
12-j	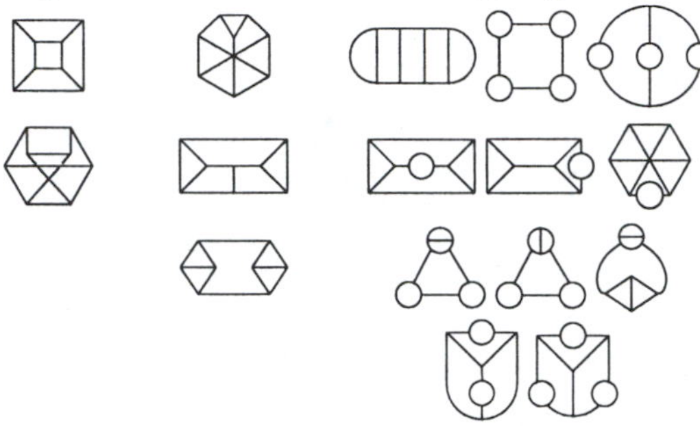			16

Table 5.1 Topologically distinct types of Wigner $3n$-j coefficients, enumerated by drawing all possible graphs, eliminating the topologically equivalent ones by hand. Lines meeting in any 3-vertex correspond to any three irreducible representations with a nonvanishing Clebsch-Gordan coefficient, so in general these graphs cannot be reduced to simpler graphs by means of such as the Lie algebra (4.47) and Jacobi identity (4.48).

5.2 WIGNER $3n$-j COEFFICIENTS

An arbitrary higher-order contribution to a 2-particle scattering process will give a complicated matrix element. The corresponding energy levels, crosssections, *etc.*, are expressed in terms of scalars obtained by contracting all tensor indices; diagrammatically they look like "vacuum bubbles," with $3n$ internal lines. The topologically distinct vacuum bubbles in low orders are given in table 5.1.

In group-theoretic literature, these diagrams are called $3n$-j symbols, and are studied in considerable detail. Fortunately, any $3n$-j symbol that contains as a subdiagram a loop with, let us say, seven vertices,

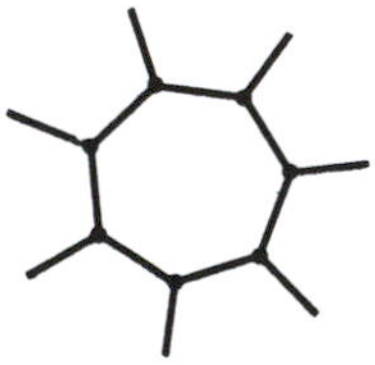

,

can be expressed in terms of 6-j coefficients. Replace the dotted pair of vertices by the cross-channel sum (5.13):

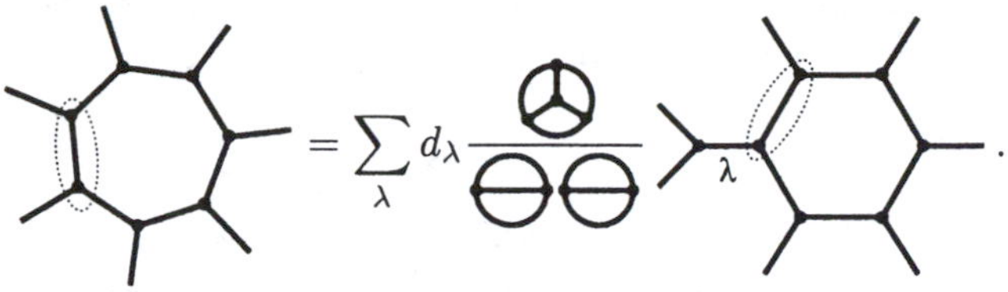

$$= \sum_{\lambda} d_\lambda \frac{\ }{\ } \quad \lambda \tag{5.16}$$

Now the loop has six vertices. Repeating the replacement for the next pair of vertices, we obtain a loop of length five:

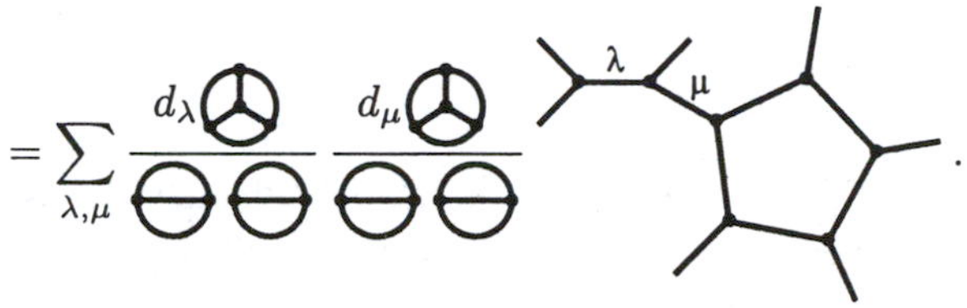

$$= \sum_{\lambda,\mu} \frac{d_\lambda}{\ } \frac{d_\mu}{\ } \quad \lambda \quad \mu \tag{5.17}$$

Repeating this process we can eliminate the loop altogether, producing 5-vertex-trees times bunches of 6-j coefficients. In this way we have expressed the original $3n$-j coefficients in terms of $3(n$-$1)$-j coefficients and 6-j coefficients. Repeating the process for the $3(n$-$1)$-j coefficients, we eventually arrive at the result that

$$(3n-j) = \sum \left(\text{products of } \right) . \tag{5.18}$$

5.3 WIGNER-ECKART THEOREM

For concreteness, consider an arbitrary invariant tensor with four indices:

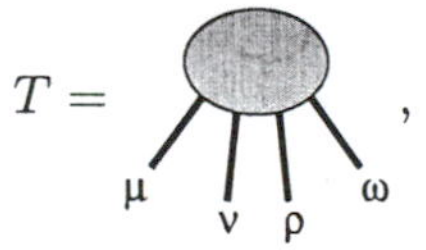

$$T = \quad \mu \ \nu \ \rho \ \omega , \tag{5.19}$$

where μ, ν, ρ and ω are rep labels, and indices and line arrows are suppressed. Now insert repeatedly the completeness relation (5.8) to obtain

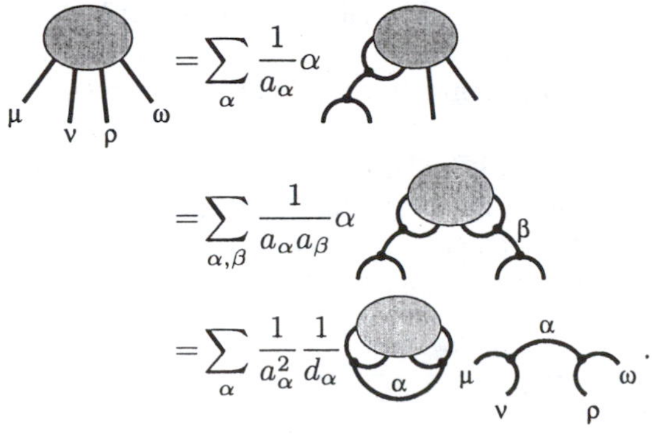

$$\mu \ \nu \ \rho \ \omega = \sum_{\alpha} \frac{1}{a_\alpha} \alpha$$

$$= \sum_{\alpha,\beta} \frac{1}{a_\alpha a_\beta} \alpha \quad \beta$$

$$= \sum_{\alpha} \frac{1}{a_\alpha^2} \frac{1}{d_\alpha} \quad \alpha \quad \mu \ \nu \ \alpha \ \rho \ \omega . \tag{5.20}$$

In the last line we have used the orthonormality of projection operators — as in (5.13) or (5.23).

In this way any invariant tensor can be reduced to a sum over clebsches (*kinematics*) weighted by *reduced matrix elements*:

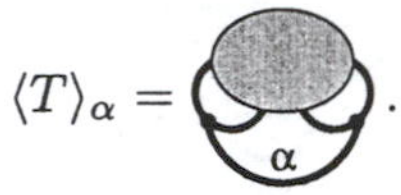

$$\langle T \rangle_\alpha = \quad (\alpha) \quad . \tag{5.21}$$

This theorem has many names, depending on how the indices are grouped. If T is a vector, then only the 1-dimensional reps (singlets) contribute

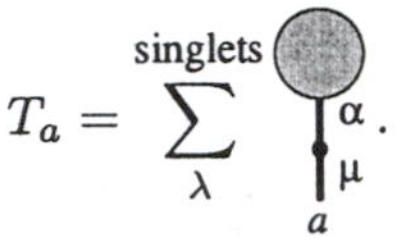

$$T_a = \sum_\lambda^{\text{singlets}} \quad \alpha,\ \mu,\ a \quad . \tag{5.22}$$

If T is a matrix, and the reps α, μ are irreducible, the theorem is called *Schur's Lemma* (for an irreducible rep an invariant matrix is either zero, or proportional to the unit matrix):

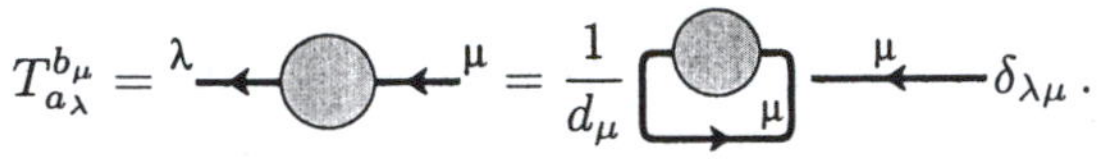

$$T^{b_\mu}_{a_\lambda} = \lambda \quad \mu = \frac{1}{d_\mu} \quad (\mu) \quad \mu \ \delta_{\lambda\mu} \, . \tag{5.23}$$

If T is an "invariant tensor operator," then the theorem is called the *Wigner-Eckart theorem* [345, 107]:

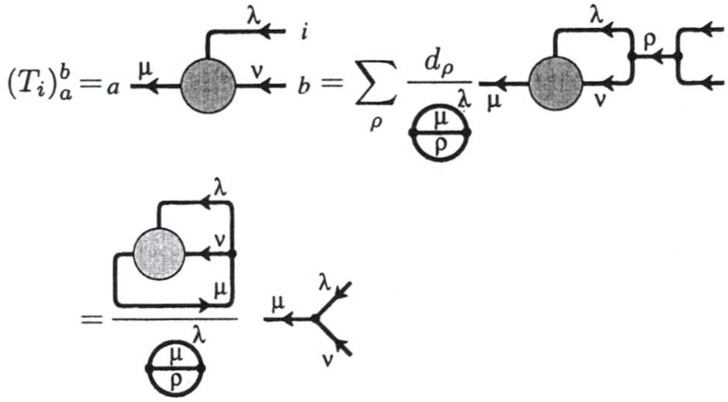

$$(T_i)^b_a = a \ \mu \quad \nu \ b \ (\lambda\ i) = \sum_\rho \frac{d_\rho}{(\mu,\rho)} \quad \lambda\ \mu\ \nu\ \rho$$

$$= \frac{(\lambda,\ \nu,\ \mu)}{(\mu,\ \lambda,\ \rho)} \quad \mu\ \lambda\ \nu \tag{5.24}$$

(assuming that μ appears only once in $\overline{\lambda} \otimes \mu$ Kronecker product). If T has many indices, as in our original example (5.19), the theorem is ascribed to Yutsis, Levinson, and Vanagas [357]. The content of all these theorems is that they reduce spectroscopic calculations to evaluation of "vacuum bubbles" or "reduced matrix elements" (5.21).

The rectangular matrices $(C_\lambda)^\alpha_\sigma$ from (3.27) do not look very much like the clebsches from the quantum mechanics textbooks; neither does the Wigner-Eckart theorem in its birdtrack version (5.24). The difference is merely a difference of notation. In the bra-ket formalism, a clebsch for $\lambda_1 \otimes \lambda_2 \to \lambda$ is written as

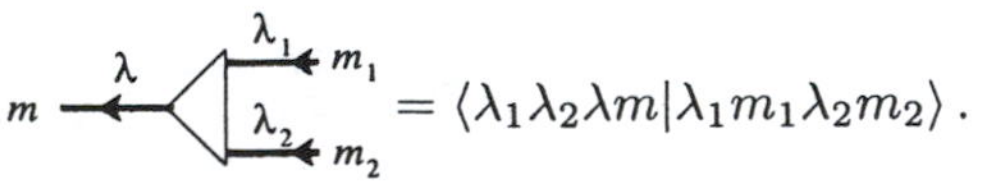

$$m \ \lambda \quad \lambda_1\ m_1 \quad \lambda_2\ m_2 = \langle \lambda_1 \lambda_2 \lambda m | \lambda_1 m_1 \lambda_2 m_2 \rangle \, . \tag{5.25}$$

Representing the $[d_\lambda \times d_\lambda]$ rep of a group element g diagrammatically by a black triangle,

$$D^{\lambda}_{m,m'}(g) = m \longleftarrow m' , \tag{5.26}$$

we can write the Clebsch-Gordan series (3.49) as

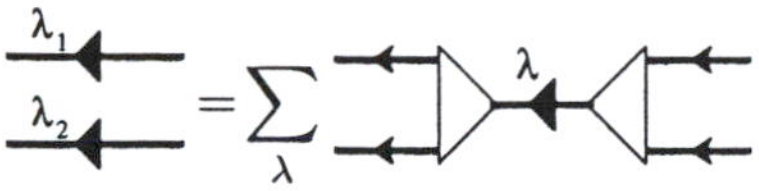

$$D^{\lambda_1}_{m_1 m_1'}(g) D^{\lambda_2}_{m_2 m_2'}(g) =$$

$$\sum_{\tilde{\lambda},\tilde{m},\tilde{m}_1} \langle \lambda_1 m_1 \lambda_2 m_2 | \lambda_1 \lambda_2 \tilde{\lambda} \tilde{m} \rangle D^{\tilde{\lambda}}_{\tilde{m}\tilde{m}_1}(g) \langle \lambda_1 \lambda_2 \tilde{\lambda} \tilde{m}_1 | \lambda_1 m_1' \lambda_2 m_2' \rangle .$$

An "invariant tensor operator" can be written as

$$\langle \lambda_2 m_2 | T^{\lambda}_{m} | \lambda_1 m_1 \rangle = \tag{5.27}$$

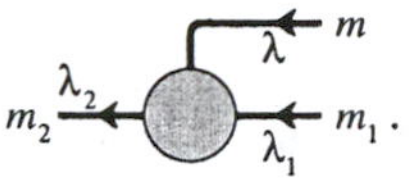

In the bra-ket formalism, the Wigner-Eckart theorem (5.24) is written as

$$\langle \lambda_2 m_2 | T^{\lambda}_{m} | \lambda_1 m_1 \rangle = \langle \lambda \lambda_1 \lambda_2 m_2 | \lambda m \lambda_1 m_1 \rangle T(\lambda, \lambda_1 \lambda_2) , \tag{5.28}$$

where the reduced matrix element is given by

$$T(\lambda, \lambda_1 \lambda_2) = \frac{1}{d_{\lambda_2}} \sum_{n_1,n_2,n} \langle \lambda n \lambda_1 n_1 | \lambda \lambda_1 \lambda_2 n_2 \rangle \langle \lambda_2 n_2 | T^{\lambda}_{n} | \lambda_1 n_1 \rangle$$

$$= \frac{1}{d_{\lambda_2}} \quad . \tag{5.29}$$

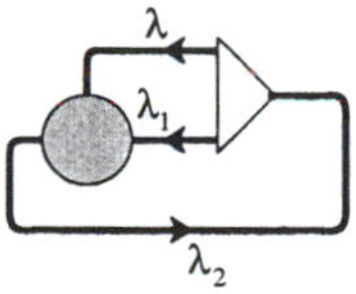

We do not find the bra-ket formalism convenient for the group-theoretic calculations that will be discussed here.

Chapter Six

Permutations

The simplest example of invariant tensors is the products of Kronecker deltas. On tensor spaces they represent index permutations. This is the way in which the symmetric group S_p, the group of permutations of p objects, enters into the theory of tensor reps. In this chapter, I introduce birdtracks notation for permutations, symmetrizations and antisymmetrizations and collect a few results that will be useful later on. These are the (anti)symmetrization expansion formulas (6.10) and (6.19), Levi-Civita tensor relations (6.28) and (6.30), the characteristic equations (6.50), and the invariance conditions (6.54) and (6.56). The theory of Young tableaux (or plethysms) is developed in chapter 9.

6.1 SYMMETRIZATION

Operation of permuting tensor indices is a linear operation, and we can represent it by a $[d \times d]$ matrix:

$$\sigma_\alpha^\beta = \sigma_{b_1 \ldots b_p}{}^{a_1 a_2 \ldots a_q}{}_{,c_q \ldots c_2 c_1}{}^{d_p \ldots d_1} \,. \tag{6.1}$$

As the covariant and contravariant indices have to be permuted separately, it is sufficient to consider permutations of purely covariant tensors.

For 2-index tensors, there are two permutations:

$$\begin{aligned} \text{identity: } \mathbf{1}_{ab,}{}^{cd} &= \delta_a^d \delta_b^c = [\text{birdtrack}] \\ \text{flip: } \sigma_{(12)ab,}{}^{cd} &= \delta_a^c \delta_b^d = [\text{birdtrack}] \,. \end{aligned} \tag{6.2}$$

For 3-index tensors, there are six permutations:

$$\begin{aligned} \mathbf{1}_{a_1 a_2 a_3,}{}^{b_3 b_2 b_1} &= \delta_{a_1}^{b_1} \delta_{a_2}^{b_2} \delta_{a_3}^{b_3} = [\text{birdtrack}] \\ \sigma_{(12)a_1 a_2 a_3,}{}^{b_3 b_2 b_1} &= \delta_{a_1}^{b_2} \delta_{a_2}^{b_1} \delta_{a_3}^{b_3} = [\text{birdtrack}] \\ \sigma_{(23)} = [\text{birdtrack}] \,, &\quad \sigma_{(13)} = [\text{birdtrack}] \\ \sigma_{(123)} = [\text{birdtrack}] \,, &\quad \sigma_{(132)} = [\text{birdtrack}] \,. \end{aligned} \tag{6.3}$$

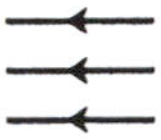

Subscripts refer to the standard permutation cycles notation. For the remainder of this chapter we shall mostly omit the arrows on the Kronecker delta lines.

The symmetric sum of all permutations,

$$S_{a_1 a_2 \ldots a_p,}{}^{b_p \ldots b_2 b_1} = \frac{1}{p!}\left\{\delta^{b_1}_{a_1}\delta^{b_2}_{a_2}\ldots\delta^{b_p}_{a_p} + \delta^{b_1}_{a_2}\delta^{b_2}_{a_1}\ldots\delta^{b_p}_{a_p} + \ldots\right\}$$

$$S = \frac{1}{p!}\left\{ \ldots + \ldots + \ldots + \ldots \right\}, \qquad (6.4)$$

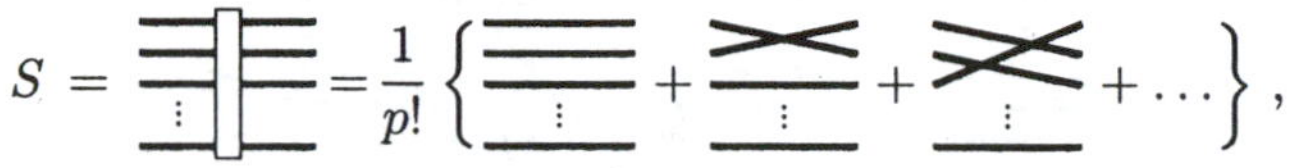

yields the symmetrization operator S. In birdtrack notation, a white bar drawn across p lines will always denote symmetrization of the lines crossed. A factor of $1/p!$ has been introduced in order for S to satisfy the projection operator normalization

$$S^2 = S \qquad (6.5)$$

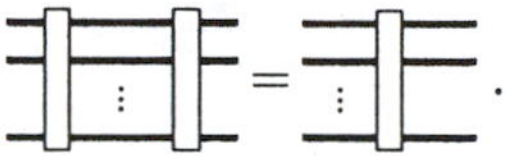

A subset of indices $a_1, a_2, \ldots a_q$, $q < p$ can be symmetrized by symmetrization matrix $S_{12\ldots q}$

$$(S_{12\ldots q})_{a_1 a_2 \ldots a_q \ldots a_p,}{}^{b_p \ldots b_q \ldots b_2 b_1} =$$
$$\frac{1}{q!}\left\{\delta^{b_1}_{a_1}\delta^{b_2}_{a_2}\ldots\delta^{b_q}_{a_q} + \delta^{b_1}_{a_2}\delta^{b_2}_{a_1}\ldots\delta^{b_q}_{a_q} + \ldots\right\}\delta^{b_q+1}_{a_q+1}\ldots\delta^{b_p}_{a_p}$$

$$S_{12\ldots q} = \qquad (6.6)$$

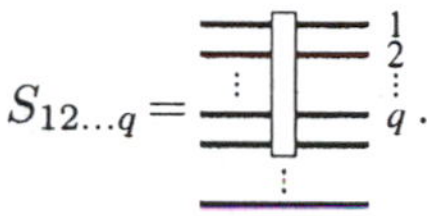

Overall symmetrization also symmetrizes any subset of indices:

$$S S_{12\ldots q} = S \qquad (6.7)$$

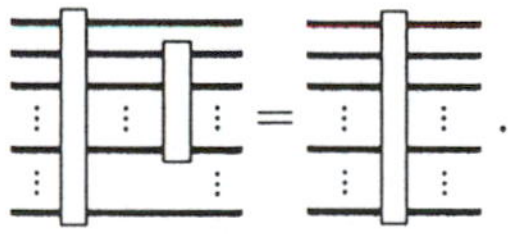

Any permutation has eigenvalue 1 on the symmetric tensor space:

$$\sigma S = S \qquad (6.8)$$

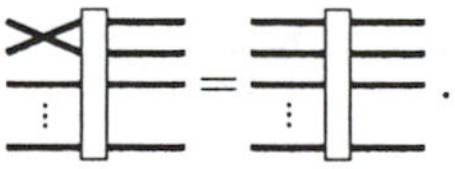

Diagrammatically this means that legs can be crossed and uncrossed at will.

The definition (6.4) of the symmetrization operator as the sum of all $p!$ permutations is inconvenient for explicit calculations; a recursive definition is more useful:

$$S_{a_1 a_2 \ldots a_p,}{}^{b_p \ldots b_2 b_1} = \frac{1}{p}\left\{\delta^{b_1}_{a_1} S_{a_2 \ldots a_p,}{}^{b_p \ldots b_2} + \delta^{b_1}_{a_2} S_{a_1 a_3 \ldots a_p,}{}^{b_p \ldots b_2} + \ldots\right\}$$

$$S = \frac{1}{p}\left(1 + \sigma_{(21)} + \sigma_{(321)} + \ldots + \sigma_{(p\ldots 321)}\right) S_{23\ldots p}$$

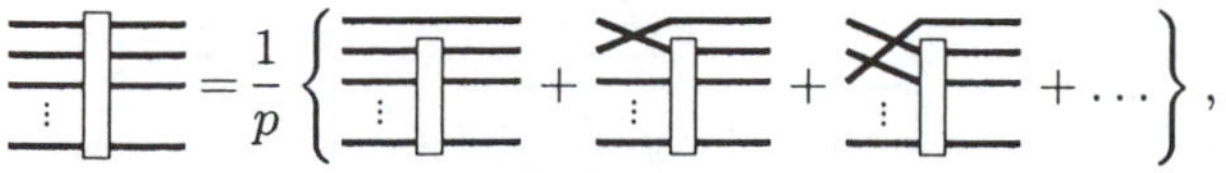

$$= \frac{1}{p}\left\{ \ldots + \ldots + \ldots + \ldots \right\}, \qquad (6.9)$$

which involves only p terms. This equation says that if we start with the first index, we end up either with the first index, or the second index and so on. The remaining indices are fully symmetric. Multiplying by $S_{23}\ldots p$ from the left, we obtain an even more compact recursion relation with two terms only:

$$= \frac{1}{p}\left(+ (p-1) \right) . \tag{6.10}$$

As a simple application, consider computation of a contraction of a single pair of indices:

$$= \frac{1}{p}\left\{ + (p-1) \right\}$$
$$= \frac{n+p-1}{p}$$
$$S_{a_p a_{p-1}\ldots a_1,}{}^{b_1\ldots b_{p-1}a_p} = \frac{n+p-1}{p} S_{a_{p-1}\ldots a_1,}{}^{b_1\ldots b_{p-1}} . \tag{6.11}$$

For a contraction in $(p-k)$ pairs of indices, we have

$$= \frac{(n+p-1)!k!}{p!(n+k-1)!} . \tag{6.12}$$

The trace of the symmetrization operator yields the number of independent components of fully symmetric tensors:

$$d_S = \operatorname{tr} S = = \frac{n+p-1}{p} = \frac{(n+p-1)!}{p!(n-1)!} . \tag{6.13}$$

For example, for 2-index symmetric tensors,

$$d_S = n(n+1)/2 . \tag{6.14}$$

6.2 ANTISYMMETRIZATION

The alternating sum of all permutations,

$$A_{a_1 a_2\ldots a_p,}{}^{b_p\ldots b_2 b_1} = \frac{1}{p!}\left\{ \delta_{a_1}^{b_1}\delta_{a_2}^{b_2}\ldots\delta_{a_p}^{b_p} - \delta_{a_2}^{b_1}\delta_{a_1}^{b_2}\ldots\delta_{a_p}^{b_p} + \ldots \right\}$$
$$A = = \frac{1}{p!}\left\{ - + - \ldots \right\} , \tag{6.15}$$

yields the antisymmetrization projection operator A. In birdtrack notation, antisymmetrization of p lines will always be denoted by a black bar drawn across the lines. As in the previous section

$$A^2 = A$$

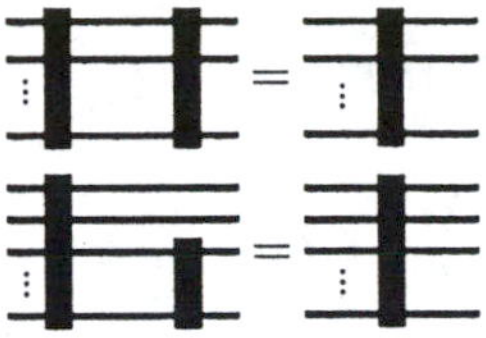

(6.16)

and in addition

$$SA = 0$$

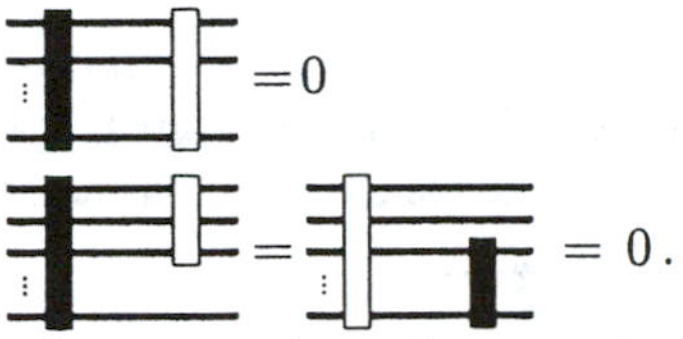

$$= 0 \,. \tag{6.17}$$

A transposition has eigenvalue -1 on the antisymmetric tensor space

$$\sigma_{(i,i+1)} A = -A$$

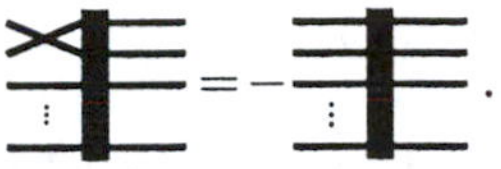

(6.18)

Diagrammatically this means that legs can be crossed and uncrossed at will, but with a factor of -1 for a transposition of any two neighboring legs.

As in the case of symmetrization operators, the recursive definition is often computationally convenient

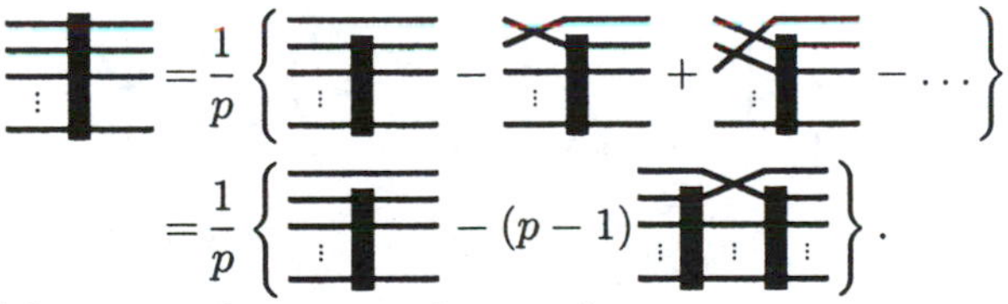

$$= \frac{1}{p}\left\{ \ldots - \ldots + \ldots - \ldots \right\}$$

$$= \frac{1}{p}\left\{ \ldots - (p-1) \ldots \right\}. \tag{6.19}$$

This is useful for computing contractions such as

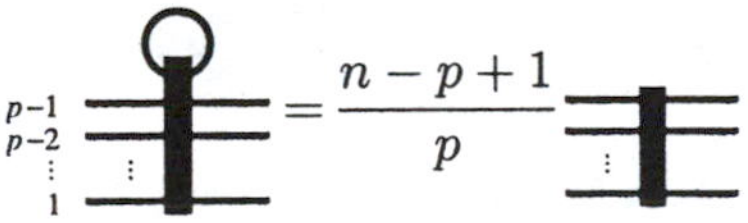

$$= \frac{n-p+1}{p}$$

$$A_{aa_{p-1}\ldots a_1},{}^{b_1\ldots b_{p-1}a} = \frac{n-p+1}{p} A_{a_{p-1}\ldots a_1},{}^{b_1\ldots b_{p-1}} \,. \tag{6.20}$$

The number of independent components of fully antisymmetric tensors is given by

$$d_A = \mathrm{tr}\, A = \quad = \frac{n-p+1}{p}\frac{n-p+2}{p-1}\cdots\frac{n}{1}$$

$$= \begin{cases} \frac{n!}{p!(n-p)!}\,, & n \geq p \\ 0\,, & n < p \end{cases} \,. \tag{6.21}$$

For example, for 2-index antisymmetric tensors the number of independent components is

$$d_A = \frac{n(n-1)}{2}\,. \tag{6.22}$$

Tracing $(p-k)$ pairs of indices yields

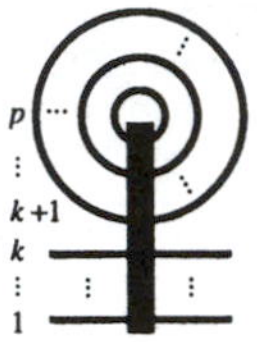

$$= \frac{k!(n-k)!}{p!(n-p)!}$$

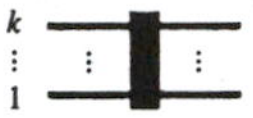

(6.23)

The antisymmetrization tensor $A_{a_1 a_2 \ldots,}{}^{b_p \ldots b_2 b_1}$ has nonvanishing components, only if all lower (or upper) indices differ from each other. If the defining dimension is smaller than the number of indices, the tensor A has no nonvanishing components:

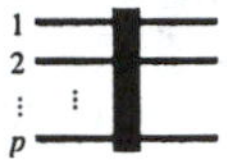

$$= 0 \qquad \text{if } p > n\,. \tag{6.24}$$

This identity implies that for $p > n$, not all combinations of p Kronecker deltas are linearly independent. A typical relation is the $p = n+1$ case

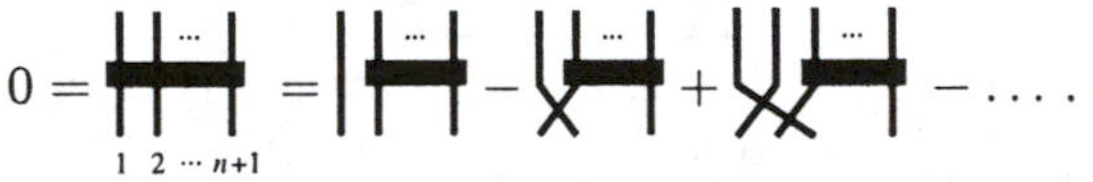

(6.25)

For example, for $n = 2$ we have

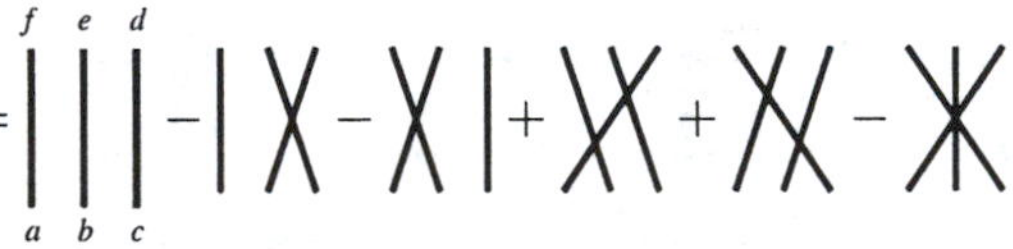

(6.26)

$$0 = \delta_a^f \delta_b^e \delta_c^d - \delta_a^f \delta_c^e \delta_b^d - \delta_b^f \delta_a^e \delta_c^d + \delta_b^f \delta_c^e \delta_a^d + \delta_c^f \delta_a^e \delta_b^d - \delta_c^f \delta_b^e \delta_a^d\,.$$

6.3 LEVI-CIVITA TENSOR

An antisymmetric tensor, with n indices in defining dimension n, has only one independent component ($d_n = 1$ by (6.21)). The clebsches (4.17) are in this case proportional to the *Levi-Civita tensor*:

$$(C_A)_1{}^{,a_n \ldots a_2 a_1} = C\epsilon^{a_n \ldots a_2 a_1} =$$

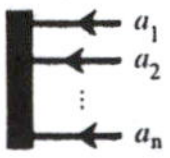

$$(C_A)_{a_1 a_2 \ldots a_n}{}^{,1} = C\epsilon_{a_1 a_2 \ldots a_n} =$$

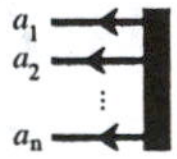

, (6.27)

with $\epsilon^{12\ldots n} = \epsilon_{12\ldots n} = 1$. This diagrammatic notation for the Levi-Civita tensor was introduced by Penrose [280]. The normalization factors C are physically irrelevant.

They adjust the phase and the overall normalization in order that the Levi-Civita tensors satisfy the projection operator (4.18) and orthonormality (4.19) conditions:

$$\frac{1}{N!}\epsilon_{b_1 b_2 \ldots b_n}\epsilon^{a_1 a_2 \ldots a_n} = A_{b_1 b_2 \ldots b_n},{}^{a_n \ldots a_2 a_1}$$

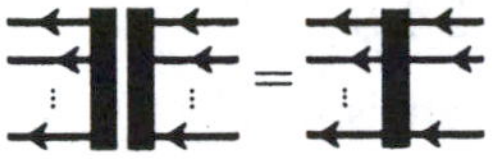

$$\frac{1}{N!}\epsilon_{a_1 a_2 \ldots a_n}\epsilon^{a_1 a_2 \ldots a_n} = \delta_{11} = 1, \qquad = 1. \tag{6.28}$$

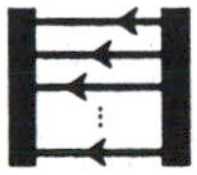

With our conventions,

$$C = \frac{i^{n(n-1)/2}}{\sqrt{n!}}. \tag{6.29}$$

The phase factor arises from the hermiticity condition (4.15) for clebsches (remember that indices are always read in the counterclockwise order around a diagram),

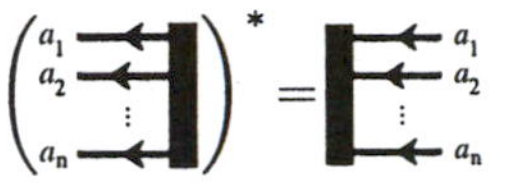

$$i^{-\phi}\epsilon_{a_1 a_2 \ldots a_n} = i^{-\phi}\epsilon_{a_n \ldots a_2 a_1}.$$

Transposing the indices

$$\epsilon_{a_1 a_2 \ldots a_n} = -\epsilon_{a_2 a_1 \ldots a_n} = \ldots = (-1)^{n(n-1)/2}\epsilon_{a_n \ldots a_2 a_1},$$

yields $\phi = n(n-1)/2$. The factor $1/\sqrt{n!}$ is needed for the projection operator normalization (3.50).

Given n dimensions we cannot label more than n indices, so Levi-Civita tensors satisfy

$$0 = \quad . \tag{6.30}$$

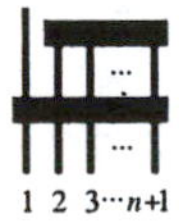

1 2 3⋯n+1

For example, for

$$n = 2: \qquad 0 = \delta_a^d \epsilon_{bc} - \delta_b^d \epsilon_{ac} + \delta_c^d \epsilon_{ab}. \tag{6.31}$$

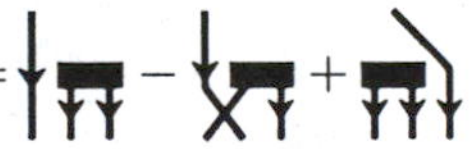

This is actually the same as the completeness relation (6.28), as can be seen by contracting (6.31) with ϵ_{cd} and using

$$n = 2: \quad = \quad = \frac{1}{2}$$

$$\epsilon_{ac}\epsilon^{bc} = \delta_a^b. \tag{6.32}$$

This relation is one of a series of relations obtained by contracting indices in the completeness relation (6.28) and substituting (6.23):

$$\epsilon_{a_n \ldots a_{k+1} b_k \ldots b_1}\epsilon^{a_n \ldots a_{k+1} a_k \ldots a_1} = k!(n-k)!\,A_{b_k \ldots b_1},{}^{a_1 \ldots a_k}$$

$$= \frac{k!(n-k)!}{n!} \quad . \tag{6.33}$$

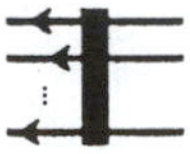

Such identities are familiar from relativistic calculations ($n = 4$):

$$\begin{aligned} \epsilon_{abcd}\epsilon^{agfe} &= \delta^{gfe}_{bcd}\,, \qquad \epsilon_{abcd}\epsilon^{abfe} = 2\delta^{fe}_{cd} \\ \epsilon_{abcd}\epsilon^{abce} &= 6\delta^{e}_{d}\,, \qquad \epsilon_{abcd}\epsilon^{abcd} = 24\,, \end{aligned} \tag{6.34}$$

where the generalized Kronecker delta is defined by

$$\frac{1}{p!}\delta^{b_1 b_2 \ldots b_p}_{a_1 a_2 \ldots a_p} = A_{a_1 a_2 \ldots a_p}{}^{b_p \ldots b_2 b_1} \,. \tag{6.35}$$

6.4 DETERMINANTS

Consider an $[n^p \times n^p]$ matrix $M_\alpha{}^\beta$ defined by a direct product of $[n \times n]$ matrices M_a^b

$$\begin{aligned} M_\alpha{}^\beta &= M_{a_1 a_2 \ldots a_p}{}^{b_p \ldots b_2 b_1} = M_{a_1}^{b_1} M_{a_2}^{b_2} \ldots M_{a_p}^{b_p} \\ M &= \quad \text{M} \quad = \quad , \end{aligned} \tag{6.36}$$

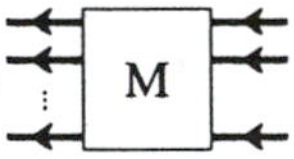

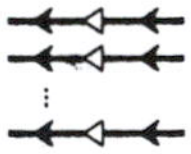

where

$$M_a^b = {}^a \; {}^b \,. \tag{6.37}$$

The trace of the antisymmetric projection of $M_\alpha{}^\beta$ is given by

$$\begin{aligned} \mathrm{tr}_p\, AM &= A_{abc\ldots d,}{}^{d'\ldots c'b'a'}\; M_{a'}^a M_{b'}^b \ldots M_{d'}^d \\ &= \quad . \end{aligned} \tag{6.38}$$

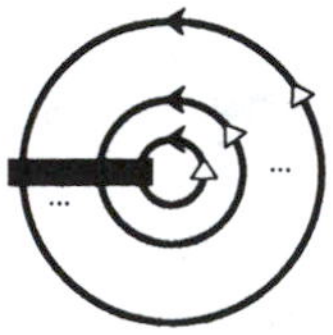

The subscript p on $\mathrm{tr}_p(\ldots)$ distinguishes the traces on $[n^p \times n^p]$ matrices M_α^β from the $[n \times n]$ matrix trace $\mathrm{tr}\, M$. To derive a recursive evaluation rule for $\mathrm{tr}_p\, AM$, use (6.19) to obtain

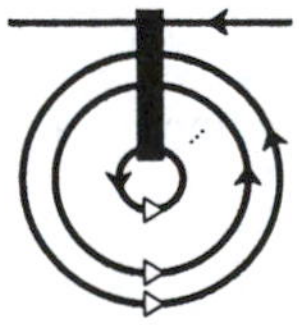

$$= \frac{1}{p}\left\{ \quad - (p-1) \quad \right\} . \tag{6.39}$$

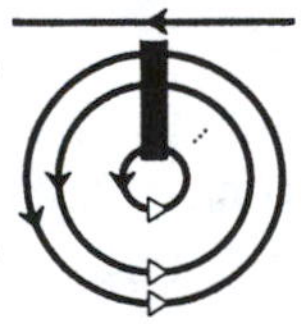

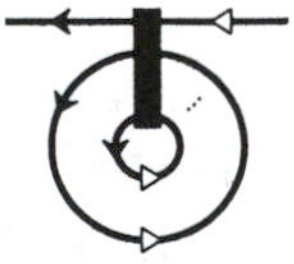

Iteration yields

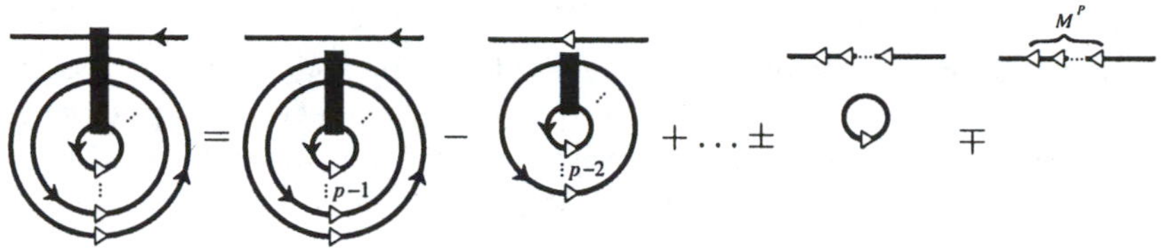

(6.40)

Contracting with M_a^b, we obtain

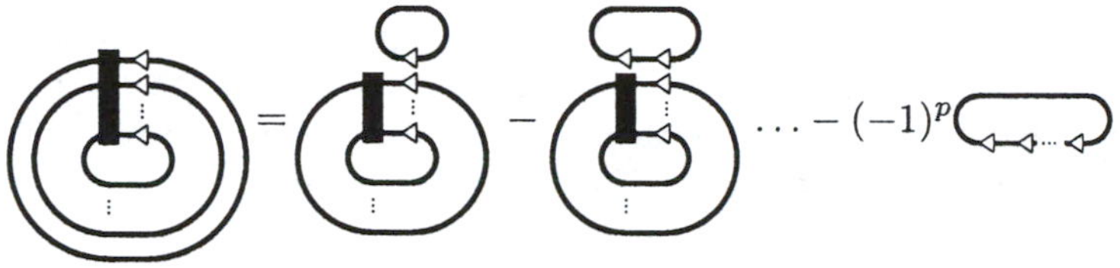

$$\mathrm{tr}_p\, AM = \frac{1}{p} \sum_{k=1}^{p} (-1)^{k-1} \left(\mathrm{tr}_{p-k}\, AM\right) \mathrm{tr}\, M^k . \tag{6.41}$$

This formula enables us to compute recursively all $\mathrm{tr}_p\, AM$ as polynomials in traces of powers of M:

$$\mathrm{tr}_0\, AM = 1 , \qquad \mathrm{tr}_1\, AM = \quad = \mathrm{tr}\, M \tag{6.42}$$

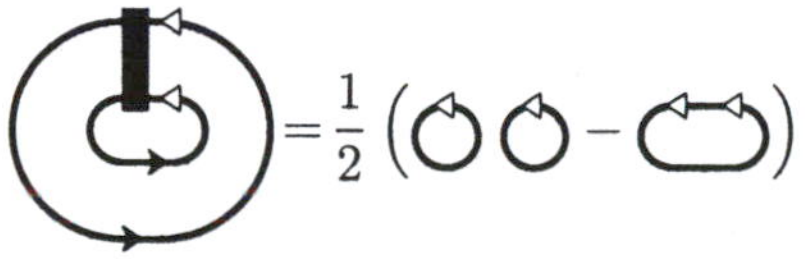

$$\mathrm{tr}_2\, AM = \frac{1}{2} \left\{ (\mathrm{tr}\, M)^2 - \mathrm{tr}\, M^2 \right\} \tag{6.43}$$

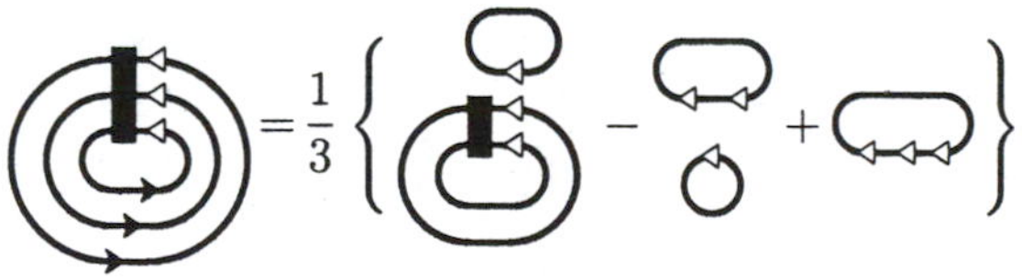

$$\mathrm{tr}_3\, AM = \frac{1}{3!} \left\{ (\mathrm{tr}\, M)^3 - 3(\mathrm{tr}\, M)(\mathrm{tr}\, M^2) + 2\ \mathrm{tr}\, M^3 \right\} \tag{6.44}$$

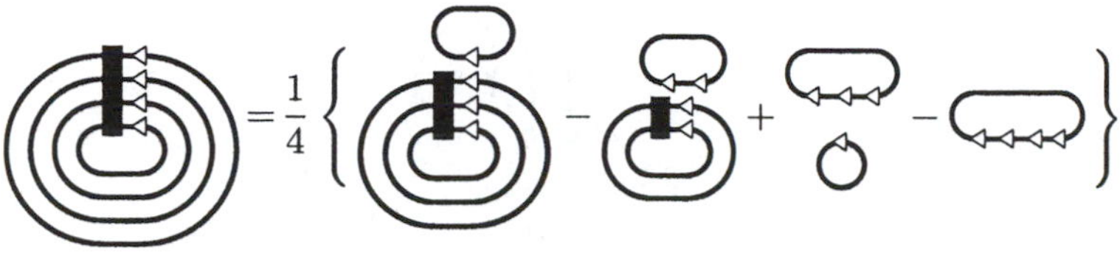

$$\begin{aligned} \mathrm{tr}_4\, AM = \frac{1}{4!} \{ & (\mathrm{tr}\, M)^4 - 6(\mathrm{tr}\, M)^2\, \mathrm{tr}\, M^2 \\ & + 3(\mathrm{tr}\, M^2)^2 + 8\ \mathrm{tr}\, M^3\, \mathrm{tr}\, M - 6\ \mathrm{tr}\, M^4 \} . \end{aligned} \tag{6.45}$$

For $p = n$ (M_a^b are $[n \times n]$ matrices) the antisymmetrized trace is the determinant

$$\det M = \mathrm{tr}_n AM = A_{a_1 a_2 \ldots a_n},{}^{b_n \ldots b_2 b_1} M_{b_1}^{a_1} M_{b_2}^{a_2} \ldots M_{b_n}^{a_n} . \tag{6.46}$$

The coefficients in the above expansions are simple combinatoric numbers. A general term for $(\mathrm{tr}\, M^{\ell_1})^{\alpha_1} \cdots (\mathrm{tr}\, M^{\ell_s})^{\alpha_s}$, with α_1 loops of length ℓ_1, α_2 loops of length ℓ_2 and so on, is divided by the number of ways in which this pattern may be obtained:

$$\ell_1^{\alpha_1} \ell_2^{\alpha_2} \ldots \ell_s^{\alpha_s} \alpha_1! \alpha_2! \ldots \alpha_s! . \tag{6.47}$$

6.5 CHARACTERISTIC EQUATIONS

We have noted that the dimension of the antisymmetric tensor space is zero for $n < p$. This is rather obvious; antisymmetrization allows each label to be used at most once, and it is impossible to label more legs than there are labels. In terms of the antisymmetrization operator this is given by the identity

$$A = 0 \qquad \text{if } p > n . \tag{6.48}$$

This trivial identity has an important consequence: it guarantees that any $[n \times n]$ matrix satisfies a characteristic (or Hamilton-Cayley or secular) equation. Take $p = n + 1$ and contract with M_a^b n index pairs of A:

$$A_{c a_1 a_2 \ldots a_n},{}^{b_n \ldots b_2 b_1 d} M_{b_1}^{a_1} M_{b_2}^{a_2} \ldots M_{b_n}^{a_n} = 0$$

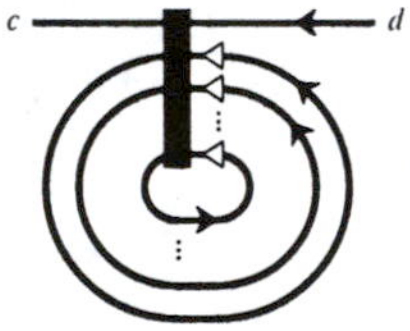

c d

$$= 0 . \tag{6.49}$$

We have already expanded this in (6.40). For $p = n+1$ we obtain the *characteristic equation*

$$0 = \sum_{k=0}^{n} (-1)^k (\mathrm{tr}_{n-k}\, AM) M^k , \tag{6.50}$$
$$= M^n - (\mathrm{tr}\, M) M^{n-1} + (\mathrm{tr}_2\, AM)\, M^{n-2} - \ldots + (-1)^n (\det M)\, \mathbf{1} .$$

6.6 FULLY (ANTI)SYMMETRIC TENSORS

We shall denote a fully *symmetric* tensor by a small circle (white dot)

$$d_{abc\ldots f} = \quad . \tag{6.51}$$

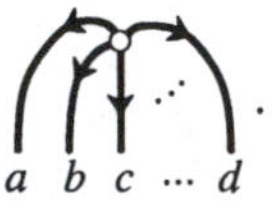

a b c ... d

A symmetric tensor $d_{abc\ldots d} = d_{bac\ldots d} = d_{acb\ldots d} = \ldots$ satisfies

$$Sd = d$$

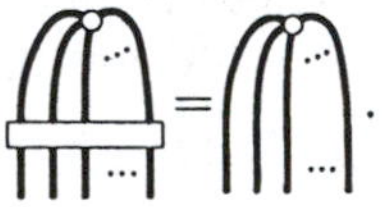

$$= \quad . \tag{6.52}$$

If this tensor is also an invariant tensor, the invariance condition (4.36) can be written as

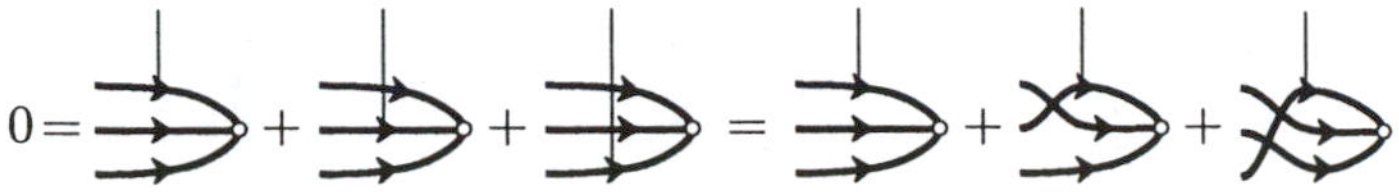

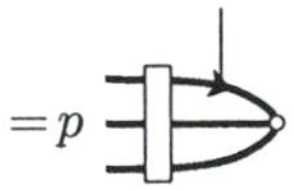

$$0 = \ldots + \ldots + \ldots = \ldots + \ldots + \ldots$$

$$= p \ldots \qquad (p = \text{number of indices})\,. \tag{6.53}$$

Hence, the invariance condition for symmetric tensors is

$$0 = \ldots\,. \tag{6.54}$$

The fully *antisymmetric* tensors with *odd* numbers of legs will be denoted by black dots

$$f_{abc\ldots d} = \ldots\,, \tag{6.55}$$

$a \; b \; c \; \cdots \; d$

with the invariance condition stated compactly as

$$0 = \ldots\,. \tag{6.56}$$

If the number of legs is *even*, an antisymmetric tensor is *anticyclic,*

$$f_{abc\ldots d} = -f_{bc\ldots da}\,, \tag{6.57}$$

and the birdtrack notation must distinguish the first leg. A black dot is inadequate for the purpose. A bar, as for the Levi-Civita tensor (6.27), or a semicircle for the symplectic invariant introduced below in (12.3), and fully skew-symmetric invariant tensors investigated in (15.27)

$$f^{ab\ldots c} = \ldots\,, \qquad f_{ab\ldots c} = \ldots \tag{6.58}$$

or a similar notation fixes the problem.

6.7 IDENTICALLY VANISHING TENSORS

Noting that a given group-theoretic weight vanishes *identically* is often an important step in a birdtrack calculation. Some examples are

$$\ldots \equiv 0\,, \qquad \ldots \equiv 0\,, \tag{6.59}$$

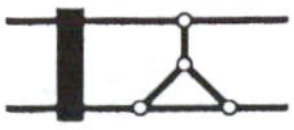

$$\ldots \equiv 0\,, \qquad \ldots \equiv 0\,. \tag{6.60}$$

In graph theory [267, 293] the left graph in (6.59) is known as the Kuratowsky graph, and the right graph in (6.60) as the Peterson graph.

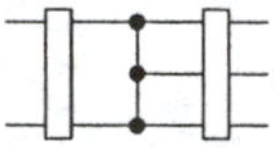

$$\equiv 0 , \quad \equiv 0 , \quad \equiv 0 , \qquad (6.61)$$

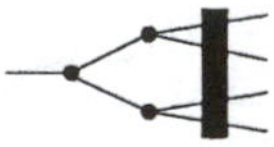

$$\equiv 0 , \quad \equiv 0 , \qquad (6.62)$$

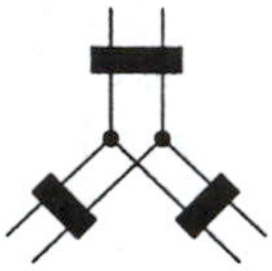

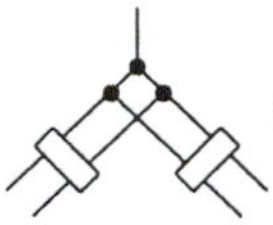

$$\equiv 0 , \quad \equiv 0 . \qquad (6.63)$$

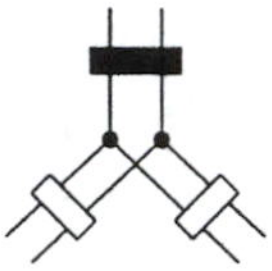

The above identities hold for any antisymmetric 3-index tensor; in particular, they hold for the Lie algebra structure constants iC_{ijk}. They are proven by mapping a diagram into itself by index transpositions. For example, interchange of the top and bottom vertices in (6.59) maps the diagram into itself, but with the $(-1)^5$ factor.

From the Lie algebra (4.47) it also follows that for any irreducible rep we have

$$= 0 , \quad = 0 . \qquad (6.64)$$

Chapter Seven

Casimir operators

The construction of invariance groups, developed elsewhere in this monograph, is self-contained, and none of the material covered in this chapter is necessary for understanding the remainder of the monograph. We have argued in section 5.2 that all relevant group-theoretic numbers are given by vacuum bubbles (reduced matrix elements, $3n$-j coefficients, *etc.*), and we have described the algorithms for their evaluation. That is all that is really needed in applications.

However, one often wants to cross-check one's calculation against the existing literature. In this chapter we discuss why and how one introduces casimirs (or Dynkin indices), we construct independent Casimir operators for the classical groups and finally we compile values of a few frequently used casimirs.

Our approach emphasizes the role of primitive invariants in constructing reps of Lie groups. Given a list of primitives, we present a systematic algorithm for constructing invariant matrices M_i and the associated projection operators (3.48).

In the canonical, Cartan-Killing approach one faces a somewhat different problem. Instead of the primitives, one is given the generators T_i explicitly and no other invariants. Hence, the invariant matrices M_i can be constructed only from contractions of generators; typical examples are matrices

$$M_2 = \quad \mu , \qquad M_4 = \quad \sigma \;\; \mu , \qquad \ldots , \tag{7.1}$$

where σ, μ could be any reps, reducible or irreducible. Such invariant matrices are called *Casimir operators*.

What is a minimal set of Casimir operators, sufficient to reduce any rep to its irreducible subspaces? Such sets can be useful, as the corresponding r Casimir operators uniquely label each irreducible rep by their eigenvalues $\lambda_1, \lambda_2, \ldots \lambda_r$.

The invariance condition for any invariant matrix (3.31) is

$$0 = [T_i, M] = \quad \mu \quad - \quad \mu$$

so all Casimir operators commute

$$M_2 M_4 = \quad \mu \quad = \quad \mu \quad = M_4 M_2, \; etc. ,$$

and, according to section 3.6, can be used to simultaneously decompose the rep μ. If $M_1, M_2, \ldots$ have been used in the construction of projection operators (3.48), any matrix polynomial $f(M_1, M_2 \ldots)$ takes value $f(\lambda_1, \lambda_2, \ldots)$ on the irreducible subspace projected by P_i, so polynomials in M_i induce no further decompositions. Hence, it is sufficient to determine the finite number of M_i's that form a polynomial basis for all Casimir operators (7.1). Furthermore, as we show in the next section, it is sufficient to restrict the consideration to the symmetrized casimirs. This observation enables us to explicitly construct, in section 7.2, a set of independent casimirs for each classical group.

Exceptional groups pose a more difficult challenge, partially met here in a piecemeal fashion in chapters on each of the exceptional groups. For a definitive, systematic calculation of all casimirs for all simple Lie groups, consult van Ritbergen, Schellekens, and Vermaseren [294].

7.1 CASIMIRS AND LIE ALGEBRA

There is no general agreement on a unique definition of a Casimir operator. We could choose to call the trace of a product of k generators

$$\mathrm{tr}(T_i T_j \ldots T_k) = \quad , \tag{7.2}$$

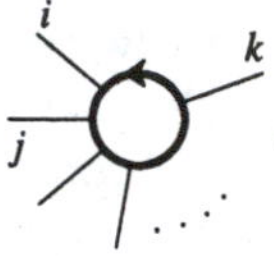

a kth order *casimir*. With such definition,

$$\mathrm{tr}(T_j T_i \ldots T_k) =$$

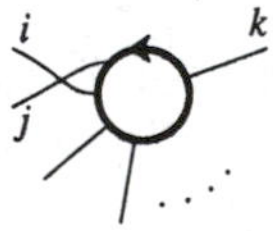

would also be a casimir, independent of the first one. However, all traces of T_i's that differ by a permutation of indices are related by Lie algebra. For example,

$$= - \;. \tag{7.3}$$

The last term involves a $(k\text{-}1)$th order casimir and is antisymmetric in the i, j indices. Only the fully symmetrized traces

$$h_{ij\ldots k} \equiv \frac{1}{p!} \sum_{perm} \mathrm{tr}(T_i T_j \ldots T_k) = \tag{7.4}$$

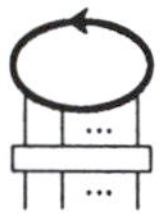

are not affected by the Lie algebra relations. Hence from now on, we shall use the term "casimir" to denote *symmetrized* traces (ref. [248] follows the same usage, for example). Any unsymmetrized trace $\mathrm{tr}(T_i T_j \ldots T_k)$ can be expressed in terms of the symmetrized traces. For example, using the symmetric group identity (see figure 9.1)

$$= + + \frac{4}{3} + \frac{4}{3} \;, \tag{7.5}$$

the Jacobi identity (4.48) and the d_{ijk} definition (9.87), we can express the trace of four generators in any rep of any semisimple Lie group in terms of the quartic and cubic casimirs:

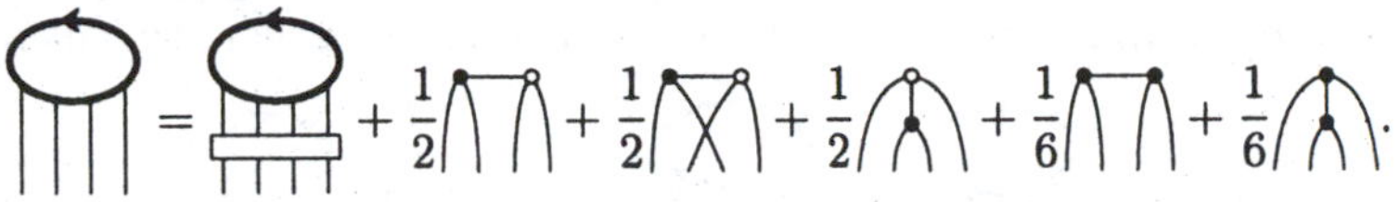

$$= + \frac{1}{2} + \frac{1}{2} + \frac{1}{2} + \frac{1}{6} + \frac{1}{6} . \quad (7.6)$$

In this way, an arbitrary kth order trace can be written as a sum over tree contractions of casimirs. The symmetrized casimirs (7.4) are conveniently manipulated as monomial coefficients:

$$\operatorname{tr} X^k = h_{ij\ldots m}\, x_i x_j \ldots x_m . \quad (7.7)$$

For a rep λ, X is a $[d_\lambda \times d_\lambda]$ matrix $X = x_i T_i$, where x_i is an arbitrary N-dimensional vector. We shall also use a birdtrack notation (6.37):

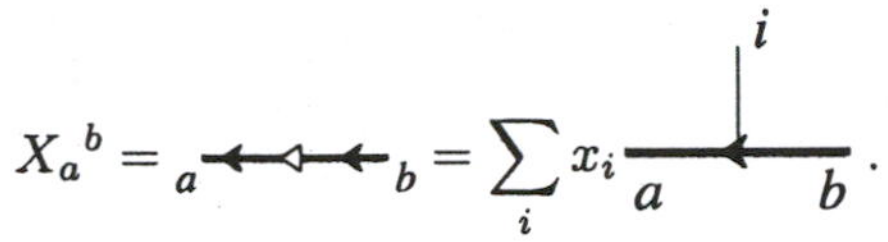

$$X_a{}^b = {}_a \longleftarrow {}_b = \sum_i x_i \; {}^{i}_{a \quad b} . \quad (7.8)$$

The symmetrization (7.4) is automatic

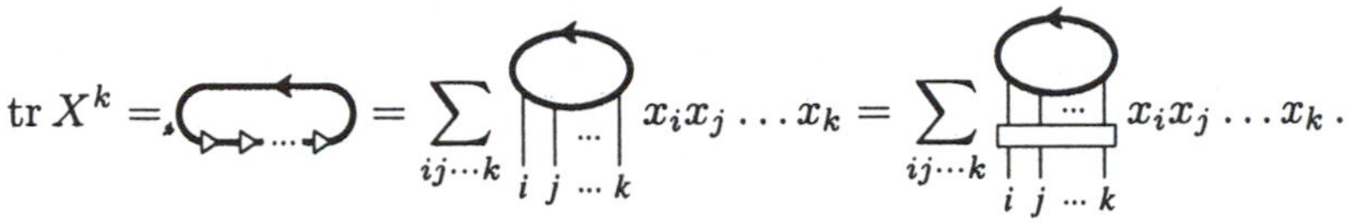

$$\operatorname{tr} X^k = = \sum_{ij\cdots k} \; x_i x_j \ldots x_k = \sum_{ij\cdots k} \; x_i x_j \ldots x_k . \quad (7.9)$$

7.2 INDEPENDENT CASIMIRS

Not all $\operatorname{tr} X^k$ are independent. For an n-dimensional rep a typical relation relating various $\operatorname{tr} X^k$ is the characteristic equation (6.50):

$$X^n = (\operatorname{tr} X) X^{n-1} - (\operatorname{tr}_2 AX) X^{n-2} + \ldots \pm (\det X). \quad (7.10)$$

Scalar coefficients $\operatorname{tr}_k AX$ are polynomials in $\operatorname{tr} X^m$, computed in section 6.4. The characteristic equation enables us to express any $X^p, p \geq n$ in terms of the matrix powers $X^k, k < n$ and the scalar coefficients $\operatorname{tr} X^k, k \leq n$. Therefore, if a group has an n-dimensional rep, it has at most n independent casimirs,

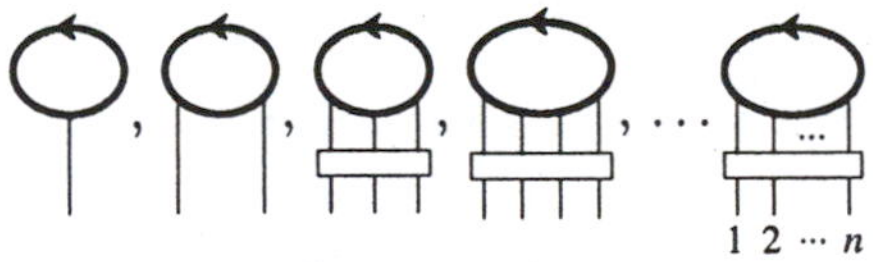

$$, \; , \; , \; , \; \cdots \quad (1\,2 \cdots n)$$

corresponding to $\operatorname{tr} X, \operatorname{tr} X^2, \operatorname{tr} X^3, \ldots \operatorname{tr} X^n$.

For a simple Lie group, the number of independent casimirs is called the *rank* of the group and is always smaller than n, the dimension of the lowest-dimensional rep. For example, for all simple groups $\operatorname{tr} T_i = 0$, the first casimir is always identically zero. For this reason, the rank of $SU(n)$ is $n - 1$, and the independent casimirs are

$$SU(n): \quad , \; , \; , \ldots , \; . \quad (7.11)$$

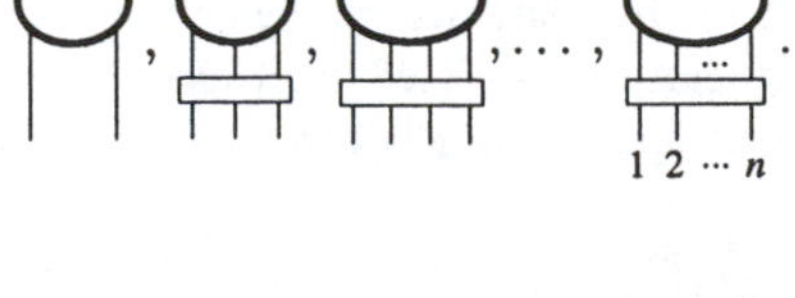

1 2 ⋯ n

The defining reps of $SO(n)$, $Sp(n)$, G_2, F_4, E_7 and E_8 groups have an invertible bilinear invariant g_{ab}, either symmetric or skew-symmetric. Inserting $\delta_a^c = g_{ab}g^{bc}$ any place in a trace of k generators, and moving the tensor g_{ab} through the generators by means of the invariance condition (10.5), we can reverse the defining rep arrow:

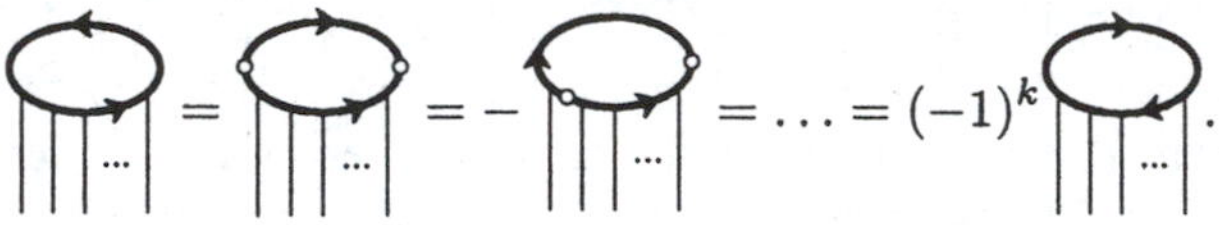

$$= \quad = - \quad = \ldots = (-1)^k \quad . \qquad (7.12)$$

Hence for the above groups, $\operatorname{tr} X^k = 0$ for k odd, and all their casimirs are of even order.

The odd and the even-dimensional orthogonal groups differ in the orders of independent casimirs. For $n = 2r + 1$, there are r independent casimirs

$$SO(2r+1): \quad , \quad , \ldots , \quad . \qquad (7.13)$$

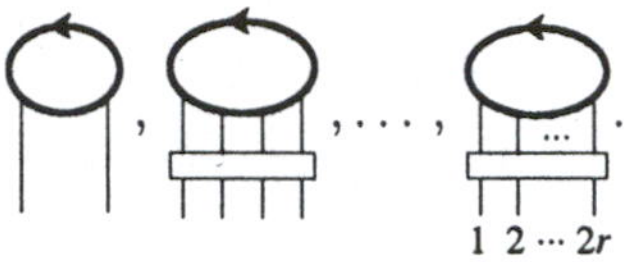

1 2 ⋯ 2r

For $n = 2r$, a symmetric invariant can be formed by contracting r defining reps with a Levi-Civita tensor (the adjoint projection operator (10.13) is antisymmetric):

$$I_r(x) = \quad . \qquad (7.14)$$

$\operatorname{tr} X^{2r}$ is not independent, as by (6.28), it is contained in the expansion of $I_r(x)^2$

$$I_r(x)^2 = \quad = \quad \simeq \quad + \ldots . \qquad (7.15)$$

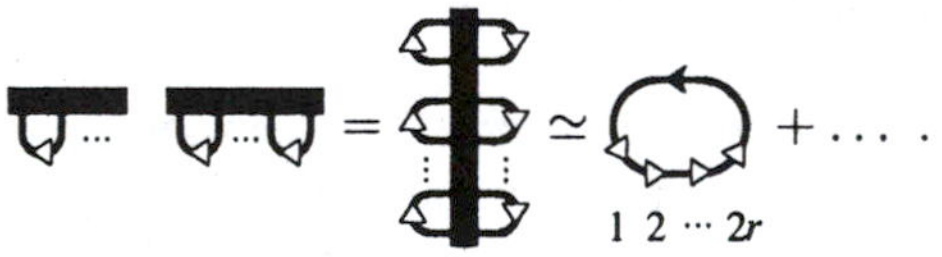

1 2 ⋯ 2r

Hence, the r independent casimirs for even-dimensional orthogonal groups are

$$SO(2r): \quad , \quad , \ldots , \quad , \quad . \qquad (7.16)$$

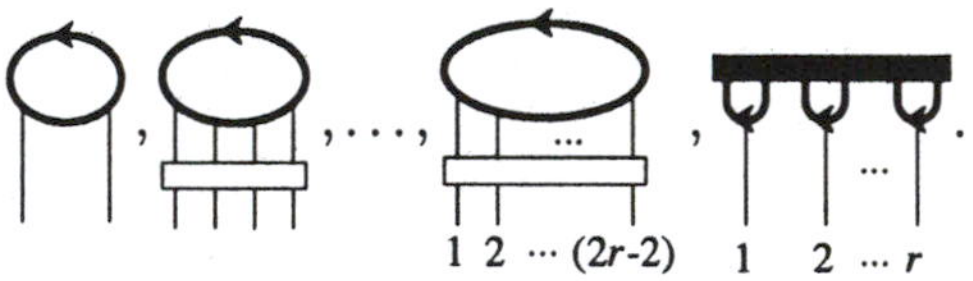

1 2 ⋯ (2r-2) 1 2 ⋯ r

For $Sp(2r)$ there are no special relations, and the r independent casimirs are $\operatorname{tr} X^{2k}, 0 < l \leq r$;

$$Sp(2r): \quad , \quad , \ldots , \quad . \qquad (7.17)$$

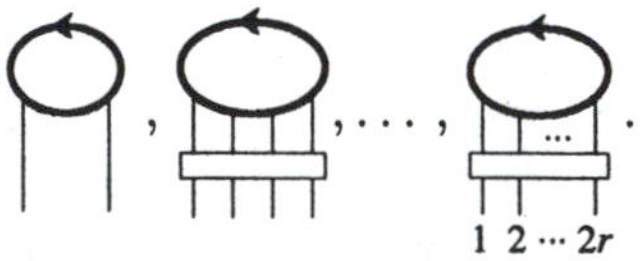

1 2 ⋯ 2r

The characteristic equation (7.10), by means of which we count the independent casimirs, applies to the lowest-dimensional rep of the group, and one might worry that other reps might be characterized by further independent casimirs. The answer is no; all casimirs can be expressed in terms of the defining rep. For $SU(n)$, $Sp(n)$ and $SO(n)$ tensor reps this is obvious from the explicit form of the generators in

A_r	2, 3, ..., $r+1$	$\sim$	$SU(r+1)$
B_r	2, 4, 6, ..., $2r$	$\sim$	$SO(2r+1)$
C_r	2, 4, 6, ..., $2r$	$\sim$	$Sp(2r)$
D_r	2, 4, ..., $2r-2$, r	$\sim$	$SO(2r)$
G_2	2, 6		
F_4	2, 6, 8, 12		
E_6	2, 5, 6, 8, 9, 12		
E_7	2, 6, 8, 10, 12, 14, 18		
E_8	2, 8, 12, 14, 18, 20, 24, 30		

Table 7.1 Betti numbers for the simple Lie groups.

higher reps (see section 9.4 and related results for $Sp(n)$ and $SO(n)$); they are tensor products of the defining rep generators and Kronecker deltas, and a higher rep casimir always reduces to sums of the defining rep casimirs, times polynomials in n (see examples of section 9.7).

For the exceptional groups, cubic and higher defining rep invariants enter, and the situation is not so trivial. We shall show below, by explicit computation, that $\mathrm{tr}\, X^3 = 0$ for E_6 and $\mathrm{tr}\, X^4 = c(\mathrm{tr}\, X^2)^2$ for all exceptional groups. We shall also prove the reduction to the 2nd- and 6th-order casimirs for G_2 in section 16.4 and partially prove the reduction for other exceptional groups in section 18.8. The orders of all independent casimirs are known [30, 288, 134, 54] as the Betti numbers, listed here in table 7.1. There are too many papers on computation of casimirs to even attempt a survey here; we recommend ref. [294].

7.3 ADJOINT REP CASIMIRS

For simple Lie algebras the Cartan-Killing bilinear form (4.41) is proportional to δ_{ij}, so by the argument of (7.12) all adjoint rep casimirs are even. In addition, the Jacobi identity (4.48) relates a loop to a symmetrized trace together with a set of tree contractions of lower casimirs, linearly indepenent under applications of the Jacobi identity. For example, we have from (7.6)

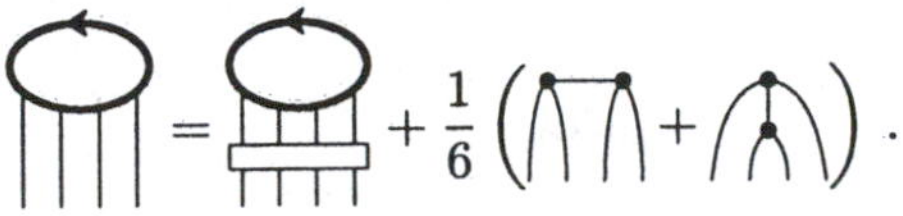

$$= \ + \frac{1}{6}\left(\ + \ \right) . \qquad (7.18)$$

The numbers of linearly independent tree contractions are discussed in ref. [73].

7.4 CASIMIR OPERATORS

Most physicists would not refer to $\operatorname{tr} X^k$ as a casimir. Casimir's [49] quadratic operator and its generalizations [288] are $[d_\mu \times d_\mu]$ matrices:

$$(I_p)_a^b = \quad = [\operatorname{tr}_\lambda(T_k \ldots T_i T_j)]\,(T_i T_j \ldots T_k)_a^b. \qquad (7.19)$$

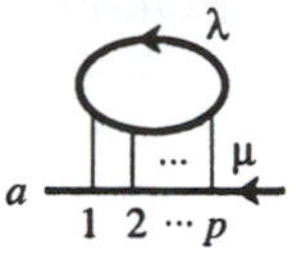

We have shown in section 5.2 that all invariants are reducible to $6j$ coefficients. I_p's are particularly easy to express in terms of $6j$'s. Define

$$M_{b\,,\,\beta}^{\alpha\;\mu} = \qquad \alpha,\beta = 1,\ldots,d_\lambda\,,\quad a,b = 1,2,\ldots,d_\mu\,. \qquad (7.20)$$

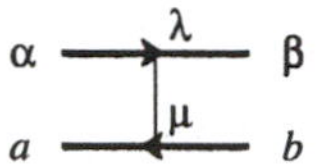

Inserting the complete Clebsch-Gordan series (5.8) for $\lambda \otimes \mu$, we obtain

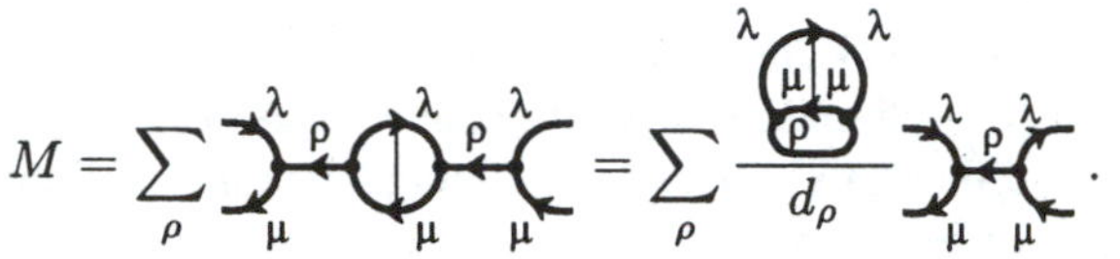

$$M = \sum_\rho \qquad = \sum_\rho \frac{\quad}{d_\rho} \qquad (7.21)$$

The eigenvalues of M are Wigner's $6j$ coefficients (5.15). It is customary to express these $6j$'s in terms of quadratic Casimir operators by using the invariance condition (4.40):

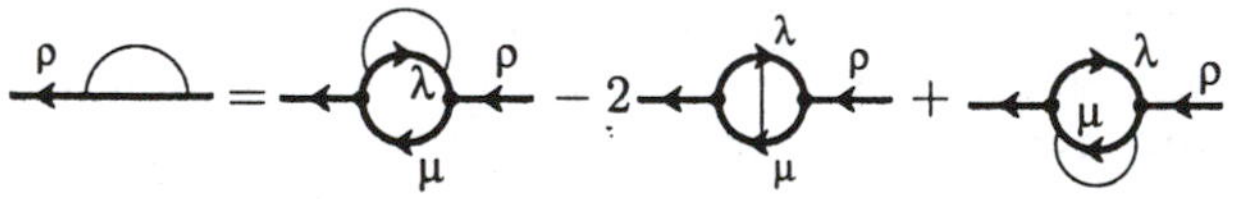

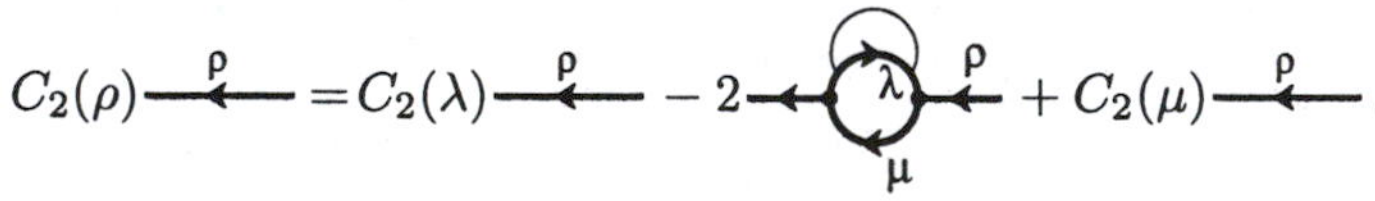

$$C_2(\rho) \quad = C_2(\lambda) \quad - 2 \quad + C_2(\mu) \quad . \qquad (7.22)$$

This is an ancient formula familiar from quantum mechanics textbooks: if the total angular momentum is $J = L + S$, then $L \cdot S = \frac{1}{2}(J^2 - L^2 - S^2)$. In the present case we trace both sides to obtain

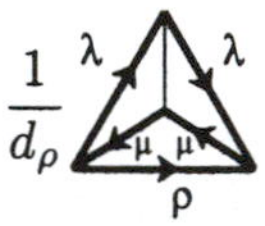

$$\frac{1}{d_\rho} \quad = -\frac{1}{2}\left\{C_2(\rho) - C_2(\lambda) - C_2(\mu)\right\}. \qquad (7.23)$$

The pth order casimir is thus [255]

$$(I_p)_a^b = (M^p)_{a\alpha}^{\alpha b} = \sum_\rho^{irreduc.} \left(\frac{C_2(\rho) - C_2(\lambda) - C_2(\mu)}{2}\right)^p \quad .$$

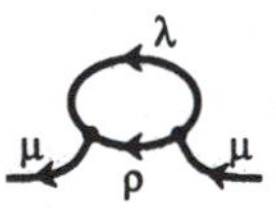

If μ is an irreducible rep, (5.23) yields

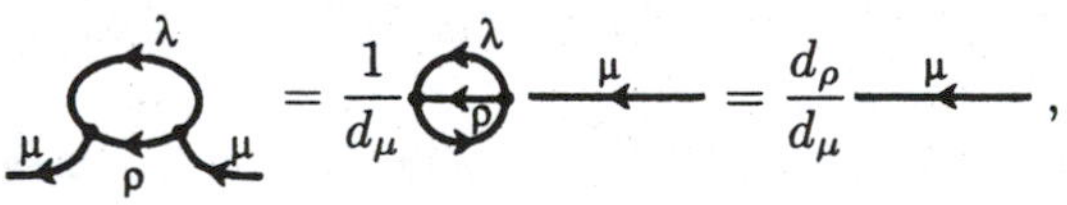

$$\quad = \frac{1}{d_\mu} \quad = \frac{d_\rho}{d_\mu} \quad ,$$

and the μ rep eigenvalue of I_p is given by

$$\sum_\rho \left(\frac{C_2(\rho) - C_2(\lambda) - C_2(\mu)}{2} \right)^p d_\rho \,. \tag{7.24}$$

Here the sum goes over all $\rho \subset \lambda \otimes \mu$, where ρ, λ and μ are irreducible reps.

Another definition of the generalized Casimir operator, in the spirit of (7.4), uses the fully symmetrized trace:

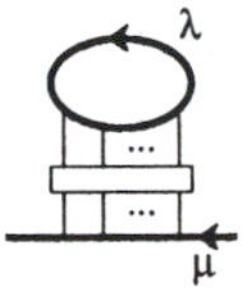

$$= h(\lambda)_{ij\ldots k} (T_i T_j \ldots T_k)_a^b \,. \tag{7.25}$$

We shall return to this definition in the next section.

7.5 DYNKIN INDICES

As we have seen so far, there are many ways of defining casimirs; in practice it is usually quicker to directly evaluate a given birdtrack diagram than to relate it to somebody's "standard" casimirs. Still, it is good to have an established convention, if for no other reason than to be able to cross-check one's calculation against the tabulations available in the literature.

Usually a rep is specified by its dimension. If the group has several inequivalent reps with the same dimensions, further numbers are needed to uniquely determine the rep. Specifying the *Dynkin index* [104],

$\ell_\lambda =$ 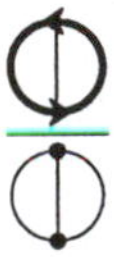 $= \dfrac{\mathrm{tr}_\lambda(T_i T_i)}{\mathrm{tr}(C_i C_i)} \,,$ (7.26)

usually (but not always) does the job. A Dynkin index is easy to evaluate by birdtrack methods. By the Lie algebra (4.47), the defining rep Dynkin index is related to a $6j$ coefficient:

$\ell^{-1} =$ 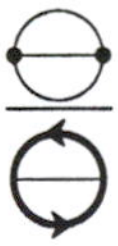 $= \dfrac{2}{a^2 N}$  $= \dfrac{2N}{n} - \dfrac{2}{N}\dfrac{1}{a^2}$  . (7.27)

The $6j$ coefficient  $= \mathrm{tr}(T_i T_j T_i T_j)$ is evaluated by the usual birdtrack tricks. For $SU(n)$, for example

 $=$  $- \dfrac{1}{n}$  $= -\dfrac{n^2 - 1}{n}.$ (7.28)

The Dynkin index of a rep ρ in the Clebsch-Gordan series for $\lambda \otimes \mu$ is related to a $6j$ coefficient by (7.23):

$\ell_\rho / d_\rho = \ell_\lambda / d_\lambda + \ell_\mu / d_\mu + 2 \dfrac{\ell}{N} \dfrac{1}{d_\rho}$  . (7.29)

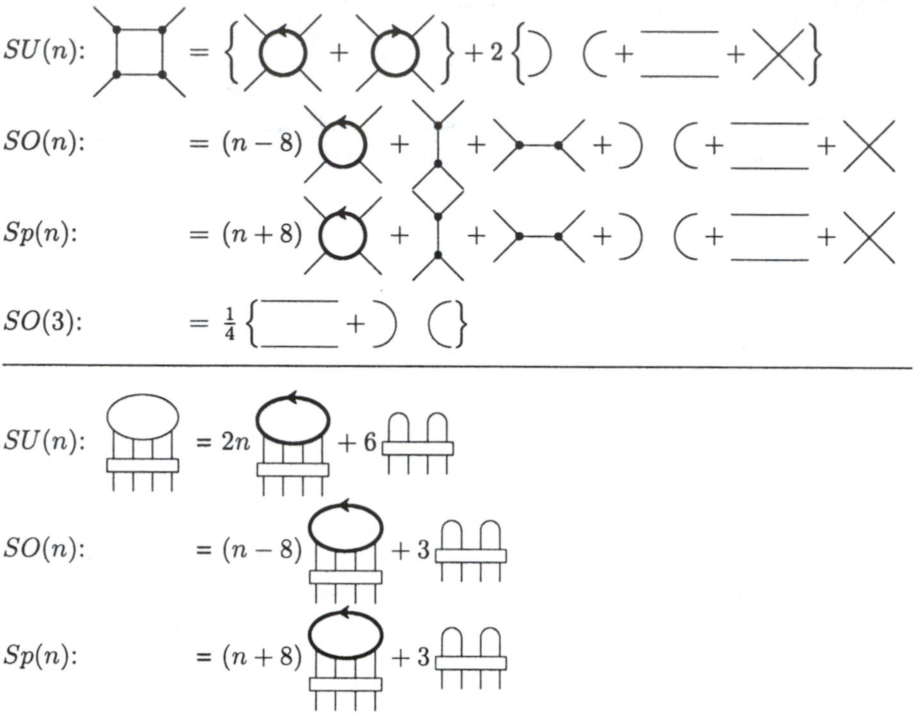

$SU(n)$: = { + } + 2{ + + }

$SO(n)$: $= (n-8)$ + + + +

$Sp(n)$: $= (n+8)$ + + + +

$SO(3)$: $= \frac{1}{4}\{$ + $\}$

$SU(n)$: $= 2n$ $+ 6$

$SO(n)$: $= (n-8)$ $+ 3$

$Sp(n)$: $= (n+8)$ $+ 3$

Table 7.2 (*Top*) Expansions of the adjoint rep quartic casimirs in terms of the defining rep, and (*bottom*) reduction of adjoint quartic casimirs to the defining rep quartic casimirs, for the classical simple Lie algebras. The normalization (7.38) is set to $a = 1$.

We shall usually evaluate Dynkin indices by this relation. Another convenient formula for evaluation of Dynkin indices for semisimple groups is

$$\ell_\lambda = \frac{\mathrm{tr}_\lambda X^2}{\mathrm{tr}_A X^2}, \tag{7.30}$$

with X defined in section 6.7. An application of this formula is given in section 9.7.

The form of the Dynkin index is motivated by a few simple considerations. First, we want an invariant number, so we trace all indices. Second, we want a pure, normalization independent number, so we take a ratio. $\mathrm{tr}(C_i C_i)$ is the natural normalization scale, as all groups have the adjoint rep. Furthermore, unlike the Casimir operators (7.19), which have single eigenvalues $I_p(\lambda)$ only for irreducible reps, the Dynkin index is a pure number for both reducible and irreducible reps. [Exercise: compute the Dynkin index for $U(n)$.]

The above criteria lead to the Dynkin index as the unique group-theoretic scalar corresponding to the quadratic Casimir operator. The choice of group-theoretic scalars corresponding to higher casimirs is rather more arbitrary. Consider the reductions of I_4 for the adjoint reps, tabulated in table 7.2. (The $SU(n)$ was evaluated as an introductory example, section 2.2. The remaining examples are evaluated by inserting the appropriate adjoint projection operators, derived below.)

Quartic casimirs contain quadratic bits, and in general, expansions of $h(\lambda)$'s in terms of the defining rep will contain lower-order casimirs. To construct the "pure" pth order casimirs, we introduce

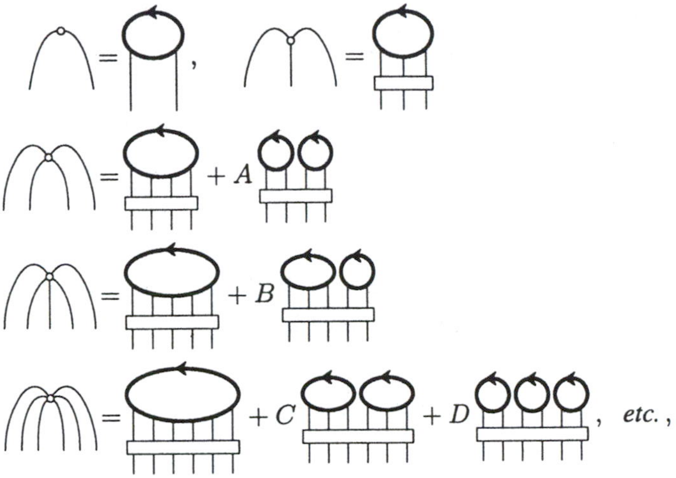

$$+A \qquad (7.31)$$

$$+B$$

$$+C \qquad +D \qquad , \quad etc.,$$

and fix the constants $A, B, C, \ldots$ by requiring that these casimirs are *orthogonal*:

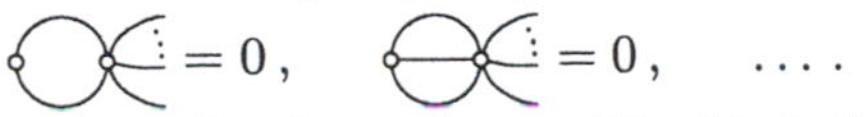

$$= 0\,, \qquad = 0\,, \qquad \ldots\,. \qquad (7.32)$$

Now we can define the *generalized* or *orthogonal* Dynkin indices [259, 294] by

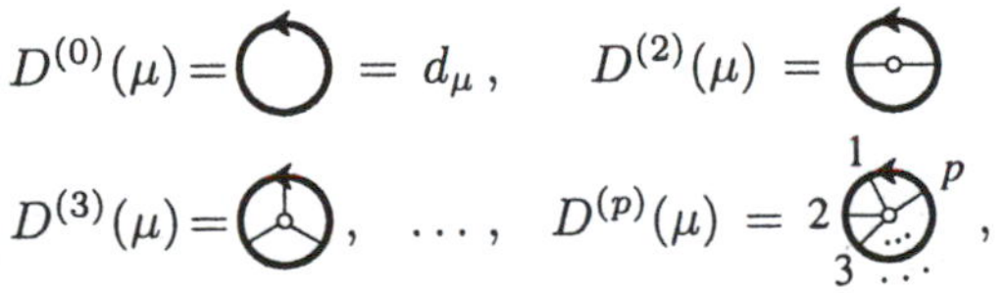

$$D^{(0)}(\mu) = \quad = d_\mu\,, \qquad D^{(2)}(\mu) =$$

$$D^{(3)}(\mu) = \quad , \quad \ldots\,, \quad D^{(p)}(\mu) = \quad , \qquad (7.33)$$

where the thick line stands for μ rep. Here we have chosen normalization $\mathrm{tr}(C_i C_i) = 1$.

The generalized Dynkin indices are not particularly convenient or natural from the computational point of view (see ref. [294] for discussion of indices in "orthogonal basis") but they do have some nice properties. For example (as we shall show later on), the exceptional groups $\mathrm{tr}\, X^4 = C(\mathrm{tr}\, X^2)^2$ are singled out by $D^{(4)} = 0$.

If μ is a Kronecker product of two reps, $\mu = \lambda \otimes \rho$, the generalized Dynkin indices satisfy

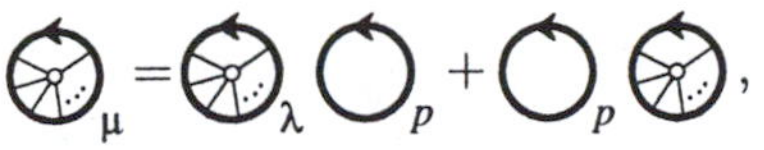

$$D^{(p)}(\mu) = D^{(p)}(\lambda) d_\rho + d_\lambda D^{(p)}(\rho) > 0\,, \qquad (7.34)$$

as the cross terms vanish by the orthonormality conditions (7.32). Substituting the completeness relation (5.7), $\lambda \otimes \rho = \sum \sigma$, we obtain a family of *sum rules* for the generalized Dynkin indices:

$$\sum_\sigma \quad = \sum_\sigma D^{(p)}(\sigma) = D^{(p)}(\lambda) d_\rho + d_\lambda D^{(p)}(\rho). \qquad (7.35)$$

For $p = 2$ this is a $\lambda \otimes \rho = \sum \sigma$ sum rule for Dynkin indices (7.28)

$$\ell_\lambda d_\rho + d_\lambda \ell_\rho = \sum_\sigma \ell_\sigma , \tag{7.36}$$

useful in checking Clebsch-Gordan decompositions.

7.6 QUADRATIC, CUBIC CASIMIRS

As the low-order Casimir operators appear so often in physics, it is useful to list them and their relations.

Given two generators T_i, T_j in $[n \times n]$ rep λ, there are only two ways to form a loop:

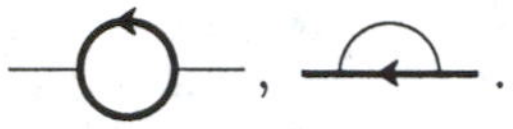

If the λ rep is irreducible, we define C_F casimir as

$$= C_F$$

$$(T_i T_i)_a^b = C_F \delta_a^b . \tag{7.37}$$

If the adjoint rep is irreducible, we define

$$= a$$

$$\mathrm{tr}\, T_i T_j = a\, \delta_{ij} . \tag{7.38}$$

Usually we take λ to be the defining rep and fix the overall normalization by taking $a = 1$. For the adjoint rep (dimension N), we use notation

$$= C_{ik\ell} C_{jk\ell} = C_A \delta_{ij} . \tag{7.39}$$

Existence of the quadratic Casimir operator C_A is a necessary and sufficient condition that the Lie algebra is semisimple [10, 104, 273]. For compact groups $C_A > 0$. C_F, a, C_A, and ℓ, the Dynkin index (7.28), are related by tracing the above expressions:

$$= nC_F = Na = NC_A \ell . \tag{7.40}$$

While the Dynkin index is normalization independent, one of C_F, a or C_A has to be fixed by a convention. The cubic invariants formed from T_i's and C_{ijk}'s are (all but one) reducible to the quadratic Casimir operators:

$$= \left(\frac{aN}{n} - \frac{C_A}{2} \right) \tag{7.41}$$

$$= \frac{C_A}{2} \tag{7.42}$$

$$= \frac{C_A}{2} . \tag{7.43}$$

This follows from the Lie algebra (4.47)

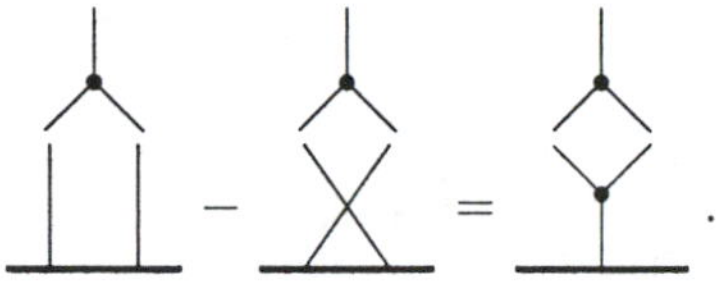

The one exception is the symmetrized third-order casimir

$$\frac{1}{2}d_{ijk} = \quad \equiv \frac{1}{2a}\left\{ \quad + \quad \right\}. \tag{7.44}$$

By (7.12) this vanishes for all groups whose defining rep is not complex. That leaves behind only $SU(n), n \geq 3$ and E_6. As we shall show in section 18.6, $d_{ijk} = 0$ for E_6, so only $SU(n)$ groups have nonvanishing cubic casimirs.

7.7 QUARTIC CASIMIRS

There is no unique definition of a quartic casimir. Any group-theoretic weight that contains a trace of four generators

(7.45)

can be called a *quartic casimir*. For example, a 4-loop contribution to the QCD β function

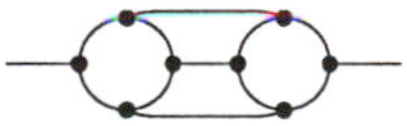

(7.46)

contains two quartic casimirs. This weight cannot be expressed as a function of quadratic casimirs and has to be computed separately for each rep and each group. For example, such quartic casimirs need to be evaluated for the purpose of classification of grand unified theories [255], weak coupling expansions in lattice gauge theories [80] and the classification of reps of simple Lie algebras [234].

Not every birdtrack diagram that contains a trace of four generators is a genuine quartic casimir. For example,

(7.47)

is reducible by (7.42) to

$$\frac{1}{4} \tag{7.48}$$

and equals $\frac{1}{4}aC_A^2$ for a simple Lie algebra. However, if all loops contain four vertices or more, Lie algebra cannot be used to reduce the diagram. For example,

$$= \quad - \quad . \tag{7.49}$$

	$\frac{1}{N}$ 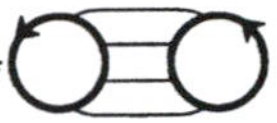	$\frac{1}{N}$ 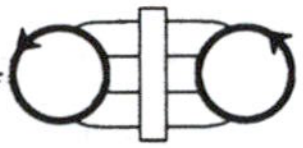	$\frac{1}{N}$ 	$\frac{1}{N}$ 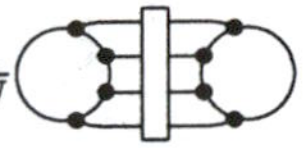	$\frac{1}{N}$ 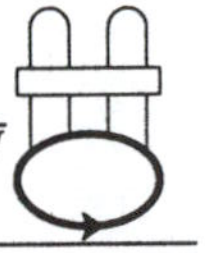
$U(n)$	n^2	$\frac{n^2+5}{6}$	$2n^2(n^2+12)$	$\frac{2n^2(n^2+36)}{3}$	$\frac{2n^2+1}{3n}$
$SU(n)$	$\frac{n^4-3n^2+3}{n^2}$	$\frac{n^4-6n^2+18}{6n^2}$	$2n^2(n^2+12)$	$\frac{2n^2(n^2+36)}{3}$	$\frac{2n^2-3}{3n}$
$SO(n)$	$\frac{n^2-3n+4}{8}$	$\frac{n^2-n+4}{24}$	$\frac{(n-2)(n^3-9n^2+54n-104)}{8}$	$\frac{(n-2)(n^3-15n^2+138n-296)}{24}$	$\frac{2n-1}{6}$
$Sp(n)$	$\frac{n^2+3n+4}{8}$	$\frac{n^2+n+4}{24}$	$\frac{(n+2)(n^3+9n^2+54n+104)}{8}$	$\frac{(n+2)(n^3+15n^2+138n+296)}{24}$	$\frac{2n+1}{6}$
$G_2(7)$	$\frac{5}{3}$	$\frac{1}{3}$	$\frac{164}{3}$	$\frac{100}{3}$	$\frac{4}{3}$
$F_4(26)$	$\frac{7}{8}$	$\frac{1}{8}$	$\frac{79}{8}$	$\frac{25}{8}$	$\frac{3}{2}$
$E_6(27)$	$\frac{41}{27}$	$\frac{5}{27}$	28	$\frac{20}{3}$	$\frac{20}{9}$
$E_7(56)$	$\frac{53}{64}$	$\frac{5}{64}$	8	$\frac{320}{81}$	$\frac{15}{8}$
$E_8(248)$	$\frac{11}{120}$	$\frac{1}{120}$	$\frac{11}{120}$	$\frac{1}{120}$	$\frac{5}{6}$

Table 7.3 Various quartic casimirs for all simple Lie algebras. The normalization in (7.38) is set to $a = 1$.

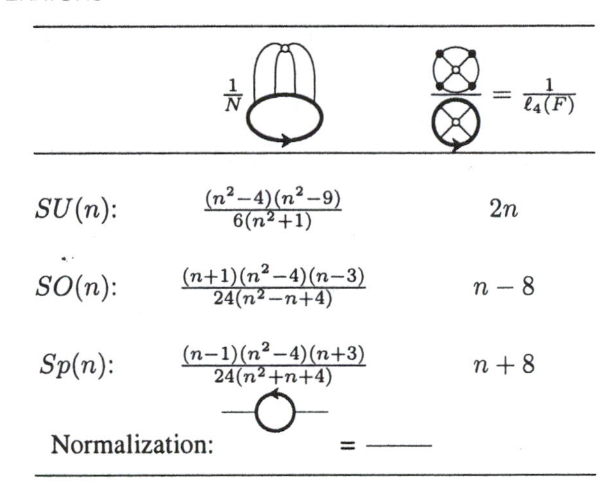

	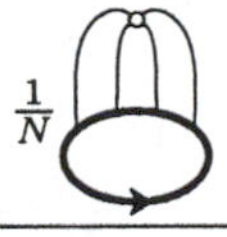	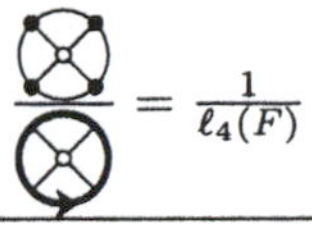
$SU(n)$:	$\frac{(n^2-4)(n^2-9)}{6(n^2+1)}$	$2n$
$SO(n)$:	$\frac{(n+1)(n^2-4)(n-3)}{24(n^2-n+4)}$	$n-8$
$Sp(n)$:	$\frac{(n-1)(n^2-4)(n+3)}{24(n^2+n+4)}$	$n+8$
Normalization:	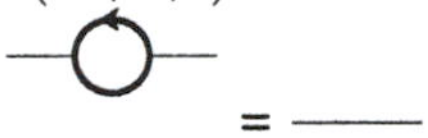	

Table 7.4 Quartic Dynkin indices (7.33) for the defining and the adjoint reps of classical groups. For the exceptional groups the quartic Dynkin indices vanish identically.

The second diagram on the right-hand side is reducible, but the first one is not. Hence, at least one quartic casimir from a family of quartic casimirs related by Lie algebra has to be evaluated directly. For the classical groups, this is a straightforward application of the birdtrack reduction algorithms. For example, for $SU(n)$ we worked this out in section 2.2.

The results for the defining and adjoint reps of all simple Lie groups are listed in table 7.3. In table 7.4 we have used the results of table 7.3 to compute the quartic Dynkin indices (7.33). These computations were carried out by the methods that will be developed in the remainder of this monograph.

7.8 SUNDRY RELATIONS BETWEEN QUARTIC CASIMIRS

In evaluations of group theory weights, the following reduction of a 2-adjoint, 2-defining indices quartic casimir is often very convenient:

(7.50)

where the constants A and B are listed in table 7.5.

For the exceptional groups, the calculation of quartic casimirs is very simple. As mentioned above, the exceptional groups have no genuine quartic casimirs, as

$$\mathrm{tr}\, X^4 = b(\mathrm{tr}\, X^2)^2$$

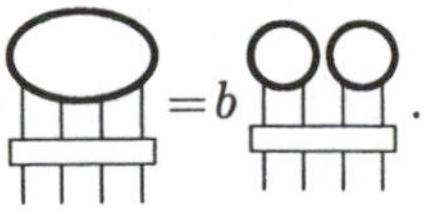

(7.51)

The constant is fixed by contracting with :

$$b = \frac{3}{N(N+2)}\frac{1}{a^2}$$ 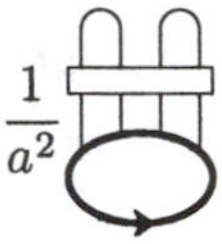 $$= \frac{3}{N(N+2)}\left(\frac{N}{n} - \frac{1}{6}\frac{C_A}{a}\right).$$

Hence, for the exceptional groups

$$\frac{1}{N}$$ 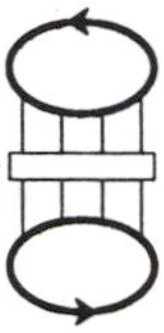 $$= \frac{3}{N+2}\left(\frac{1}{N}\right.$$ 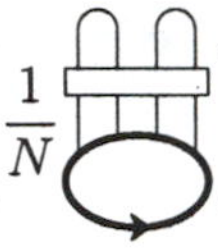 $$\left.\right)^2 = \frac{3a^4}{N+2}\left(\frac{N}{n} - \frac{C_A}{6a}\right)^2, \qquad (7.52)$$

$$\frac{1}{N}$$ 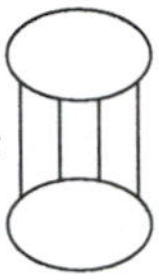 $$= C_A^4 \frac{25}{12(N+2)}, \qquad (7.53)$$

$$\frac{1}{N}$$ 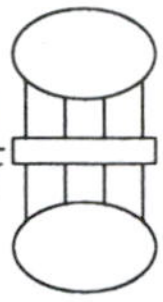 $$= C_A^4 \frac{N+27}{12(N+2)}. \qquad (7.54)$$

Here the third relation follows from the second by the Lie algebra.

To facilitate such computations, we list a selection of relations between various quartic casimirs (using normalization  $= a$ ——) for irreducible reps

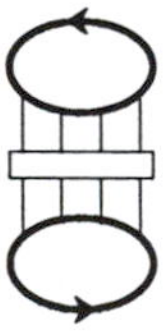 $$= \frac{1}{2}\Bigg\{$$ 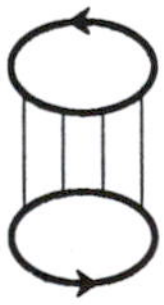 $$+$$ 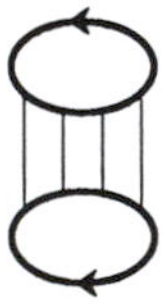 $$\Bigg\} - \frac{NC_A^2}{12}a^2 \qquad (7.55)$$

$$=$$ 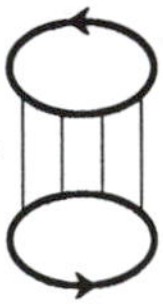 $$- \frac{C_A}{2}$$ 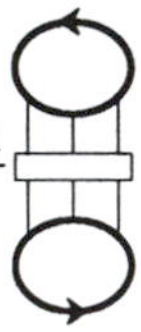 $$- \frac{NC_A^2}{12}a^2. \qquad (7.56)$$

The cubic casimir 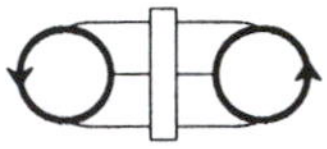 is nonvanishing only for $SU(n), n \geq 3$.

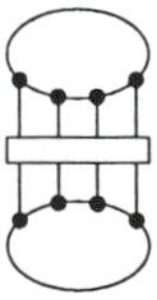 $$=$$  $$- \frac{NC_A^4}{12} \qquad (7.57)$$

$$a$$ 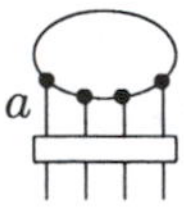 $$= C_A$$ 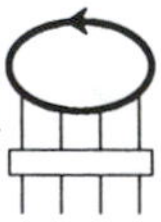 $$- 6$$ 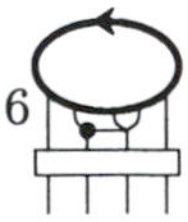 $$\qquad (7.58)$$

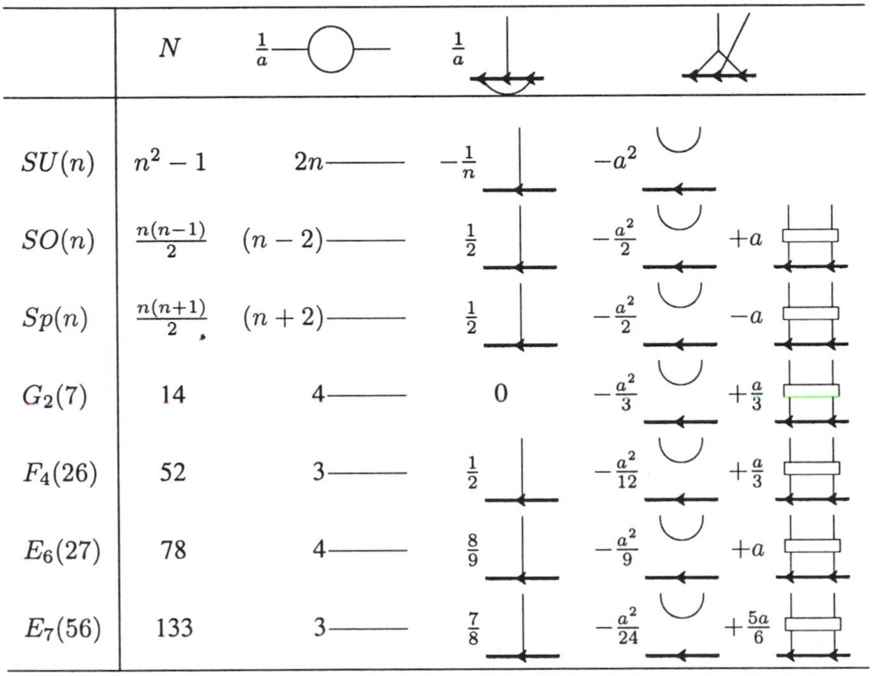

	N	$\frac{1}{a}$	$\frac{1}{a}$	
$SU(n)$	n^2-1	$2n$	$-\frac{1}{n}$	$-a^2$
$SO(n)$	$\frac{n(n-1)}{2}$	$(n-2)$	$\frac{1}{2}$	$-\frac{a^2}{2}$ $+a$
$Sp(n)$	$\frac{n(n+1)}{2}$	$(n+2)$	$\frac{1}{2}$	$-\frac{a^2}{2}$ $-a$
$G_2(7)$	14	4	0	$-\frac{a^2}{3}$ $+\frac{a}{3}$
$F_4(26)$	52	3	$\frac{1}{2}$	$-\frac{a^2}{12}$ $+\frac{a}{3}$
$E_6(27)$	78	4	$\frac{8}{9}$	$-\frac{a^2}{9}$ $+a$
$E_7(56)$	133	3	$\frac{7}{8}$	$-\frac{a^2}{24}$ $+\frac{5a}{6}$

Table 7.5 The dimension N of the adjoint rep, the quadratic casimir of the adjoint rep $1/\ell$, the vertex casimir C_v and the quartic casimir (7.50) for the defining rep of all simple Lie algebras.

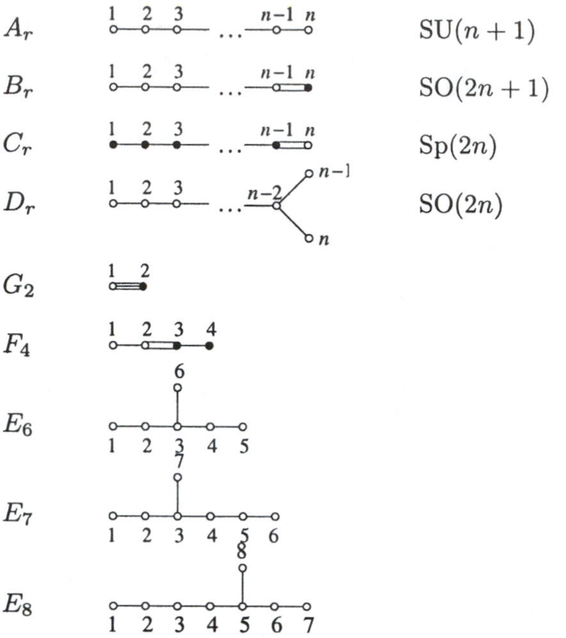

A_r	1 2 3 … $n-1$ n	$\mathrm{SU}(n+1)$
B_r	1 2 3 … $n-1$ n	$\mathrm{SO}(2n+1)$
C_r	1 2 3 … $n-1$ n	$\mathrm{Sp}(2n)$
D_r	1 2 3 … $n-2$ $n-1$ n	$\mathrm{SO}(2n)$
G_2	1 2	
F_4	1 2 3 4	
E_6	6; 1 2 3 4 5	
E_7	7; 1 2 3 4 5 6	
E_8	8; 1 2 3 4 5 6 7	

Table 7.6 Dynkin diagrams for the simple Lie groups.

$$\frac{1}{a^2N} \quad = \frac{1}{3a}(2C_F + C_V) = \frac{N}{n} - \frac{C_A}{6a} \tag{7.59}$$

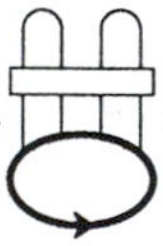

$$\frac{1}{N} \quad = \frac{5}{6}C_A^2 \tag{7.60}$$

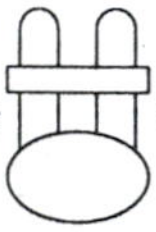

$$\frac{1}{a^3N} \quad = \frac{1}{3}(C_F^2 + C_F C_V + C_V^2)\,. \tag{7.61}$$

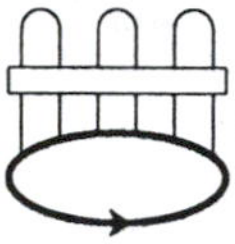

7.9 DYNKIN LABELS

"Why are they called Dynkin diagrams?"
H. S. M. Coxeter [67]

It is standard to identify a rep of a simple group of rank r by its Dynkin labels, a set of r integers $(a_1 a_2 \ldots a_r)$ assigned to the simple roots of the group by the Dynkin diagrams. The Dynkin diagrams (table 7.6) are the most concise summary of the Cartan-Killing construction of semisimple Lie algebras. We list them here only to facilitate the identification of the reps and do not attempt to derive or explain them. In this monograph, we emphasize the tensorial techniques for constructing irreps. Dynkin's canonical construction is described in refs. [312, 126]. However, in order to help the reader connect the two approaches, we will state the correspondence between the tensor reps (identified by the Young tableaux) and the canonical reps (identified by the Dynkin labels) for each group separately, in the appropriate chapters.

Chapter Eight

Group integrals

In this chapter we discuss evaluation of group integrals of form

$$\int dg\, G_a{}^b G_c{}^d \dots G^e{}_f G^g{}_h\,, \tag{8.1}$$

where $G_a{}^b$ is a $[n \times n]$ defining matrix rep of $g \in \mathcal{G}$, $G^\dagger$ is the matrix rep of the action of g on the conjugate vector space, which we write as in (3.12),

$$G^a{}_b = (G^\dagger)_b{}^a\,,$$

and the integration is over the entire range of g. As always, we assume that $\mathcal{G}$ is a compact Lie group, and $G_a{}^b$ is unitary. Such integrals are of import for certain quantum field theory calculations, and the chapter should probably be skipped by a reader not interested in such applications. The integral (8.1) is defined by two requirements:

1. Normalization:

$$\int dg = 1\,. \tag{8.2}$$

2. The action of $g \in \mathcal{G}$ is to rotate a vector x_a into $x'_a = G_a{}^b x_b$:

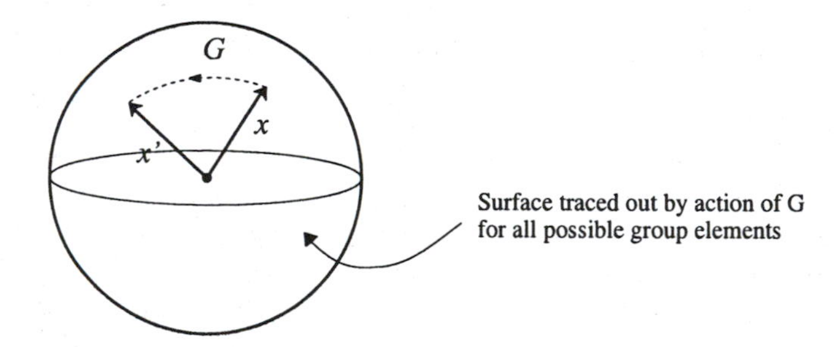

The averaging smears x in all directions, hence the second integration rule,

$$\int dg\, G_a{}^b = 0\,, \qquad G \text{ is a } \textit{nontrivial} \text{ rep of } g\,, \tag{8.3}$$

simply states that the average of a vector is zero.

A rep is trivial if $G = 1$ for all group elements g. In this case no averaging is taking place, and the first integration rule (8.2) applies.

What happens if we average a pair of vectors x, y? There is no reason why a pair should average to zero; for example, we know that $|x|^2 = \sum_a x_a x_a^* = x_a x^a$ is invariant, so it cannot have a vanishing average. Therefore, in general,

$$\int dg\, G_a{}^b G^c{}_d \neq 0\,. \tag{8.4}$$

8.1 GROUP INTEGRALS FOR ARBITRARY REPS

To get a feeling of what the right-hand side of (8.4) looks like, let us work out an $SU(n)$ example.

Let $G_a{}^b$ be the defining $[n \times n]$ matrix rep of $SU(n)$. The defining rep is nontrivial, so it averages to zero by (8.3). The first nonvanishing average involves $G^\dagger$, the matrix rep of the action of g on the conjugate vector space. As we shall soon have to face a lot of indices, we immediately resort to birdtracks. In the birdtracks notation of section 4.1,

$$G_a{}^b = a \quad b\,, \qquad G^a{}_b = a \quad b\,. \tag{8.5}$$

For G the arrows and the triangle point the same way, while for $G^\dagger$ they point the opposite way. Unitarity $G^\dagger G = 1$ is given by

$$= \quad = \quad . \tag{8.6}$$

In this notation, the $GG^\dagger$ integral to be evaluated is

$$\int dg \; {}^{a}_{b} \quad {}^{d}_{c}\,. \tag{8.7}$$

As in the $SU(n)$ example of section 2.2, the $V \otimes \overline{V}$ tensors decompose into the singlet and the adjoint rep

$$\begin{aligned} &= \frac{1}{n} \quad + \\ \delta_a^d \delta_c^b &= \frac{1}{n}\delta_a^b \delta_c^d + \frac{1}{a}(T_i)_a^b (T_i)_c^d\,. \end{aligned} \tag{8.8}$$

We multiply (8.7) with the above decomposition of the identity. The unitarity relation (8.7) eliminates G's from the singlet:

$$= \frac{1}{n} \quad + \quad . \tag{8.9}$$

The generators T_i are invariant (see (4.47)):

$$(T_i)_b^a = G^a{}_{a'} G_b{}^{b'} G_{ii'} (T_{i'})_{b'}^{a'}\,, \tag{8.10}$$

where G_{ij} is the adjoint $[N \times N]$ matrix rep of $g \in \mathcal{G}$. Multiplying by $(G^{-1})_{ij}$, we obtain

$$= \quad . \tag{8.11}$$

Hence, the pair $GG^\dagger$ in the defining rep can be traded in for a single G in the adjoint rep,

$$= \frac{1}{n} \quad + \quad . \tag{8.12}$$

The adjoint rep G_{ij} is nontrivial, so it gets averaged to zero by (8.3). Only the singlet survives:

$$\begin{aligned} \int dg\, G_a{}^d G^b{}_c &= \frac{1}{d}\delta_c^d \delta_a^b \\ \int dg \quad &= \frac{1}{d} \quad . \end{aligned} \tag{8.13}$$

Now let G be any $[d \times d]$ irrep of a compact semisimple Lie group. Irreducibility means that any $[d \times d]$ invariant tensor A^a_b is proportional to δ^a_b (otherwise one could use A to construct projection operators of section 3.5 and decompose the d-dimensional rep). As the only bilinear invariant is δ^a_b, the Clebsch-Gordan

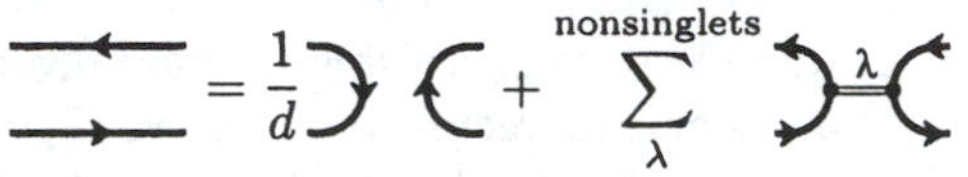

$$= \frac{1}{d} \quad + \sum_{\lambda}^{\text{nonsinglets}} \quad \lambda \qquad (8.14)$$

series contains one and only one singlet. Only the singlet survives the group averaging, and (8.13) is true for any $[d \times d]$ *irreducible* rep (with $n \to d$). If we take $G^{(\mu)}{}_\alpha{}^\beta$ and $G^{(\lambda)}{}_d{}^c$ in inequivalent reps λ, μ (there is no matrix K such that $G^{(\lambda)} = KG^{(\mu)}K^{-1}$ for all $g \in \mathcal{G}$), then there is no way of forming a singlet, and

$$\int dg\, G^{(\lambda)}{}_a{}^d G_{(\mu)}{}^\beta{}_\alpha = 0 \qquad \text{if} \qquad \lambda \neq \mu\,. \qquad (8.15)$$

What happens if G is a reducible rep? In the compact index notation of section 3.2, the group integral (8.1) that we want to evaluate is given by

$$\int dg\, G_\alpha{}^\beta\,. \qquad (8.16)$$

A reducible rep can be expanded in a Clebsch-Gordan series (3.60)

$$\int dg\, G = \sum_\lambda C_\lambda^\dagger \int dg\, G_\lambda C_\lambda\,. \qquad (8.17)$$

By the second integration rule (8.3), all nonsinglet reps average to zero, and one is left with a sum over singlet projection operators:

$$\int dg\, G = \sum_{\text{singlets}} C_\lambda^\dagger C_\lambda = \sum_{\text{singlets}} \mathbf{P}_\lambda\,. \qquad (8.18)$$

Group integration amounts to projecting out all singlets in a given Kronecker product. We now flesh out the logic that led to (8.18) with a few details. For concreteness, consider the Clebsch-Gordan series (5.8) for $\mu \times \nu = \sum \lambda$. Each clebsch

$$(C_\lambda)_{ac}{}^i = \quad {}^a_c \quad \lambda \quad i \qquad (8.19)$$

is an invariant tensor (see (4.39)):

$$C_{ac}{}^i = G_a{}^{a'} G_c{}^{c'} G^i{}_{i'} C_{a'c'}{}^{i'}$$

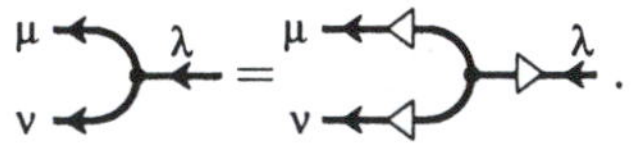

$$\mu, \nu \quad \lambda \quad = \quad \mu, \nu \quad \lambda\,. \qquad (8.20)$$

Multiplying with $G^{(\lambda)}$ from right, we obtain the rule for the "propagation" of g through the "vertex" C:

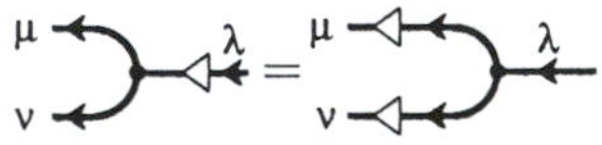

$$\mu, \nu \quad \lambda \quad = \quad \mu, \nu \quad \lambda$$

$$C_{ac}{}^{i'} G_{i'}{}^i = G_a{}^{a'} G_c{}^{c'} C_{a'c'}{}^i\,. \qquad (8.21)$$

In this way, $G^{(\mu)}G^{(\nu)}$ can be written as a Clebsch-Gordan series, each term with a single matrix $G^{(\lambda)}$ (see (5.8)):

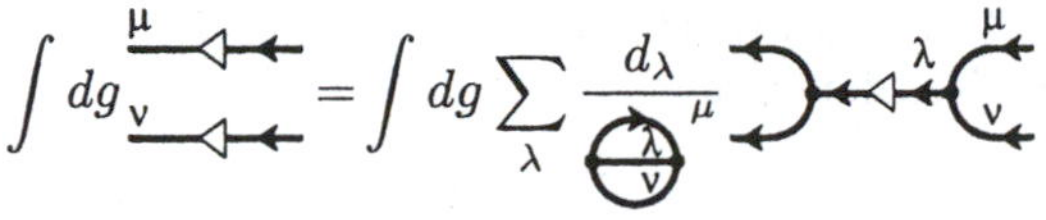

$$= \sum_\lambda (C^\lambda)_{ab}{}^i (C_\lambda)_j{}^{cd} \int dg\, G^{(\lambda)}{}_i{}^j \,. \qquad (8.22)$$

Clebsches are invariant tensors, so they are untouched by group integration. Integral over $G^{(\mu)}G^{(\nu)}$ reduces to clebsches times integrals:

$$\int dg\, G^{(\lambda)}{}_i{}^j = \begin{cases} 1 \text{ for } \lambda \text{ singlet} \\ 0 \text{ for } \lambda \text{ nonsinglet} \end{cases} . \qquad (8.23)$$

Nontrivial reps average to zero, yielding (8.18). We have gone into considerable detail in deriving (8.22) in order to motivate the sum-over-the-singlets projection operators rule (8.18). Clebsches were used in the above derivations for purely pedagogical reasons; all that is actually needed are the singlet projection operators.

8.2 CHARACTERS

Physics calculations (such as lattice gauge theories) often involve group-invariant quantities formed by contracting G with invariant tensors. Such invariants are of the form $\mathrm{tr}(hG) = h_b{}^a G_a{}^b$, where h stands for any invariant tensor. The trace of an irreducible $[d \times d]$ matrix rep λ of g is called the *character* of the rep:

$$\chi_\lambda(g) = \mathrm{tr}\, G^{(\lambda)} = G^{(\lambda)}{}_a{}^a \,. \qquad (8.24)$$

The character of the conjugate rep is

$$\chi^\lambda(g) = \mathrm{tr}\, G^{(\lambda)\dagger} = G^{(\lambda)a}{}_a = \chi_\lambda(g)^* \,. \qquad (8.25)$$

Contracting (8.14) with two arbitrary invariant $[d \times d]$ tensors $h_d{}^a$ and $(f^\dagger)_b{}^c$, we obtain the *character orthonormality relation*:

$$\int dg\, \chi_\lambda(hg)\chi^\mu(gf) = \delta^\mu_\lambda \frac{1}{d_\lambda} \chi_\lambda(hf^\dagger) \qquad (8.26)$$

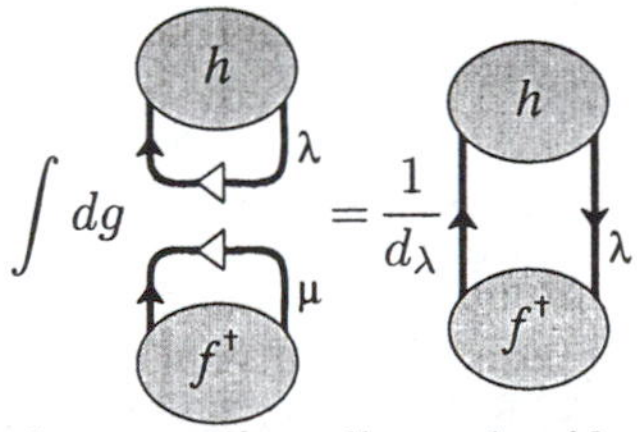

$$\int dg \ldots = \frac{1}{d_\lambda} \ldots \qquad \left(\begin{array}{l} \lambda, \mu \text{ irreducible} \\ \text{reps} \end{array} \right) .$$

The character orthonormality tells us that if two group-invariant quantities share a $GG^\dagger$ pair, the group averaging sews them into a single group-invariant quantity. The replacement of $G_a{}^b$ by the character $\chi_\lambda(h^\dagger g)$ does not mean that any of the tensor index structure is lost; $G_a{}^b$ can be recovered by differentiating

$$G_a{}^b = \frac{d}{dh_b{}^a} \chi_\lambda(h^\dagger g) \,. \qquad (8.27)$$

The birdtracks and the characters are two equivalent notations for evaluating group integrals.

8.3 EXAMPLES OF GROUP INTEGRALS

We will illustrate (8.18) by two examples: $SU(n)$ integrals over GG and $GGG^\dagger G^\dagger$. A product of two G's is drawn as

$$G_a{}^b G_c{}^d = \quad . \tag{8.28}$$

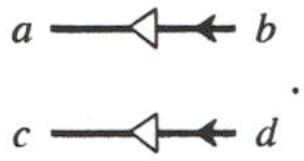

G's are acting on $\otimes V^2$ tensor space, which is decomposable by (9.4) into the symmetric and the antisymmetric subspace

$$\delta_a^b \delta_c^d = (P_s)_{ac},{}^{db} + (P_A)_{ac},{}^{db}\,, \tag{8.29}$$

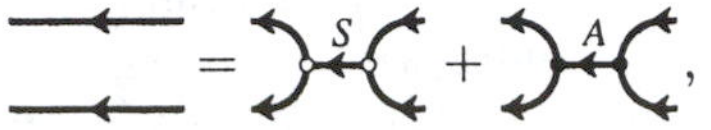

$$(P_s)_{ac},{}^{db} = \frac{1}{2}\left(\delta_a^b\delta_c^d + \delta_a^d\delta_c^b\right)$$

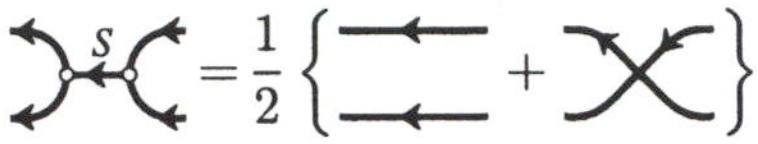

$$(P_A)_{ac},{}^{db} = \frac{1}{2}\left(\delta_a^b\delta_c^d + \delta_a^d\delta_c^b\right) \tag{8.30}$$

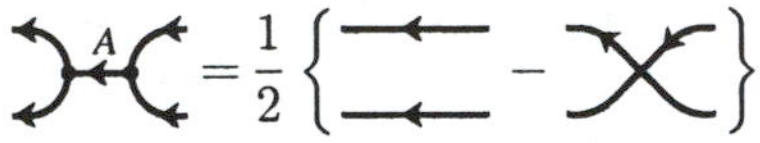

$$d_s = \frac{n(n+1)}{2}\,, \qquad d_A = \frac{n(n-1)}{2}\,. \tag{8.31}$$

The transposition of indices b and d is explained in section 4.1; it ensures a simple correspondence between tensors and birdtracks.

For $SU(2)$ the antisymmetric subspace has dimension $d_A = 1$. We shall return to this case in section 15.1. For $n \geq 3$, both subspaces are nonsinglets, and by the second integration rule,

$$SU(n) : \int dg\, G_a{}^b G_c{}^d = 0\,, \qquad n > 2\,. \tag{8.32}$$

As the second example, consider the group integral over $GGG^\dagger G^\dagger$. This rep acts on $V^2 \otimes \overline{V}^2$ tensor space. There are various ways of constructing the singlet projectors; we shall give two.

We can treat the $V^2 \otimes \overline{V}^2$ space as a Kronecker product of spaces $\otimes V^2$ and $\otimes \overline{V}^2$. We first reduce the particle and antiparticle spaces separately by (8.29):

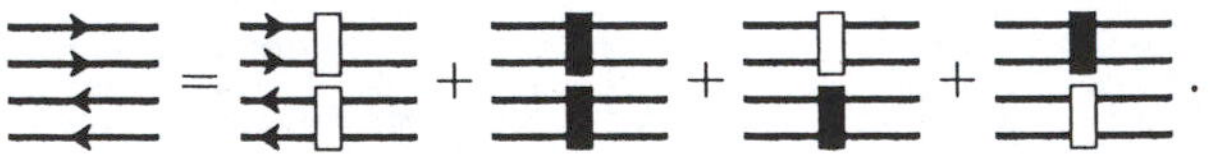

$$\tag{8.33}$$

The only invariant tensors that can project singlets out of this space (for $n \geq 3$) are index contraction with no intermediate lines:

$$\tag{8.34}$$

Contracted with the last two reps in (8.33), they yield zero. Only the first two reps yield singlets

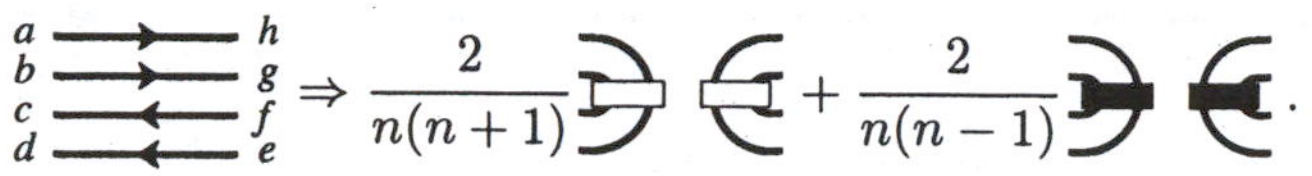

(8.35)

The projector normalization factors are the dimensions of the associated reps (3.24). The $GGG^\dagger G^\dagger$ group integral written out in tensor notation is

$$\int dg\, G^a{}_h G^b{}_g G_c{}^f G_d{}^e = \frac{1}{2n(n+1)} \left(\delta^a_d \delta^b_c + \delta^a_c \delta^b_d\right)\left(\delta^e_h \delta^f_g + \delta^e_g \delta^f_h\right) + \frac{1}{2n(n-1)} \left(\delta^a_d \delta^b_c - \delta^a_c \delta^b_d\right)\left(\delta^e_h \delta^f_g - \delta^e_g \delta^f_h\right). \quad (8.36)$$

We have obtained this result by first reducing $\otimes V^2$ and $\otimes \overline{V}^2$. What happens if we reduce $V^2 \otimes \overline{V}^2$ as $(V \otimes \overline{V})^2$? We first decompose the two $V \otimes \overline{V}$ tensor subspaces into singlets and adjoint reps (see section 2.2):

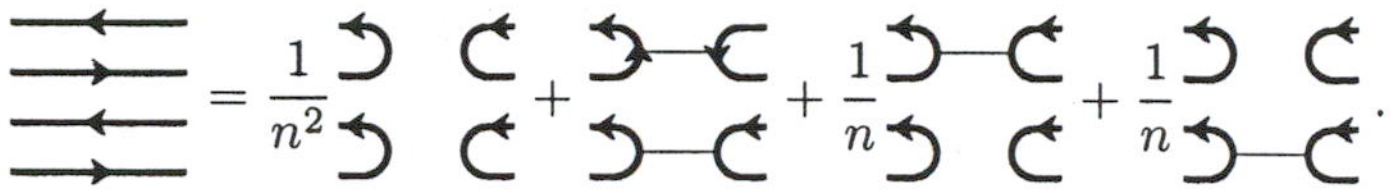

(8.37)

The two cross terms with one intermediate adjoint line cannot be reduced further. The 2-index adjoint intermediate state contains only one singlet in the Clebsch-Gordan series (15.25), so that the final result [69] is

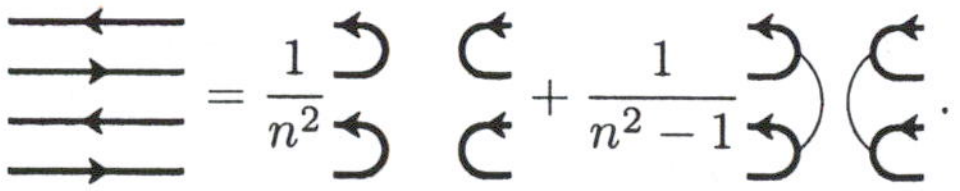

(8.38)

By substituting adjoint rep projection operators (9.54), one can check that this is the same combination of Kronecker deltas as (8.36).

To summarize, the projection operators constructed in this monograph are all that is needed for evaluation of group integrals; the group integral for an arbitrary rep is given by the sum over all singlets (8.18) contained in the rep.

Chapter Nine

Unitary groups

P. Cvitanović, H. Elvang, and A. D. Kennedy

$U(n)$ is the group of all transformations that leave invariant the norm $\overline{q}q = \delta_b^a q^b q_a$ of a complex vector q. For $U(n)$ there are no other invariant tensors beyond those constructed of products of Kronecker deltas. They can be used to decompose the tensor reps of $U(n)$. For purely covariant or contravariant tensors, the symmetric group can be used to construct the Young projection operators. In sections. 9.1–9.2 we show how to do this for 2- and 3-index tensors by constructing the appropriate characteristic equations.

For tensors with more indices it is easier to construct the Young projection operators directly from the Young tableaux. In section 9.3 we review the Young tableaux, and in section 9.4 we show how to construct Young projection operators for tensors with any number of indices. As examples, 3- and 4-index tensors are decomposed in section 9.5. We use the projection operators to evaluate $3n$-j coefficients and characters of $U(n)$ in sections. 9.6–9.9, and we derive new sum rules for $U(n)$ 3-j and 6-j symbols in section 9.7. In section 9.8 we consider the consequences of the Levi-Civita tensor being an extra invariant for $SU(n)$.

For mixed tensors the reduction also involves index contractions and the symmetric group methods alone do not suffice. In sections. 9.10–9.12 the mixed $SU(n)$ tensors are decomposed by the projection operator techniques introduced in chapter 3. $SU(2)$, $SU(3)$, $SU(4)$, and $SU(n)$ are discussed from the "invariance group" perspective in chapter 15.

9.1 TWO-INDEX TENSORS

Consider 2-index tensors $q^{(1)} \otimes q^{(2)} \in \otimes V^2$. According to (6.1), all permutations are represented by invariant matrices. Here there are only two permutations, the identity and the flip (6.2),

$$\sigma = \text{[diagram]} .$$

The flip satisfies

$$\sigma^2 = \text{[diagram]} = \mathbf{1} ,$$

$$(\sigma + \mathbf{1})(\sigma - \mathbf{1}) = 0 . \qquad (9.1)$$

The eigenvalues are $\lambda_1 = 1, \lambda_2 = -1$, and the corresponding projection operators (3.48) are

$$\mathbf{P}_1 = \frac{\sigma - (-1)\mathbf{1}}{1-(-1)} = \frac{1}{2}(\mathbf{1}+\sigma) = \frac{1}{2}\left\{ \quad + \quad \right\}, \tag{9.2}$$

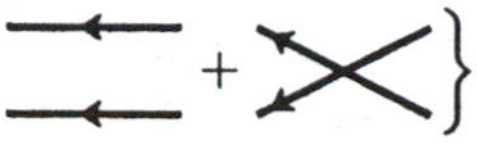

$$\mathbf{P}_2 = \frac{\sigma - \mathbf{1}}{-1-1} = \frac{1}{2}(\mathbf{1}-\sigma) = \frac{1}{2}\left\{ \quad - \quad \right\}. \tag{9.3}$$

We recognize the symmetrization, antisymmetrization operators (6.4), (6.15); $\mathbf{P}_1 = \mathbf{S}, \mathbf{P}_2 = \mathbf{A}$, with subspace dimensions $d_1 = n(n+1)/2, d_2 = n(n-1)/2$. In other words, under general linear transformations the symmetric and the antisymmetric parts of a tensor x_{ab} transform separately:

$$x = \mathbf{S}x + \mathbf{A}x\,,$$
$$x_{ab} = \frac{1}{2}(x_{ab}+x_{ba}) + \frac{1}{2}(x_{ab}-x_{ba})$$ 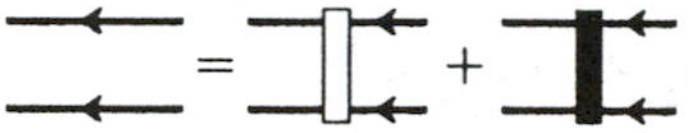
$$= \quad + \quad . \tag{9.4}$$

The Dynkin indices for the two reps follow by (7.29) from $6j$'s:

$$= \frac{1}{2}(0) + \frac{1}{2} \quad = \frac{N}{2}$$

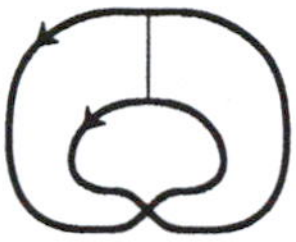

$$\ell_1 = \frac{2\ell}{n}\cdot d_1 + \frac{2\ell}{N}\cdot\frac{N}{2}$$
$$= \ell(n+2)\,. \tag{9.5}$$

Substituting the defining rep Dynkin index $\ell^{-1} = C_A = 2n$, computed in section 2.2, we obtain the two Dynkin indices

$$\ell_1 = \frac{n+2}{2n}, \qquad \ell_2 = \frac{n-2}{2n}. \tag{9.6}$$

9.2 THREE-INDEX TENSORS

Three-index tensors can be reduced to irreducible subspaces by adding the third index to each of the 2-index subspaces, the symmetric and the antisymmetric. The results of this section are summarized in figure 9.1 and table 9.1. We mix the third index into the symmetric 2-index subspace using the invariant matrix

$$\mathbf{Q} = \mathbf{S}_{12}\sigma_{(23)}\mathbf{S}_{12} = \quad . \tag{9.7}$$

Here projection operators $\mathbf{S}_{12}$ ensure the restriction to the 2-index symmetric subspace, and the transposition $\sigma_{(23)}$ mixes in the third index. To find the characteristic equation for $\mathbf{Q}$, we compute $\mathbf{Q}^2$:

$$\mathbf{Q}^2 = \mathbf{S}_{12}\sigma_{(23)}\mathbf{S}_{12}\sigma_{(23)}\mathbf{S}_{12} = \frac{1}{2}\left\{\mathbf{S}_{12} + \mathbf{S}_{12}\sigma_{(23)}\mathbf{S}_{12}\right\} = \frac{1}{2}\mathbf{S}_{12} + \frac{1}{2}\mathbf{Q}$$
$$= \quad = \frac{1}{2}\left\{ \quad + \quad \right\}.$$

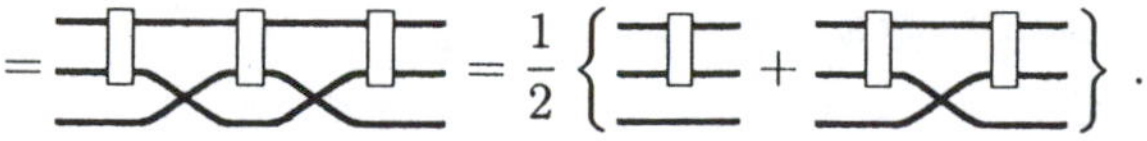

Hence, $\mathbf{Q}$ satisfies

$$(\mathbf{Q}-1)(\mathbf{Q}+1/2)\mathbf{S}_{12}=0\,, \tag{9.8}$$

and the corresponding projection operators (3.48) are

$$\mathbf{P}_1 = \frac{\mathbf{Q}+\frac{1}{2}\mathbf{1}}{1+\frac{1}{2}}\mathbf{S}_{12} = \frac{1}{3}\left\{\sigma_{(23)}+\sigma_{(123)}+\mathbf{1}\right\}\mathbf{S}_{12} = \mathbf{S}$$

$$= \frac{1}{3}\left\{ \cdots + \cdots + \cdots \right\} \cdots = \cdots \tag{9.9}$$

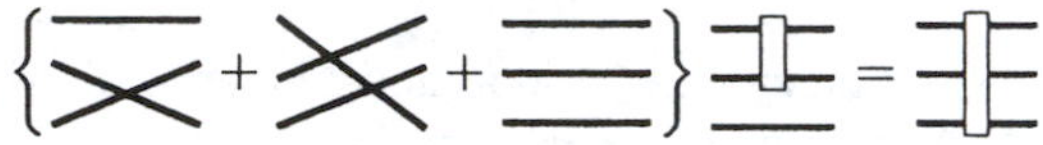

$$\mathbf{P}_2 = \frac{\mathbf{Q}-\mathbf{1}}{-\frac{1}{2}-1}\mathbf{S}_{12} = \frac{4}{3}\mathbf{S}_{12}A_{23}\mathbf{S}_{12} = \frac{4}{3}\cdots\,. \tag{9.10}$$

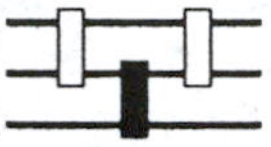

Hence, the symmetric 2-index subspace combines with the third index into a symmetric 3-index subspace (6.13) and a mixed symmetry subspace with dimensions

$$d_1 = \mathrm{tr}\,\mathbf{P}_1 = n(n+1)(n+2)/3! \tag{9.11}$$

$$d_2 = \mathrm{tr}\,\mathbf{P}_2 = \frac{4}{3}\cdots = n(n^2-1)/3\,. \tag{9.12}$$

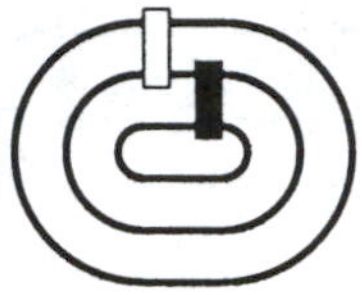

The antisymmetric 2-index subspace can be treated in the same way using the invariant matrix

$$\mathbf{Q} = \mathbf{A}_{12}\sigma_{(23)}\mathbf{A}_{12} = \cdots\,. \tag{9.13}$$

The resulting projection operators for the antisymmetric and mixed symmetry 3-index tensors are given in figure 9.1. Symmetries of the subspace are indicated by the corresponding Young tableaux, table 9.2. For example, we have just constructed

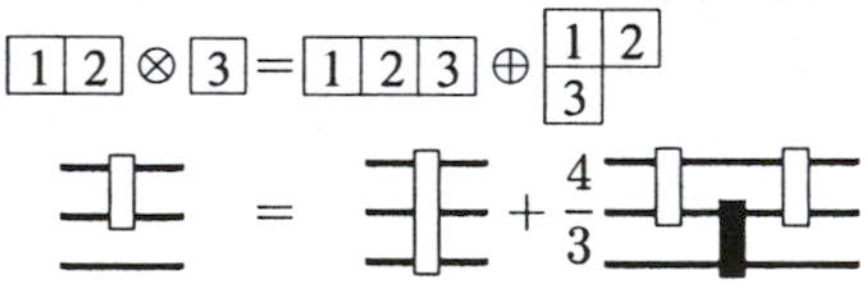

$$\boxed{1\,|\,2} \otimes \boxed{3} = \boxed{1\,|\,2\,|\,3} \oplus \begin{array}{|c|c|}\hline 1 & 2 \\ \hline 3 \\ \cline{1-1}\end{array}$$

$$\cdots = \cdots + \frac{4}{3}\cdots$$

$$\frac{n^2(n+1)}{2} = \frac{n(n+1)(n+2)}{3!} + \frac{n(n^2-1)}{3}\,. \tag{9.14}$$

The projection operators for tensors with up to 4 indices are shown in figure 9.1, and in figure 9.2 the corresponding stepwise reduction of the irreps is given in terms of Young standard tableaux (defined in section 9.3.1).

9.3 YOUNG TABLEAUX

We have seen in the examples of sections. 9.1–9.2 that the projection operators for 2-index and 3-index tensors can be constructed using characteristic equations. For tensors with more than three indices this method is cumbersome, and it is much simpler to construct the projection operators directly from the Young tableaux. In this section we review the Young tableaux and some aspects of symmetric group representations that will be important for our construction of the projection operators in section 9.4.

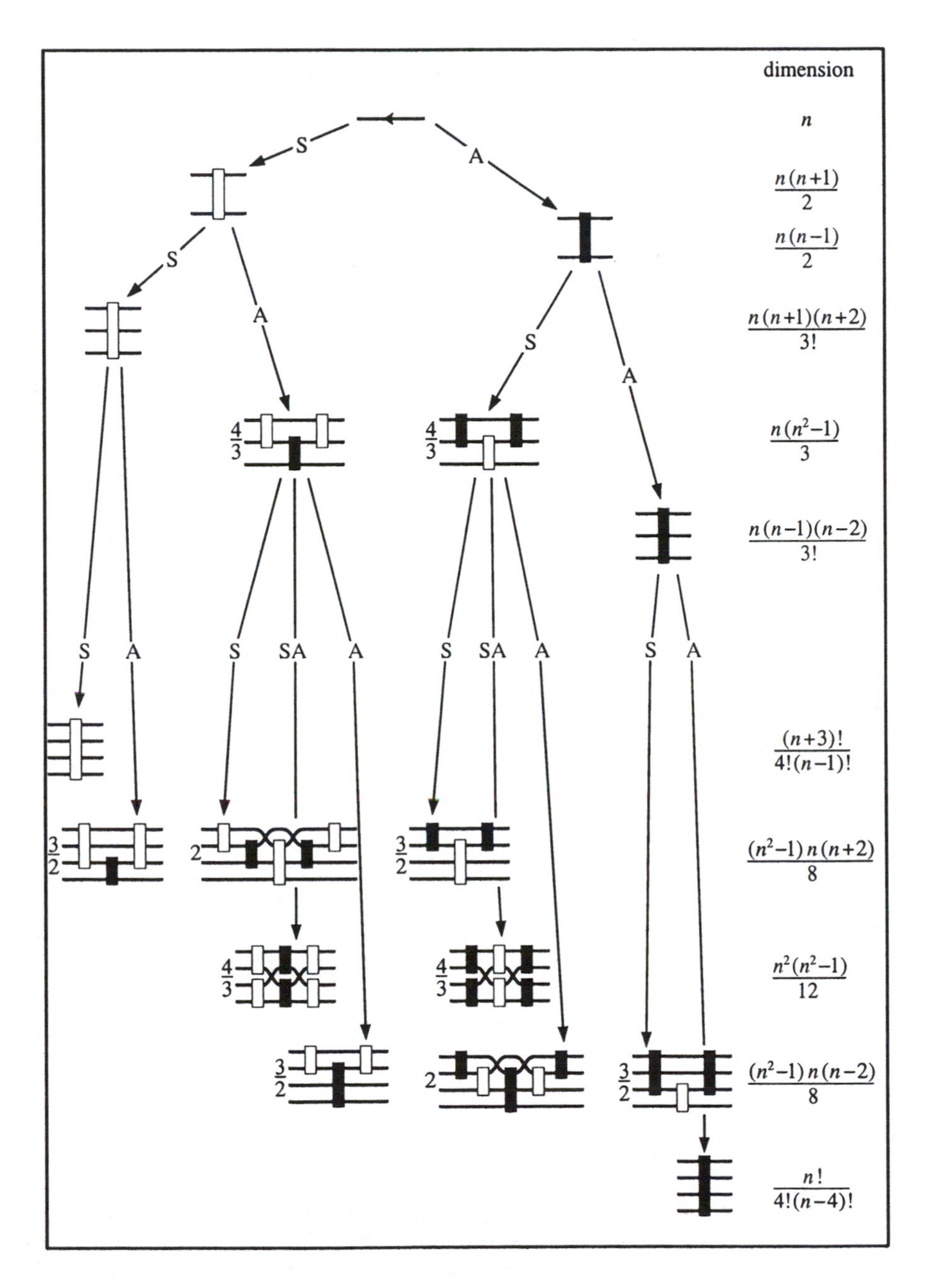

Figure 9.1 Projection operators for 2-, 3-, and 4-index tensors in $U(n)$, $SU(n)$, $n \geq p =$ number of indices.

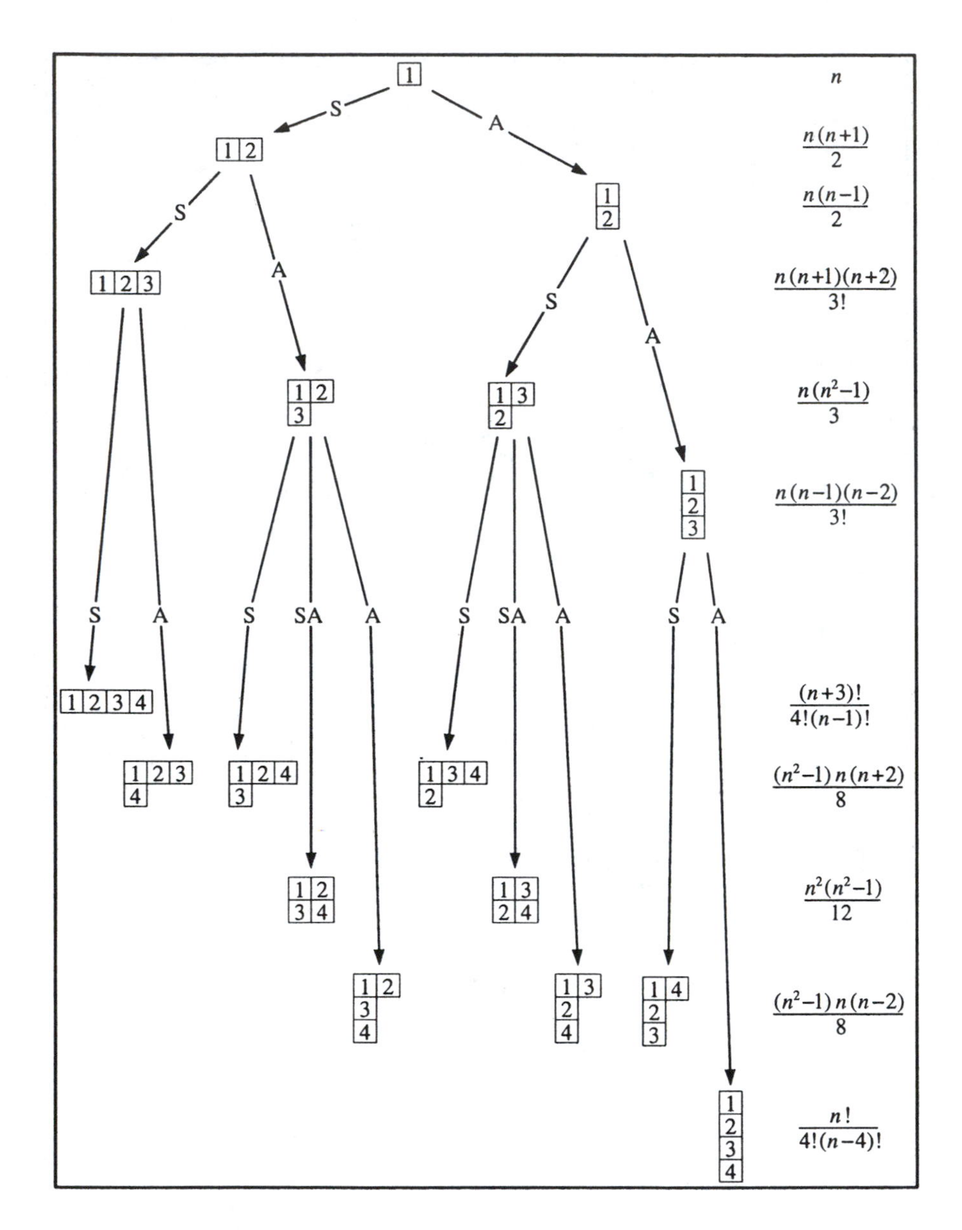

Figure 9.2 Young tableaux for the irreps of the symmetric group for 2-, 3-, and 4-index tensors. Rows correspond to symmetrizations, columns to antisymmetrizations. The reduction procedure is not unique, as it depends on the order in which the indices are combined; this order is indicated by labels 1, 2, 3 , ..., p in the boxes of Young tableaux.

9.3.1 Definitions

Partition k identical boxes into D subsets, and let λ_m, $m = 1, 2, \ldots, D$, be the number of boxes in the subsets ordered so that $\lambda_1 \geq \lambda_2 \geq \ldots \geq \lambda_D \geq 1$. Then the partition $\lambda = [\lambda_1, \lambda_2, \ldots, \lambda_D]$ fulfills $\sum_{m=1}^{D} \lambda_m = k$. The diagram obtained by drawing the D rows of boxes on top of each other, left aligned, starting with λ_1 at the top, is called a *Young diagram* Y.

Examples:
The ordered partitions for $k = 4$ are $[4], [3, 1], [2, 2], [2, 1, 1]$ and $[1, 1, 1, 1]$. The corresponding Young diagrams are

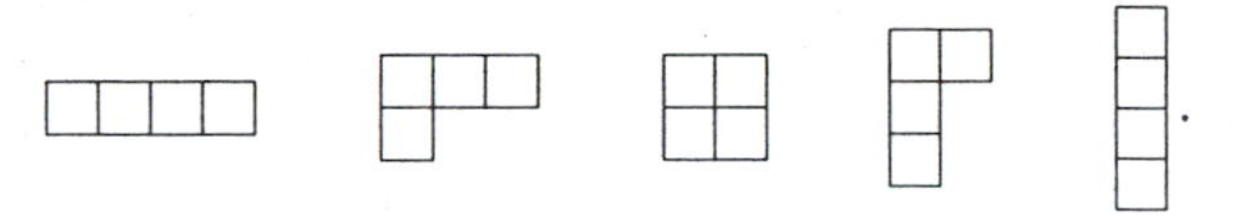

Inserting a number from the set $\{1, \ldots, n\}$ into every box of a Young diagram Y_λ in such a way that numbers increase when reading a column from top to bottom, and numbers do not decrease when reading a row from left to right, yields a *Young tableau* Y_a. The subscript a labels different tableaux derived from a given Young diagram, *i.e.*, different admissible ways of inserting the numbers into the boxes.

A *standard tableau* is a k-box Young tableau constructed by inserting the numbers $1, \ldots, k$ according to the above rules, but using each number exactly once. For example, the 4-box Young diagram with partition $\lambda = [2, 1, 1]$ yields three distinct standard tableaux:

$$\begin{array}{|c|c|}\hline 1 & 2 \\ \hline 3 \\ \cline{1-1} 4 \\ \cline{1-1}\end{array}\;, \quad \begin{array}{|c|c|}\hline 1 & 3 \\ \hline 2 \\ \cline{1-1} 4 \\ \cline{1-1}\end{array}\;, \quad \begin{array}{|c|c|}\hline 1 & 4 \\ \hline 2 \\ \cline{1-1} 3 \\ \cline{1-1}\end{array}\;. \tag{9.15}$$

An alternative labeling of a Young diagram are Dynkin labels, the list of numbers b_m of columns with m boxes: $(b_1 b_2 \ldots)$. Having k boxes we must have $\sum_{m=1}^{k} m b_m = k$. For example, the partition $[4, 2, 1]$ and the labels $(21100\cdots)$ give rise to the same Young diagram, and so do the partition $[2, 2]$ and the labels $(020\cdots)$.

We define the *transpose* diagram Y^t as the Young diagram obtained from Y by interchanging rows and columns. For example, the transpose of $[3, 1]$ is $[2, 1, 1]$,

$$\begin{array}{|c|c|c|}\hline 1 & 2 & 4 \\ \hline 3 \\ \cline{1-1}\end{array}^{\,t} = \begin{array}{|c|c|}\hline 1 & 3 \\ \hline 2 \\ \cline{1-1} 4 \\ \cline{1-1}\end{array}\;,$$

or, in terms of Dynkin labels, the transpose of $(210\ldots)$ is $(1010\ldots)$.

The Young tableaux are useful for labeling irreps of various groups. We shall use the following facts (see for instance ref. [153]):

1. The k-box *Young diagrams* label all irreps of the symmetric group S_k.

2. The *standard tableaux* of k-box Young diagrams with no more than n rows label the irreps of $GL(n)$, in particular they label the irreps of $U(n)$.

3. The *standard tableaux* of k-box Young diagrams with no more than $n-1$ rows label the irreps of $SL(n)$, in particular they label the irreps of $SU(n)$.

In this section, we consider the Young tableaux for reps of S_k and $U(n)$, while the case of $SU(n)$ is postponed to section 9.8.

9.3.2 Symmetric group S_k

The irreps of the symmetric group S_k are labeled by the k-box Young diagrams. For a given Young diagram, the basis vectors of the corresponding irrep can be labeled by the standard tableaux of Y; consequently the dimension $\Delta_{\rm Y}$ of the irrep is the number of standard tableaux that can be constructed from the Young diagram Y. The example (9.15) shows that the irrep $\lambda = [2, 1, 1]$ of S_4 is 3-dimensional.

As an alternative to counting standard tableaux, the dimension $\Delta_{\rm Y}$ of the irrep of S_k corresponding to the Young diagram Y can be computed easily as

$$\Delta_{\rm Y} = \frac{k!}{|{\rm Y}|}, \tag{9.16}$$

where the number $|{\rm Y}|$ is computed using a "hook" rule: Enter into each box of the Young diagram the number of boxes below and to the right of the box, including the box itself. Then $|{\rm Y}|$ is the product of the numbers in all the boxes. For instance,

$${\rm Y} = \begin{array}{|c|c|c|c|}\hline \; & \; & \; & \; \\ \hline \; & \; & \; \\ \cline{1-3} \; & \; \\ \cline{1-2}\end{array} \quad \longrightarrow \quad |{\rm Y}| = \begin{array}{|c|c|c|c|}\hline 6 & 5 & 3 & 1 \\ \hline 4 & 3 & 1 \\ \cline{1-3} 2 & 1 \\ \cline{1-2}\end{array} = 6!\,3\,. \tag{9.17}$$

The hook rule (9.16) was first proven by Frame, de B. Robinson, and Thrall [123]. Various proofs can be found in the literature [295, 170, 133, 142, 21]; see also Sagan [302] and references therein.

We now discuss the regular representation of the symmetric group. The elements $\sigma \in S_k$ of the symmetric group S_k form a basis of a $k!$-dimensional vector space V of elements

$$s = \sum_{\sigma \in S_k} s_\sigma \, \sigma \in V, \tag{9.18}$$

where s_σ are the components of a vector s in the given basis. If $s \in V$ has components (s_σ) and $\tau \in S_k$, then τs is an element in V with components $(\tau s)_\sigma = s_{\tau^{-1}\sigma}$. This action of the group elements on the vector space V defines an $k!$-dimensional matrix representation of the group S_k, the *regular representation*.

The regular representation is reducible, and each irrep λ appears Δ_λ times in the reduction; Δ_λ is the dimension of the subspace V_λ corresponding to the irrep λ. This gives the well-known relation between the order of the symmetric group $|S_k| = k!$ (the dimension of the regular representation) and the dimensions of the irreps,

$$|S_k| = \sum_{\text{all irreps } \lambda} \Delta_\lambda^2 \,.$$

Using (9.16) and the fact that the Young diagrams label the irreps of S_k, we have

$$1 = k! \sum_{(k)} \frac{1}{|{\rm Y}|^2}, \tag{9.19}$$

where the sum is over all Young diagrams with k boxes. We shall use this relation to determine the normalization of Young projection operators in appendix B.3.

The reduction of the regular representation of S_k gives a completeness relation,

$$\mathbf{1} = \sum_{(k)} \mathbf{P}_{\mathrm{Y}} ,$$

in terms of projection operators

$$\mathbf{P}_{\mathrm{Y}} = \sum_{\mathrm{Y}_a \in \mathrm{Y}} \mathbf{P}_{\mathrm{Y}_a} .$$

The sum is over all standard tableaux derived from the Young diagram Y. Each $\mathbf{P}_{\mathrm{Y}_a}$ projects onto a corresponding invariant subspace V_{Y_a}: for each Y there are Δ_{Y} such projection operators (corresponding to the Δ_{Y} possible standard tableaux of the diagram), and each of these project onto one of the Δ_{Y} invariant subspaces V_{Y} of the reduction of the regular representation. It follows that the projection operators are orthogonal and that they constitute a complete set.

9.3.3 Unitary group $U(n)$

The irreps of $U(n)$ are labeled by the k-box Young standard tableaux with no more than n rows. A k-index tensor is represented by a Young diagram with k boxes — one typically thinks of this as a k-particle state. For $U(n)$, a 1-index tensor has n-components, so there are n 1-particle states available, and this corresponds to the n-dimensional fundamental rep labeled by a 1-box Young diagram. There are n^2 2-particle states for $U(n)$, and as we have seen in section 9.1 these split into two irreps: the symmetric and the antisymmetric. Using Young diagrams, we write the reduction of the 2-particle system as

$$\square \otimes \square = \square\!\square \oplus \begin{array}{c}\square\\[-4pt]\square\end{array} . \tag{9.20}$$

Except for the fully symmetric and the fully antisymmetric irreps, the irreps of the k-index tensors of $U(n)$ have mixed symmetry. Boxes in a row correspond to indices that are symmetric under interchanges (symmetric multiparticle states), and boxes in a column correspond to indices antisymmetric under interchanges (antisymmetric multiparticle states). Since there are only n labels for the particles, no more than n particles can be antisymmetrized, and hence only standard tableaux with up to n rows correspond to irreps of $U(n)$.

The number of standard tableaux Δ_{Y} derived from a Young diagram Y is given in (9.16). In terms of irreducible tensors, the Young diagram determines the symmetries of the indices, and the Δ_{Y} distinct standard tableaux correspond to the independent ways of combining the indices under these symmetries. This is illustrated in figure 9.2.

For a given $U(n)$ irrep labeled by some standard tableau of the Young diagram Y, the basis vectors are labeled by the Young tableaux Y_a obtained by inserting the numbers $1, 2, \ldots, n$ into Y in the manner described in section 9.3.1. Thus the dimension of an irrep of $U(n)$ equals the number of such Young tableaux, and we note that all irreps with the same Young diagram have the same dimension. For $U(2)$, the $k = 2$ Young tableaux of the symmetric and antisymmetric irreps are

$$\begin{array}{|c|c|}\hline 1 & 1 \\ \hline \end{array}\,, \quad \begin{array}{|c|c|}\hline 1 & 2 \\ \hline \end{array}\,, \quad \begin{array}{|c|c|}\hline 2 & 2 \\ \hline \end{array}\,, \quad \text{and} \quad \begin{array}{|c|}\hline 1 \\ \hline 2 \\ \hline \end{array}\,,$$

so the symmetric state of $U(2)$ is 3-dimensional and the antisymmetric state is 1-dimensional, in agreement with the formulas (6.4) and (6.15) for the dimensions of the symmetry operators. For $U(3)$, the counting of Young tableaux shows that the symmetric 2-particle irrep is 6-dimensional and the antisymmetric 2-particle irrep is 3-dimensional, again in agreement with (6.4) and (6.15). In section 9.4.3 we state and prove a dimension formula for a general irrep of $U(n)$.

9.4 YOUNG PROJECTION OPERATORS

Given an irrep of $U(n)$ labeled by a k-box standard tableaux Y, we construct the corresponding Young projection operator $\mathbf{P}_{\mathrm{Y}}$ in birdtrack notation by identifying each box in the diagram with a directed line. The operator $\mathbf{P}_{\mathrm{Y}}$ is a block of symmetrizers to the left of a block of antisymmetrizers, all imposed on the k lines. The blocks of symmetry operators are dictated by the Young *diagram*, whereas the attachment of lines to these operators is specified by the particular standard tableau.

The Kronecker delta is invariant under unitary transformations: for $U \in U(n)$, we have $(U^\dagger)_a{}^{a'} \delta^{b'}_{a'} U_{b'}{}^b = \delta^b_a$. Consequently, any combination of Kronecker deltas, such as a symmetrizer, is invariant under unitary transformations. The symmetry operators constitute a complete set, so any $U(n)$ invariant tensor built from Kronecker deltas can be expressed in terms of symmetrizers and antisymmetrizers. In particular, the invariance of the Kronecker delta under $U(n)$ transformations implies that the same symmetry group operators that project the irreps of S_k also yield the irreps of $U(n)$.

The simplest examples of Young projection operators are those associated with the Young tableaux consisting of either one row or one column. The corresponding Young projection operators are simply the symmetrizers or the antisymmetrizers respectively. As projection operators for S_k, the symmetrizer projects onto the 1-dimensional subspace corresponding to the fully symmetric representation, and the antisymmetrizer projects onto the fully antisymmetric representation (the alternating representation).

A Young projection operator for a mixed symmetry Young tableau will here be constructed by first antisymmetrizing subsets of indices, and then symmetrizing other subsets of indices; the Young tableau determines which subsets, as will be explained shortly. Schematically,

$$\mathbf{P}_{\mathrm{Y}_a} = \alpha_{\mathrm{Y}} \qquad , \tag{9.21}$$

where the white (black) blob symbolizes a set of (anti)symmetrizers. The normalization constant α_{Y} (defined below) ensures that the operators are idempotent, $\mathbf{P}_{\mathrm{Y}_a}\mathbf{P}_{\mathrm{Y}_b} = \delta_{ab}\mathbf{P}_{\mathrm{Y}_a}$.

This particular form of projection operators is not unique: in section 9.2 we built 3-index tensor Young projection operators that were symmetric under transposition.

The Young projection operators constructed in this section are particularly convenient for explicit $U(n)$ computations, and another virtue is that we can write down the projectors explicitly from the standard tableaux, without having to solve a characteristic equation. For multiparticle irreps, the Young projection operators of this section will generally be different from the ones constructed from characteristic equations (see sections. 9.1–9.2); however, the operators are equivalent, since the difference amounts to a choice of basis.

9.4.1 Construction of projection operators

Let Y_a be a k-box standard tableau. Arrange a set of symmetrizers corresponding to the rows in Y_a, and to the right of this arrange a set of antisymmetrizers corresponding to the columns in Y_a. For a Young diagram Y with s rows and t columns we label the rows $\mathrm{S}_1, \mathrm{S}_2, \ldots, \mathrm{S}_s$ and to the columns $\mathrm{A}_1, \mathrm{A}_2, \ldots, \mathrm{A}_t$. Each symmetry operator in $\mathbf{P}_\mathrm{Y}$ is associated to a row/column in Y, hence we label a symmetry operator after the corresponding row/column, for example,

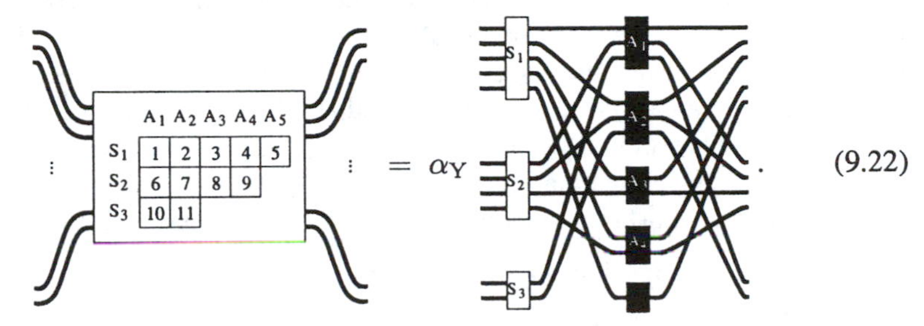

(9.22)

Let the lines numbered 1 to k enter the symmetrizers as described by the numbers in the boxes in the standard tableau and connect the set of symmetrizers to the set of antisymmetrizers in a nonvanishing way, avoiding multiple intermediate lines prohibited by (6.17). Finally, arrange the lines coming out of the antisymmetrizers such that if the lines all passed straight through the symmetry operators, they would exit in the same order as they entered. This ensures that upon expansion of all the symmetry operators, the identity appears exactly once.

We denote by $|\mathrm{S}_i|$ or $|\mathrm{A}_i|$ the *length* of a row or column, respectively, that is the number of boxes it contains. Thus $|\mathrm{A}_i|$ also denotes the number of lines entering the antisymmetrizer A_i. In the above example we have $|\mathrm{S}_1| = 5$, $|\mathrm{A}_2| = 3$, *etc.*

The normalization α_Y is given by

$$\alpha_\mathrm{Y} = \frac{\left(\prod_{i=1}^{s} |\mathrm{S}_i|!\right)\left(\prod_{j=1}^{t} |\mathrm{A}_j|!\right)}{|\mathrm{Y}|}, \tag{9.23}$$

where $|\mathrm{Y}|$ is related through (9.16) to Δ_Y, the dimension of irrep Y of S_k, and is a hook rule S_k combinatoric number. The normalization depends only on the shape of the Young diagram, not the particular tableau.

Example: The Young diagram ⊞ tells us to use one symmetrizer of length three, one of length one, one antisymmetrizer of length two, and two of length one.

There are three distinct k-standard arrangements, each corresponding to a projection operator

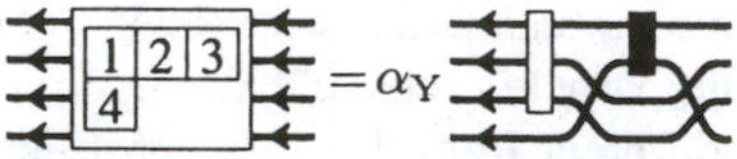

$$=\alpha_Y \qquad (9.24)$$

$$=\alpha_Y \qquad (9.25)$$

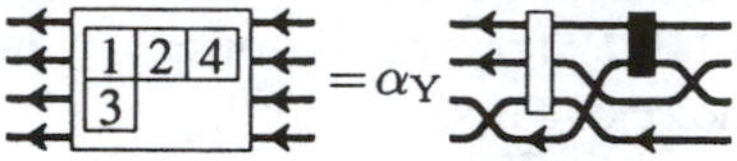

$$=\alpha_Y \quad , \qquad (9.26)$$

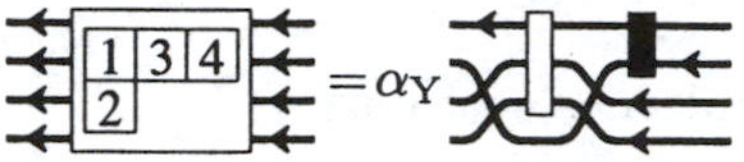

where the normalization constant is $\alpha_Y = 3/2$ by (9.23). More examples of Young projection operators are given in section 9.5.

9.4.2 Properties

We prove in appendix B that the above construction yields well defined projection operators. In particular, the internal connection between the symmetrizers and antisymmetrizers is unique up to an overall sign (proof in appendix B.1). We fix the overall sign by requiring that when all symmetry operators are expanded, the identity appears with a positive coefficient. Note that by construction (the lines exit in the same order as they enter) the identity appears exactly once in the full expansion of any of the Young projection operators.

We list here the most important properties of the Young projection operators:

1. The Young projection operators are *orthogonal*: If Y and Z are two distinct standard tableaux, then $\mathbf{P}_Y \mathbf{P}_Z = 0 = \mathbf{P}_Z \mathbf{P}_Y$.

2. With the normalization (9.23), the Young projection operators are indeed *projection operators*, *i.e.*, they are idempotent: $\mathbf{P}_Y^2 = \mathbf{P}_Y$.

3. For a given k the Young projection operators constitute a complete set such that $\mathbf{1} = \sum \mathbf{P}_Y$, where the sum is over all standard tableaux Y with k boxes.

The proofs of these properties are given in appendix B.

9.4.3 Dimensions of $U(n)$ irreps

The dimension d_Y of a $U(n)$ irrep Y can be computed diagrammatically as the trace of the corresponding Young projection operator, $d_Y = \operatorname{tr} \mathbf{P}_Y$. Expanding the symmetry operators yields a weighted sum of closed-loop diagrams. Each loop is worth n, and since the identity appears precisely once in the expansion, the dimension d_Y of a irrep with a k-box Young tableau Y is a degree k polynomial in n.

Example: We compute we dimension of the $U(n)$ irrep $\begin{array}{|c|c|}\hline 1 & 2 \\ \hline 3 \\ \cline{1-1}\end{array}$:

$$d_Y = \quad = \frac{4}{3}$$

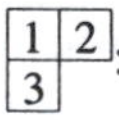

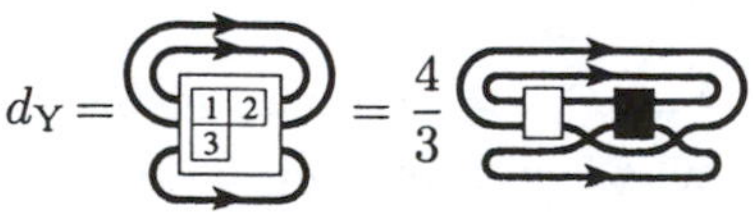

$$= \frac{4}{3}\left(\frac{1}{2!}\right)^2 \Big\{ \quad + \quad - \quad - \quad \Big\}$$

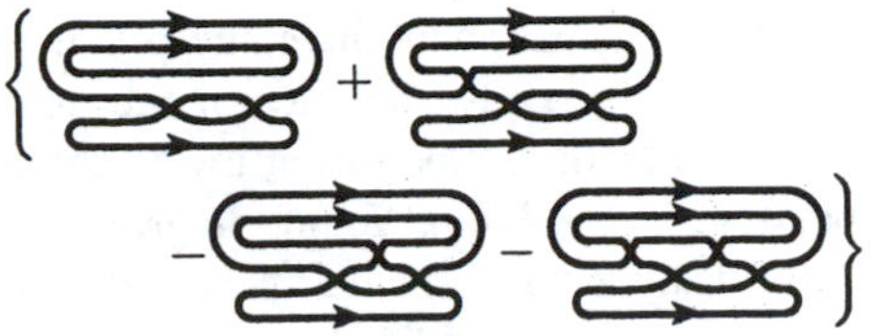

$$= \frac{1}{3}(n^3 + n^2 - n^2 - n) = \frac{n(n^2-1)}{3} . \tag{9.27}$$

In practice, this is unnecessarily laborious. The dimension of a $U(n)$ irrep Y is given by

$$d_Y = \frac{f_Y(n)}{|Y|} . \tag{9.28}$$

Here $f_Y(n)$ is a polynomial in n obtained from the Young diagram Y by multiplying the numbers written in the boxes of Y, according to the following rules:

1. The upper left box contains an n.
2. The numbers in a row increase by one when reading from left to right.
3. The numbers in a column decrease by one when reading from top to bottom.

Hence, if k is the number of boxes in Y, $f_Y(n)$ is a polynomial in n of degree k. The dimension formula (9.28) is well known (see for instance ref. [138]).

Example: In the above example with the irrep $\begin{array}{|c|c|}\hline 1 & 2 \\ \hline 3 \\ \cline{1-1}\end{array}$, we have

$$d_Y = \frac{f_Y(n)}{|Y|} = \frac{n(n^2-1)}{3}$$

in agreement with the diagrammatic trace calculation (9.27).

Example: With Y = [4,2,1], we have

$$f_Y(n) = \begin{array}{|c|c|c|c|}\hline n & n+1 & n+2 & n+3 \\ \hline n-1 & n \\ \cline{1-2} n-2 \\ \cline{1-1}\end{array} = n^2(n^2-1)(n^2-4)(n+3),$$

$$|Y| = \begin{array}{|c|c|c|c|}\hline 6 & 4 & 2 & 1 \\ \hline 3 & 1 \\ \cline{1-2} 1 \\ \cline{1-1}\end{array} = 144, \tag{9.29}$$

hence,

$$d_Y = \frac{n^2(n^2-1)(n^2-4)(n+3)}{144} . \tag{9.30}$$

Using $d_Y = \text{tr}\, \mathbf{P}_Y$, the dimension formula (9.28) can be proven diagrammatically by induction on the number of boxes in the irrep Y. The proof is given in appendix B.4.

The polynomial $f_Y(n)$ has an intuitive interpretation in terms of strand colorings of the diagram for $\operatorname{tr} \mathbf{P}_Y$. Draw the trace of the Young projection operator. Each line is a strand, a closed line, which we draw as passing straight through all of the symmetry operators. For a k-box Young diagram, there are k strands. Given the following set of rules, we count the number of ways to color the k strands using n colors. The top strand (corresponding to the leftmost box in the first row of Y) may be colored in n ways. Color the rest of the strands according to the following rules:

1. If a path, which could be colored in m ways, enters an antisymmetrizer, the lines below it can be colored in $m-1, m-2, \ldots$ ways.

2. If a path, which could be colored in m ways, enters a symmetrizer, the lines below it can be colored in $m+1, m+2, \ldots$ ways.

Using this coloring algorithm, the number of ways to color the strands of the diagram is $f_Y(n)$.

Example: For $Y = \begin{array}{|c|c|c|c|} \hline 1&2&3&6 \\ \hline 4&5&7 \\ \cline{1-3} 8 \\ \cline{1-1} \end{array}$, the strand diagram is

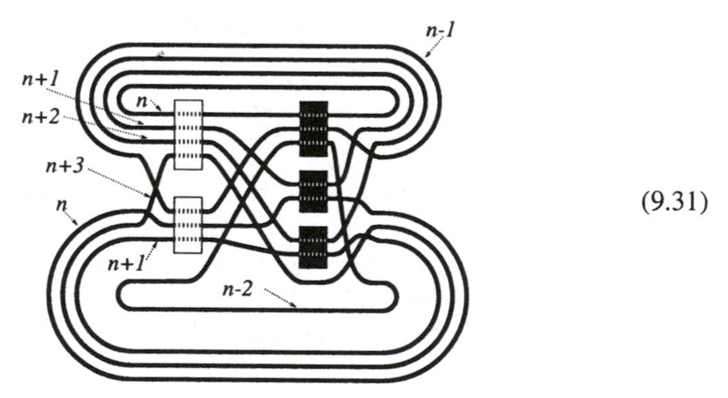

$$(9.31)$$

Each strand is labeled by the number of admissible colorings. Multiplying these numbers and including the factor $1/|Y|$, we find

$$\begin{aligned} d_Y &= (n-2)\,(n-1)\,n^2\,(n+1)^2\,(n+2)\,(n+3) \Big/ \begin{array}{|c|c|c|c|} \hline 6&4&3&1 \\ \hline 4&2&1 \\ \cline{1-3} 1 \\ \cline{1-1} \end{array} \\ &= \frac{n\,(n+1)\,(n+3)!}{2^6\,3^2\,(n-3)!}\,, \end{aligned}$$

in agreement with (9.28).

9.5 REDUCTION OF TENSOR PRODUCTS

We now work out several explicit examples of decomposition of direct products of Young diagrams/tableaux in order to motivate the general rules for decomposition

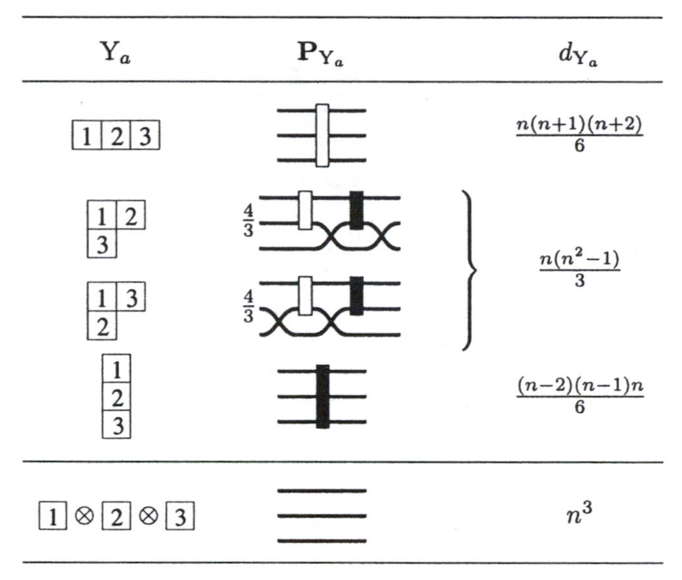

Y_a	P_{Y_a}	d_{Y_a}
$\begin{array}{|c|c|c|}\hline 1&2&3\\\hline\end{array}$		$\frac{n(n+1)(n+2)}{6}$
$\begin{array}{|c|c|}\hline 1&2\\\hline 3\\\cline{1-1}\end{array}$	$\frac{4}{3}$	$\frac{n(n^2-1)}{3}$
$\begin{array}{|c|c|}\hline 1&3\\\hline 2\\\cline{1-1}\end{array}$	$\frac{4}{3}$	
$\begin{array}{|c|}\hline 1\\\hline 2\\\hline 3\\\hline\end{array}$		$\frac{(n-2)(n-1)n}{6}$
$\boxed{1}\otimes\boxed{2}\otimes\boxed{3}$		n^3

Table 9.1 Reduction of 3-index tensor. The last row shows the direct sum of the Young tableaux, the sum of the dimensions of the irreps adding up to n^3, and the sum of the projection operators adding up to the identity as verification of completeness (3.51).

of direct products stated below, in section 9.5.1. We have already treated the decomposition of the 2-index tensor into the symmetric and the antisymmetric tensors, but we shall reconsider the 3-index tensor, since the projection operators are different from those derived from the characteristic equations in section 9.2.

The 3-index tensor reduces to

$$\boxed{1}\otimes\boxed{2}\otimes\boxed{3} = \left(\begin{array}{|c|c|}\hline 1&2\\\hline\end{array} \oplus \begin{array}{|c|}\hline 1\\\hline 2\\\hline\end{array}\right)\otimes\boxed{3}$$

$$= \begin{array}{|c|c|c|}\hline 1&2&3\\\hline\end{array} \oplus \begin{array}{|c|c|}\hline 1&2\\\hline 3\\\cline{1-1}\end{array} \oplus \begin{array}{|c|c|}\hline 1&3\\\hline 2\\\cline{1-1}\end{array} \oplus \begin{array}{|c|}\hline 1\\\hline 2\\\hline 3\\\hline\end{array}\,. \qquad (9.32)$$

The corresponding dimensions and Young projection operators are given in table 9.1. For simplicity, we neglect the arrows on the lines where this leads to no confusion.

The Young projection operators are orthogonal by inspection. We check completeness by a computation. In the sum of the fully symmetric and the fully antisymmetric tensors, all the odd permutations cancel, and we are left with

Expanding the two tensors of mixed symmetry, we obtain

Adding the two equations we get

$$+\frac{4}{3}\quad+\frac{4}{3}\quad+\quad=\quad,\tag{9.33}$$

verifying the completeness relation.

For 4-index tensors the decomposition is performed as in the 3-index case, resulting in table 9.2.

Acting with any permutation on the fully symmetric or antisymmetric projection operators gives ± 1 times the projection operator (see (6.8) and (6.18)). For projection operators of mixed symmetry the action of a permutation is not as simple, because the permutations will mix the spaces corresponding to the distinct tableaux. Here we shall need only the action of a permutation within a $3n$-j symbol, and, as we shall show below, in this case the result will again be simple, a factor ± 1 or 0.

9.5.1 Reduction of direct products

We state the rules for general decompositions of direct products such as (9.20) in terms of Young diagrams:

Draw the two diagrams next to each other and place in each box of the second diagram an a_i, $i = 1, \ldots, k$, such that the boxes in the first row all have a_1 in them, second row boxes have a_2 in them, *etc.* The boxes of the second diagram are now added to the first diagram to create new diagrams according to the following rules:

1. Each diagram must be a Young diagram.
2. The number of boxes in the new diagram must be equal to the sum of the number of boxes in the two initial diagrams.
3. For U(n) no diagram has more than n rows.
4. Making a journey through the diagram starting with the top row and entering each row from the right, at any point the number of a_i's encountered in any of the attached boxes must not exceed the number of previously encountered a_{i-1}'s.
5. The numbers must not increase when reading across a row from left to right.
6. The numbers must decrease when reading a column from top to bottom.

Rules 4–6 ensure that states that were previously symmetrized are not antisymmetrized in the product, and vice versa. Also, the rules prevent counting the same state twice.

For example, consider the direct product of the partitions [3] and [2, 1]. For $U(n)$ with $n \geq 3$ we have

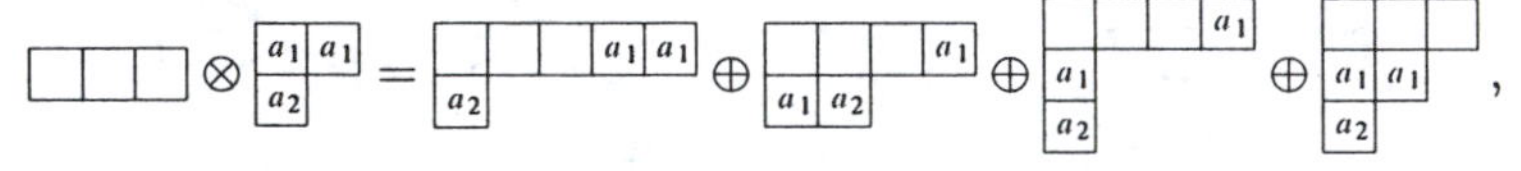

while for $n = 2$ we have

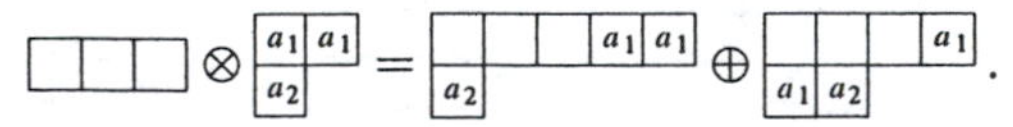

Y_a	$\mathbf{P}_{Y_a}$	d_{Y_a}
1 2 3 4	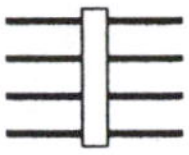	$\frac{n(n+1)(n+2)(n+3)}{24}$
1 2 3 4	 $\frac{3}{2}$	$\frac{(n-1)n(n+1)(n+2)}{8}$
1 2 4 3	 $\frac{3}{2}$	
1 3 4 2	 $\frac{3}{2}$	
1 2 3 4	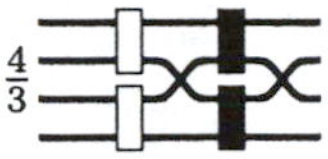 $\frac{4}{3}$	$\frac{n^2(n^2-1)}{12}$
1 3 2 4	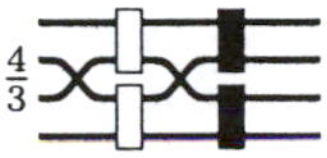 $\frac{4}{3}$	
1 2 3 4	 $\frac{3}{2}$	$\frac{(n-2)(n-1)n(n+1)}{8}$
1 3 2 4	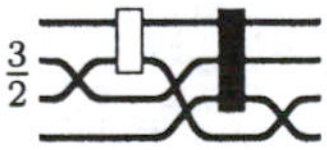 $\frac{3}{2}$	
1 4 2 3	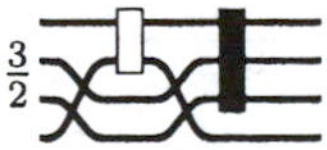 $\frac{3}{2}$	
1 2 3 4		$\frac{n(n-1)(n-2)(n-3)}{24}$
1 ⊗ 2 ⊗ 3 ⊗ 4	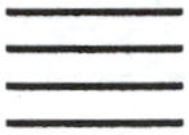	n^4

Table 9.2 Reduction of 4-index tensors. Note the symmetry under $n \leftrightarrow -n$.

As a check that a decomposition is correct, one can compute the dimensions for the product of irreps on the LHS and the sums of the irreps on the RHS to see that they match. Methods for calculating the dimension of a $U(n)$ irreps are discussed in section 9.4.3.

9.6 $U(n)$ RECOUPLING RELATIONS

For $U(n)$ (as opposed to $SU(n)$; see section 9.8) we have no antiparticles, so in recoupling relations the total particle number is conserved. Consider as an example the step-by-step reduction of a 5-particle state in terms of the Young projection operators:

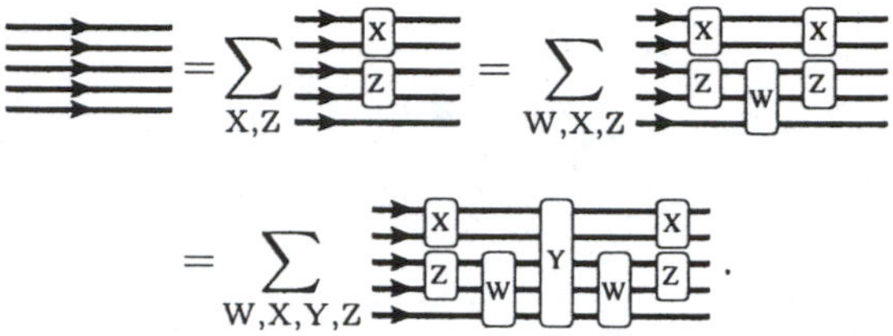

$$= \sum_{X,Z} \quad = \sum_{W,X,Z}$$

$$= \sum_{W,X,Y,Z} \quad .$$

More generally, we can visualize any sequence of $U(n)$ pairwise Clebsch-Gordan reductions as a flow with lines joining into thicker and thicker projection operators, always ending in a maximal $\mathbf{P}_{\mathrm{Y}}$ that spans across all lines. In the clebsches notation of section 5.1, this can be redrawn more compactly as

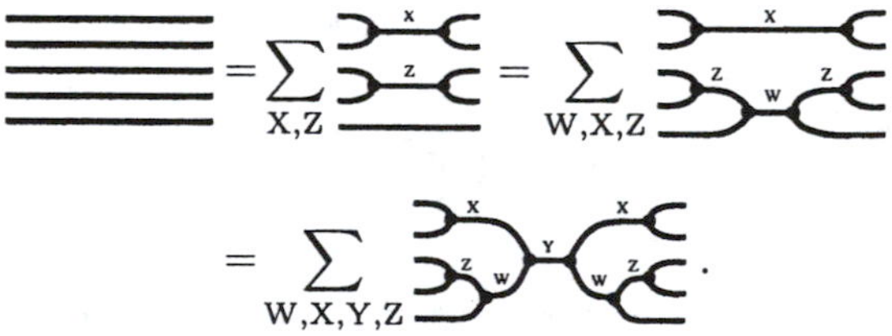

$$= \sum_{X,Z} \quad = \sum_{W,X,Z}$$

$$= \sum_{W,X,Y,Z} \quad .$$

The trace of each term in the final sum of the 5-particle state is a 12-j symbol of the form

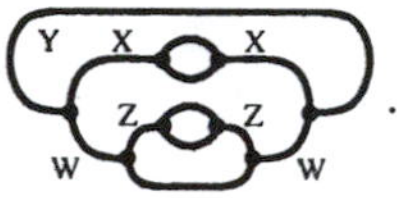

$$. \qquad (9.34)$$

In the trace (9.34) we can use the idempotency of the projection operators to double the maximal Young projection operator $\mathbf{P}_{\mathrm{Y}}$, and sandwich by it all smaller projection operators:

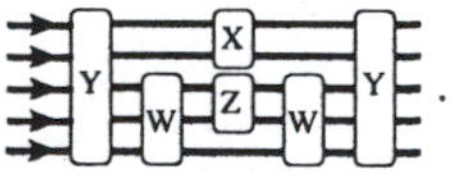

$$. \qquad (9.35)$$

From uniqueness of connection between the symmetry operators (see appendix B.1), we have for any permutation $\sigma \in S_k$:

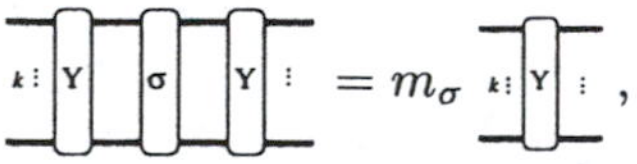

$$= m_\sigma \quad , \qquad (9.36)$$

where $m_\sigma = 0, \pm 1$. Expressions such as (9.35) can be evaluated by expanding the projection operators $\mathbf{P}_W$, $\mathbf{P}_X$, $\mathbf{P}_Z$ and determining the value of m_σ of (9.36) for each permutation σ of the expansion. The result is

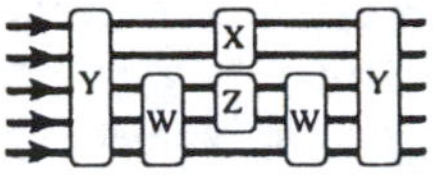

$= M(\mathrm{Y}; \mathrm{W}, \mathrm{X}, \mathrm{Z})$
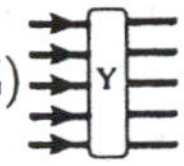

, (9.37)

where the factor $M(\mathrm{Y}; \mathrm{W}, \mathrm{X}, \mathrm{Z})$ *does not depend* on n and is determined by a purely symmetric group calculation. Examples follow.

9.7 $U(n)$ $3n$-j SYMBOLS

In this section, we construct $U(n)$ 3-j and 6-j symbols using the Young projection operators, and we give explicit examples of their evaluation. Sum rules for 3-j's and 6-j's are derived in section 9.7.3.

9.7.1 3-j symbols

Let X, Y, and Z be irreps of $U(n)$. In terms of the Young projection operators $\mathbf{P}_X$, $\mathbf{P}_Y$, and $\mathbf{P}_Z$, a $U(n)$ 3-vertex (5.4) is obtained by tying together the three Young projection operators,

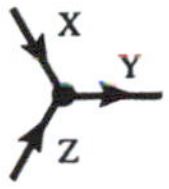

$=$
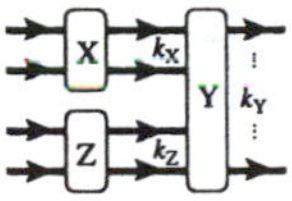

. (9.38)

Since there are no antiparticles, the construction requires $k_X + k_Z = k_Y$.

A 3-j coefficient constructed from the vertex (9.38) is then

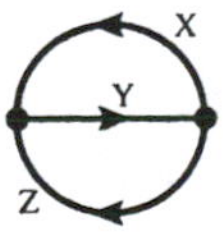

$=$
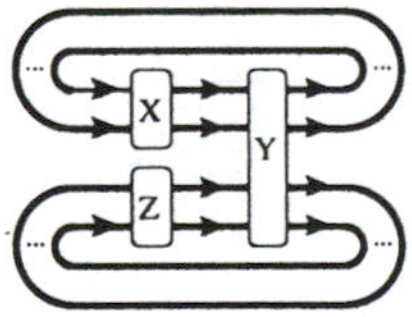

. (9.39)

As an example, take

$$\mathrm{X} = \begin{array}{|c|c|}\hline 1 & 2 \\ \hline 3 \\ \cline{1-1}\end{array}, \quad \mathrm{Y} = \begin{array}{|c|c|c|}\hline 1 & 2 & 4 \\ \hline 3 & 5 & 6 \\ \hline\end{array}, \quad \text{and} \quad \mathrm{Z} = \begin{array}{|c|c|}\hline 4 & 5 \\ \hline 6 \\ \cline{1-1}\end{array}.$$

Then

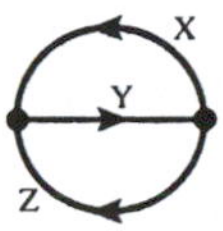

$= \frac{4}{3} \cdot 2 \cdot \frac{4}{3}$
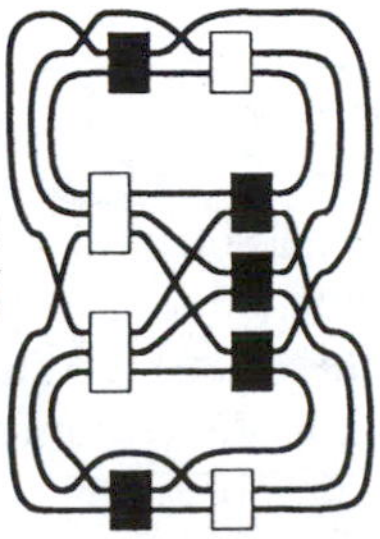
$= M \cdot d_Y$, (9.40)

where $M = 1$ here. Below we derive that d_Y (the dimension of the irrep Y) is indeed the value of this 3-j symbol.

In principle the value of a 3-j symbol (9.39) can be computed by expanding out all symmetry operators, but that is not recommended as the number of terms in such expansions grows combinatorially with the total number of boxes in the Young diagram Y. One can do a little better by carefully selecting certain symmetry operators to expand. Then one simplifies the resulting diagrams using rules such as (6.7), (6.8), (6.17), and (6.18) before expanding more symmetry operators. However, a much simpler method exploits (9.36) and leads to the answer — in the case of (9.40) it is $d_Y = (n^2 - 1)n^2(n+1)(n+2)/144$ – much faster.

The idea for evaluating a 3-j symbol (9.39) using (9.36) is to expand the projections $\mathbf{P}_X$ and $\mathbf{P}_Z$ and determine the value of m_σ in (9.36) for each permutation σ of the expansion. As an example, consider the 3-j symbol (9.40). With $\mathbf{P}_Y$ as in (9.40) we find

σ				
$\sigma \otimes 1$				
$m_{\sigma\otimes 1}$	1	0	1	-1
$1 \otimes \sigma$				
$m_{1\otimes\sigma}$	1	-1	0	-1

so

$$\mathbf{P}_U = \quad = \frac{1}{4}\{\quad - \quad + \quad - \quad\}$$

$$\mathbf{P}_X = \mathbf{P}_U \otimes 1 = \frac{1}{4}\{\quad - \quad + \quad - \quad\}$$

$$M(\mathbf{P}_Y; \mathbf{P}_X) = \frac{1}{4}\{1 - 0 + 1 - (-1)\}$$

$$\mathbf{P}_Z = 1 \otimes \mathbf{P}_U = \frac{1}{4}\{\quad - \quad + \quad - \quad\}$$

$$M(\mathbf{P}_Y; \mathbf{P}_Z) = \frac{1}{4}\{1 - (-1) + 0 - (-1)\};$$

and hence

$$\quad = \left(\frac{3}{4}\right)^2 \alpha_X \alpha_Z \quad = \quad ,$$

and the value of the 3-j is d_Y as claimed in (9.40). That the eigenvalue happens to be 1 is an accident — in tabulations of 3-j symbols [112] it takes a range of values.

The relation (9.36) implies that the value of any $U(n)$ 3-j symbol (9.39) is $M(\mathrm{Y}; \mathrm{X}, \mathrm{Z})d_Y$, where d_Y is the dimension of the maximal irrep Y. Again we remark that $M(\mathrm{Y}; \mathrm{X}, \mathrm{Z})$ is *independent* of n.

9.7.2 6-j symbols

A general $U(n)$ 6-j symbol has form

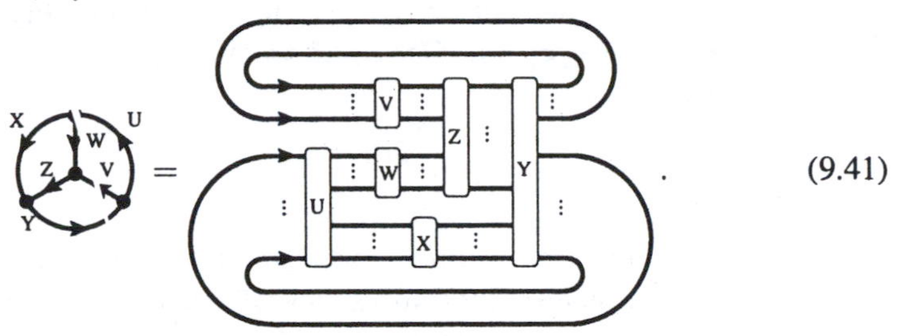

$$. \qquad (9.41)$$

Using the relation (9.36) we immediately see that

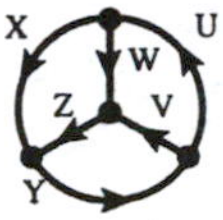

$$= M\, d_{\mathrm{Y}} \,, \qquad (9.42)$$

where M is a pure symmetric group $S_{k_{\mathrm{Y}}}$ number, independent of $U(n)$; it is surprising that the only vestige of $U(n)$ is the fact that the value of a 6-j symbol is proportional to the dimension d_{Y} of its largest projection operator.

Example: Consider the 6-j constructed from the Young tableaux

$$\mathrm{U} = \begin{array}{|c|c|}\hline 2 & 3 \\ \hline 4 \\ \cline{1-1}\end{array}\,, \quad \mathrm{V} = \boxed{1}\,, \quad \mathrm{W} = \boxed{2}\,,$$

$$\mathrm{X} = \begin{array}{|c|}\hline 3 \\ \hline 4 \\ \hline\end{array}\,, \quad \mathrm{Y} = \begin{array}{|c|c|}\hline 1 & 3 \\ \hline 2 \\ \cline{1-1} 4 \\ \cline{1-1}\end{array}\,, \quad \mathrm{Z} = \begin{array}{|c|}\hline 1 \\ \hline 2 \\ \hline\end{array}\,.$$

Using the idempotency we can double the projection $\mathbf{P}_{\mathrm{Y}}$ and sandwich the other operators, as in (9.35). Several terms cancel in the expansion of the sandwiched operator, and we are left with

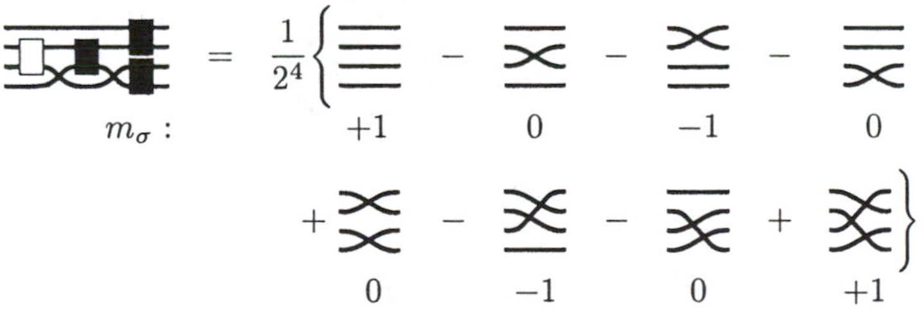

$$= \frac{1}{2^4}\Big\{ \quad - \quad - \quad - \quad + \quad - \quad - \quad + \quad \Big\} .$$

m_σ : $+1$, 0, -1, 0, 0, -1, 0, $+1$

We have listed the symmetry factors m_σ of (9.36) for each of the permutations σ sandwiched between the projection operators $\mathbf{P}_{\mathrm{Y}}$. We find that in this example the symmetric group factor M of (9.42) is

$$M = \frac{4}{2^4}\, \alpha_{\mathrm{U}}\, \alpha_{\mathrm{V}}\, \alpha_{\mathrm{W}}\, \alpha_{\mathrm{X}}\, \alpha_{\mathrm{Z}} = \frac{1}{3}\,,$$

so the value of the 6-j is

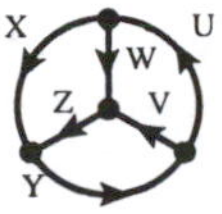

$$= \frac{1}{3} d_{\mathrm{Y}} = \frac{n\,(n^2-1)\,(n-2)}{4!}\,.$$

The method generalizes to evaluations of any $3n$-j symbol of $U(n)$.

Challenge: We have seen that there is a coloring algorithm for the dimensionality of the Young projection operators. *Open question:* Find a coloring algorithm for the 3-j's and 6-j's of $SU(n)$.

9.7.3 Sum rules

Let Y be a standard tableau with k_Y boxes, and let Λ be the set of all standard tableaux with one or more boxes (this excludes the trivial $k = 0$ representation). Then the 3-j symbols obey the sum rule

$$\sum_{X,Z\in\Lambda}$$
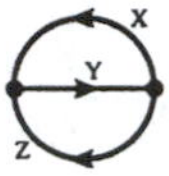

$$= (k_Y - 1)d_Y. \qquad (9.43)$$

The sum is finite, because the 3-j is nonvanishing only if the number of boxes in X and Z add up to k_Y, and this happens only for a finite number of tableaux.

To prove the 3-j sum rule (9.43), recall that the Young projection operators constitute a complete set, $\sum_{X\in\Lambda_k} \mathbf{P}_X = \mathbf{1}$, where $\mathbf{1}$ is the $[k \times k]$ unit matrix and Λ_k the set of all standard tableaux of Young diagrams with k boxes. Hence:

$$\sum_{X,Z\in\Lambda}$$
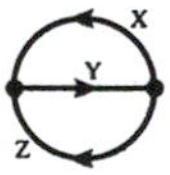

$$= \sum_{k_X=1}^{k_Y-1} \sum_{\substack{X\in\Lambda_{k_X}\\ Z\in\Lambda_{k_Y-k_X}}}$$
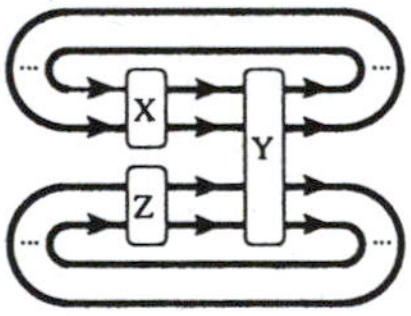

$$= \sum_{k_X=1}^{k_Y-1}$$
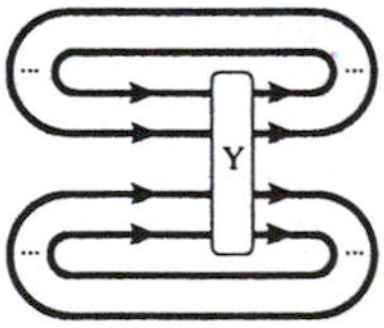

$$= \sum_{k_X=1}^{k_Y-1} d_Y = (k_Y - 1)d_Y .$$

The sum rule offers a useful cross-check on tabulations of 3-j values.

There is a similar sum rule for the 6-j symbols:

$$\sum_{X,Z,U,V,W\in\Lambda}$$

$$= \frac{1}{2}(k_Y - 1)(k_Y - 2)\, d_Y . \qquad (9.44)$$

Referring to the 6-j (9.41), let k_U be the number of boxes in the Young diagram U, k_X be the number of boxes in X, *etc.*

Let k_Y be given. From (9.41) we see that k_X takes values between 1 and $k_Y - 2$, and k_Z takes values between 2 and $k_Y - 1$, subject to the constraint $k_X + k_Z = k_Y$. We now sum over all tableaux U, V, and W keeping k_Y, k_X, and k_Z fixed. Note that

k_V can take values $1, \ldots, k_Z - 1$. Using completeness, we find

$$\sum_{U,V,W\in\Lambda} = \sum_{k_V=1}^{k_Z-1} \sum_{V\in\Lambda_{k_V}} \sum_{W\in\Lambda_{k_Z-k_V}} \sum_{U\in\Lambda_{k_Y-k_V}}$$

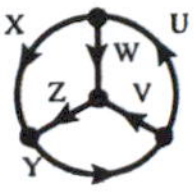

$$= \sum_{k_V=1}^{k_Z-1}$$

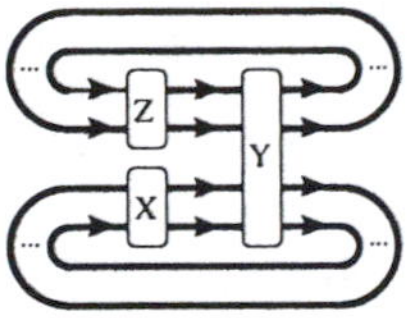

$$= (k_Z - 1) \, .$$

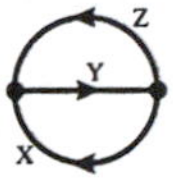

Now sum over all tableaux X and Z to find

$$\sum_{X,Z,U,V,W\in\Lambda} = \sum_{k_Z=2}^{k_Y-1} (k_Z - 1) \sum_{Z\in\Lambda_{k_Z}} \sum_{X\in\Lambda_{k_Y-k_Z}}$$

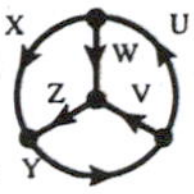

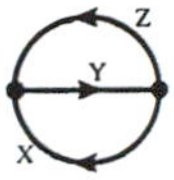

$$= \frac{1}{2}(k_Y - 1)(k_Y - 2)\, d_Y \, ,$$

verifying the sum rule (9.44) for 6-j symbols.

9.8 $SU(n)$ AND THE ADJOINT REP

The $SU(n)$ group elements satisfy det $G = 1$, so $SU(n)$ has an additional invariant, the Levi-Civita tensor $\varepsilon_{a_1 a_2 \ldots a_n} = G_{a_1}{}^{a'_1} G_{a_2}{}^{a'_2} \cdots G_{a_n}{}^{a'_n} \varepsilon_{a'_1 a'_2 \ldots a'_n}$. The diagrammatic notation for the Levi-Civita tensors was introduced in (6.27).

While the irreps of $U(n)$ are labeled by the standard tableaux with no more than n rows (see section 9.3), the standard tableaux with a maximum of $n - 1$ rows label the irreps of $SU(n)$. The reason is that in $SU(n)$, a column of length n can be removed from any diagram by contraction with the Levi-Civita tensor (6.27). For example, for $SU(4)$

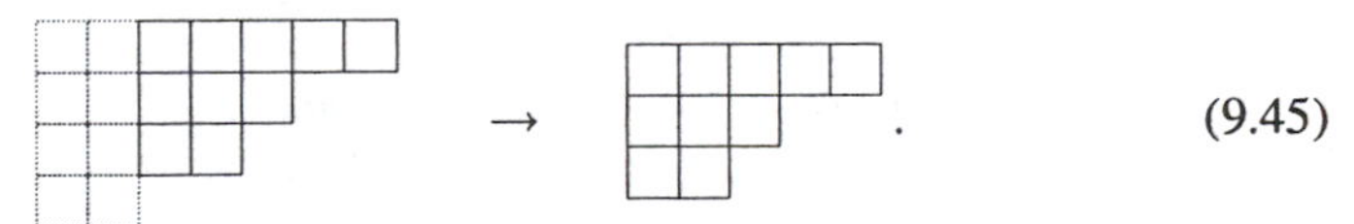

(9.45)

Standard tableaux that differ only by columns of length n correspond to equivalent irreps. Hence, for the standard tableaux labeling irreps of $SU(n)$, the highest column is of height $n - 1$, which is also the rank of $SU(n)$. A rep of $SU(n)$, or A_{n-1} in the Cartan classification (table 7.6) is characterized by $n - 1$ *Dynkin labels* $b_1 b_2 \ldots b_{n-1}$. The corresponding Young diagram (defined in section 9.3.1) is then given by $(b_1 b_2 \ldots b_{n-1} 00 \ldots)$, or $(b_1 b_2 \ldots b_{n-1})$ for short.

For $SU(n)$ a column with k boxes (antisymmetrization of k covariant indices) can be converted by contraction with the Levi-Civita tensor into a column of $(n - k)$ boxes (corresponding to $(n - k)$ contravariant indices). This operation associates with each diagram a conjugate diagram. Thus the *conjugate* of a $SU(n)$ Young diagram Y is constructed from the missing pieces needed to complete the rectangle of n rows,

$SU(5)$: 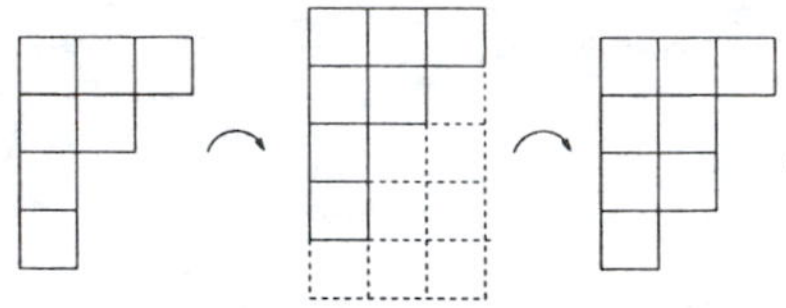 . (9.46)

To find the conjugate diagram, add squares below the diagram of Y such that the resulting figure is a rectangle with height n and width of the top row in Y. Remove the squares corresponding to Y and rotate the rest by 180 degrees. The result is the conjugate diagram of Y. For example, for $SU(6)$ the irrep (20110) has (01102) as its conjugate rep:

$SU(6)$: 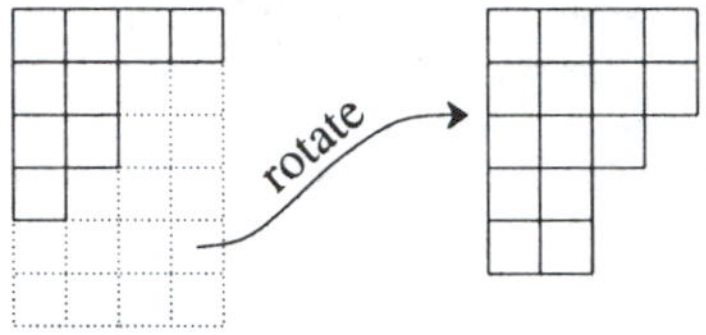

. (9.47)

In general, the $SU(n)$ reps $(b_1 b_2 \ldots b_{n-1})$ and $(b_{n-1} \ldots b_2 b_1)$ are conjugate. For example, $(10\ldots 0)$ stands for the defining rep, and its conjugate is $(00\ldots 01)$, *i.e.*, a column of $n-1$ boxes.

The Levi-Civita tensor converts an antisymmetrized collection of n–1 "in"-indices into 1 "out"-index, or, in other words, it converts an (n–1)-particle state into a single antiparticle state. We use $\bar{\square}$ to denote the single antiparticle state; it is the conjugate of the fundamental representation $\square$ single particle state. For example, for $SU(3)$ we have

(10) =  $= 3$ (20) = 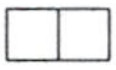 $= 6$

(01) = 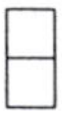 $= \bar{3}$ (02) = 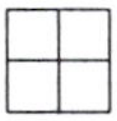 $= \bar{6}$ (9.48)

(11) = 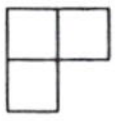 $= 8$ (21) = 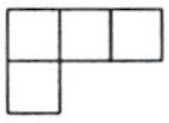 $= 15$.

The product of the fundamental rep $\square$ and the conjugate rep $\bar{\square}$ of $SU(n)$ decomposes into a singlet and the *adjoint representation*:

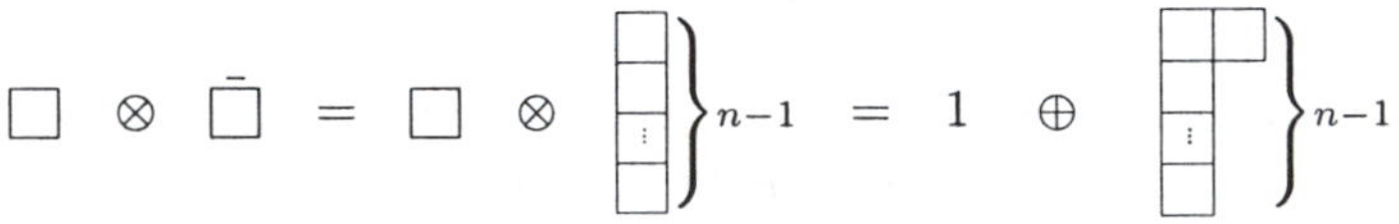

$$n \cdot n = n \cdot n = 1 + (n^2-1)\,.$$

Note that the conjugate of the diagram for the adjoint is again the adjoint.

Using the construction of section 9.4, the birdtrack Young projection operator for the adjoint representation A can be written

$$\mathbf{P}_A = \frac{2(n-1)}{n}$$ 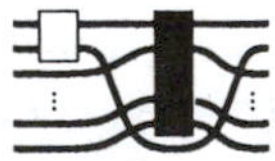 .

Using $\mathbf{P}_A$ and the definition (9.38) of the 3-vertex, $SU(n)$ group theory weights involving quarks, antiquarks, and gluons can be calculated by expansion of the symmetry operators or by application of the recoupling relation. For this reason, we prefer to keep the conjugate reps conjugate, rather than replacing them by columns of $(n-1)$ defining reps, as this will give us $SU(n)$ expressions valid for any n.

9.9 AN APPLICATION OF THE NEGATIVE DIMENSIONALITY THEOREM

An $SU(n)$ invariant scalar is a fully contracted object (vacuum bubble) consisting of Kronecker deltas and Levi-Civita symbols. Since there are no external legs, the Levi-Civitas appear only in pairs, making it possible to combine them into antisymmetrizers. In the birdtrack notation, an $SU(n)$ invariant scalar is therefore a vacuum bubble graph built only from symmetrizers and antisymmetrizers.

The negative dimensionality theorem for $SU(n)$ states that for any $SU(n)$ invariant scalar exchanging symmetrizers and antisymmetrizers is equivalent to replacing n by $-n$:

$$\mathrm{SU}(n) = \overline{\mathrm{SU}}(-n) \,, \tag{9.49}$$

where the bar on $\overline{\mathrm{SU}}$ indicates transposition, *i.e.*, exchange of symmetrizations and antisymmetrizations. The theorem also applies to $U(n)$ invariant scalars, since the only difference between $U(n)$ and $SU(n)$ is the invariance of the Levi-Civita tensor in $SU(n)$. The proof of this theorem is given in chapter 13.

We can apply the negative dimensionality theorem to computations of the dimensions of the $U(n)$ irreps, $d_{\mathrm{Y}} = \mathrm{tr}\,\mathbf{P}_{\mathrm{Y}}$. Taking the transpose of a Young diagram interchanges rows and columns, and it is therefore equivalent to interchanging the symmetrizers and antisymmetrizers in $\mathrm{tr}\,\mathbf{P}_{\mathrm{Y}}$. The dimension of the irrep corresponding to the transpose Young diagram Y^t can then be related to the dimension of the irrep labeled by Y as $d_{\mathrm{Y}^t}(n) = d_{\mathrm{Y}}(-n)$ by the negative dimensionality theorem.

Example: $[3,1]$ is the transpose of $[2,1,1]$,

$$\left(\begin{array}{|c|c|c|}\hline 1 & 2 & 3 \\ \hline 4 & \multicolumn{2}{c}{} \\ \cline{1-1}\end{array}\right)^t = \begin{array}{|c|c|}\hline 1 & 2 \\ \hline 3 & \multicolumn{1}{c}{} \\ \cline{1-1} 4 & \multicolumn{1}{c}{} \\ \cline{1-1}\end{array} \,.$$

Note the $n \to -n$ duality in the dimension formulas for these and other tableaux (table 9.2).

Now for standard tableaux X, Y, and Z, compare the diagram of the 3-j constructed from X, Y, and Z to that constructed from X^t, Z^t, and Y^t. The diagrams are related by a reflection in a vertical line, reversal of all the arrows on the lines, and interchange of symmetrizers and antisymmetrizers. The first two operations do not change the value of the diagram, and by the negative dimensionality theorem the values of two 3-j's are related by $n \leftrightarrow -n$ (and possibly an overall sign; this sign is fixed by requiring that the highest power of n comes with a positive coefficient). In tabulations, it suffices to calculate approximately half of all 3-j's. Furthermore, the 3-j sum rule (9.43) provides a cross-check.

The two 6-j symbols

(9.50)

are related by a reflection in a vertical line, reversal of all the arrows on the lines, and interchange of symmetrizers and antisymmetrizers — this can be seen by writing out the 6-j symbols in terms of the Young projection operators as in (9.41). By the negative dimensionality theorem, the values of the two 6-j symbols are therefore related by $n \leftrightarrow -n$.

9.10 $SU(n)$ MIXED TWO-INDEX TENSORS

We now return to the construction of projection operators from characteristic equations. Consider mixed tensors $q^{(1)} \otimes \overline{q}^{(2)} \in V \otimes \overline{V}$. The Kronecker delta invariants are the same as in section 9.1, but now they are drawn differently (we are looking at a "cross channel"):

$$\text{identity:} \quad \mathbf{1} = 1_{a,d}^{b\ c} = \delta_a^c \delta_d^b = \ ,$$

$$\text{trace:} \quad \mathbf{T} = T_{a,d}^{b\ c} = \delta_a^b \delta_d^c = \ . \tag{9.51}$$

The $\mathbf{T}$ matrix satisfies a trivial characteristic equation

$$\mathbf{T}^2 = \ = n\mathbf{T}\,, \tag{9.52}$$

i.e., $\mathbf{T}(\mathbf{T} - n\mathbf{1}) = 0$, with roots $\lambda_1 = 0$, $\lambda_2 = n$. The corresponding projection operators (3.48) are

$$\mathbf{P}_1 = \frac{1}{n}\mathbf{T} = \frac{1}{n}\ , \tag{9.53}$$

$$\mathbf{P}_2 = \mathbf{1} - \frac{1}{n}\mathbf{T} = \ - \frac{1}{n}\ = \ , \tag{9.54}$$

with dimensions $d_1 = \mathrm{tr}\,\mathbf{P}_1 = 1$, $d_2 = \mathrm{tr}\,\mathbf{P}_2 = n^2 - 1$. $\mathbf{P}_2$ is the projection operator for the adjoint rep of $SU(n)$. In this way, the invariant matrix $\mathbf{T}$ has resolved the space of tensors $x_b^a \in V \otimes \overline{V}$ into a singlet and a traceless part,

$$\mathbf{P}_1 x = \frac{1}{n} x_c^c \delta_a^b\,, \qquad \mathbf{P}_2 x = x_a^b - \left(\frac{1}{n} x_c^c\right) \delta_a^b\,. \tag{9.55}$$

Both projection operators leave δ_b^a invariant, so the generators of the unitary transformations are given by their sum

$$U(n): \quad \frac{1}{a}\ = \ , \tag{9.56}$$

and the dimension of the $U(n)$ adjoint rep is $N = \mathrm{tr}\,\mathbf{P}_A = \delta_a^a \delta_b^b = n^2$. If we extend the list of primitive invariants from the Kronecker delta to the Kronecker delta and the Levi-Civita tensor (6.27), the singlet subspace does not satisfy the invariance condition (6.56)

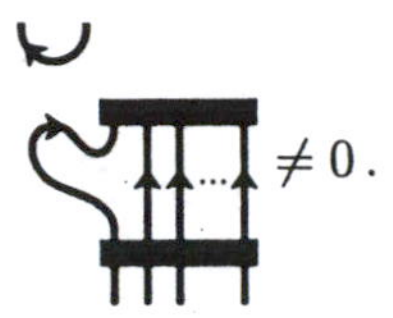

For the traceless subspace (9.54), the invariance condition is

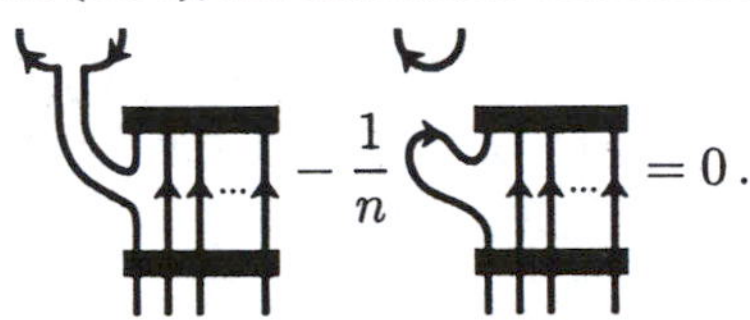

This is the same relation as (6.25), as can be shown by expanding the antisymmetrization operator using (6.19), so the invariance condition is satisfied. The adjoint rep is given by

$$SU(n): \quad \frac{1}{a}(T_i)^a_b\,(T_i)^d_c = \delta^a_c\delta^d_b - \frac{1}{n}\delta^a_b\delta^d_c\,. \tag{9.57}$$

The special unitary group $SU(n)$ is, by definition, the invariance group of the Levi-Civita tensor (hence "special") and the Kronecker delta (hence "unitary"), and its dimension is $N = n^2 - 1$. The defining rep Dynkin index follows from (7.27) and (7.28)

$$\ell^{-1} = 2n \tag{9.58}$$

(This was evaluated in the example of section 2.2.) The Dynkin index for the singlet rep (9.55) vanishes identically, as it does for any singlet rep.

9.11 $SU(n)$ MIXED DEFINING ⊗ ADJOINT TENSORS

In this and the following section we generalize the reduction by invariant matrices to spaces other than the defining rep. Such techniques will be very useful later on, in our construction of the exceptional Lie groups. We consider the defining ⊗ adjoint tensor space as a projection from $V \otimes V \otimes \overline{V}$ space:

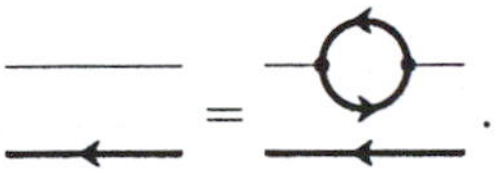

$$(9.59)$$

The following two invariant matrices acting on $V^2 \otimes \overline{V}$ space contract or interchange defining rep indices:

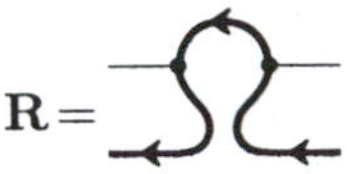

$$\mathbf{R} = \tag{9.60}$$

$$\mathbf{Q} = \tag{9.61}$$

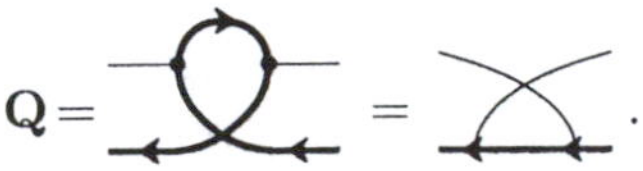

	$\underline{A} \otimes \underline{q}$	$=$	$\underline{V_1}$	$\oplus$	$\underline{V_2}$	$\oplus$	$\underline{V_3}$
Dynkin labels	$(10\ldots1)\otimes(10\ldots)$	$=$	$(10\ldots)$	$\oplus$	$(200\ldots01)$	$\oplus$	$(010\ldots01)$
Dimensions:	$(n^2-1)n$	$=$	n	$+$	$\frac{n(n-1)(n+2)}{2}$	$+$	$\frac{n(n+1)(n-2)}{2}$
Indices:	$n+\frac{n^2-1}{2n}$	$=$	$\frac{1}{2n}$	$+$	$\frac{(n+2)(3n-1)}{4n}$	$+$	$\frac{(n-2)(3n+1)}{4n}$
SU(3) example:							
Dimensions:	$8\cdot3$	$=$	3	$+$	15	$+$	6
Indices:	$13/3$	$=$	$1/6$	$+$	$10/3$	$+$	$5/6$
SU(4) example:							
Dimensions:	$15\cdot4$	$=$	4	$+$	36	$+$	20
Indices:	$47/8$	$=$	$1/8$	$+$	$33/8$	$+$	$13/8$

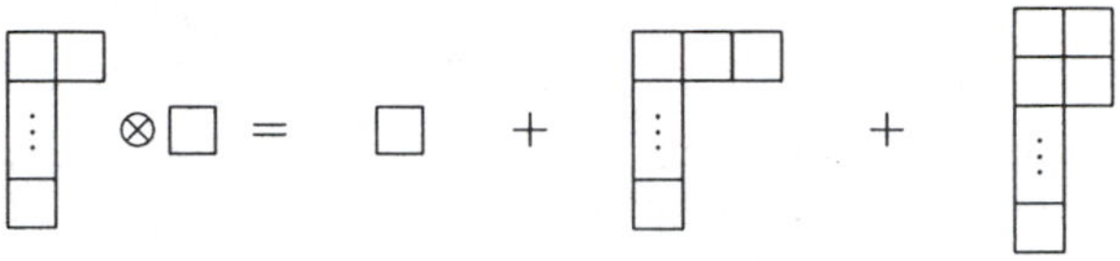

Projection operators:

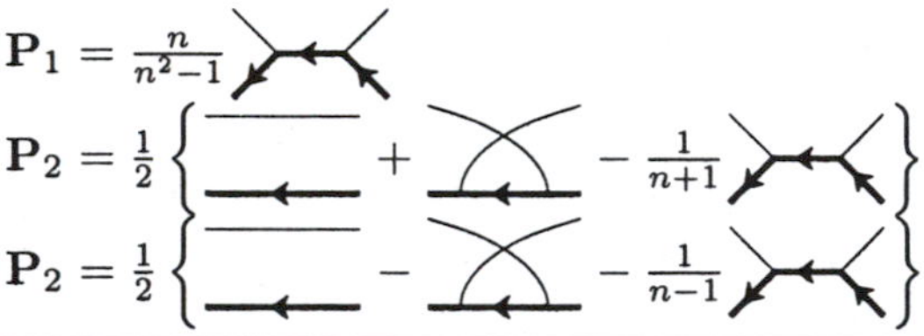

$\mathbf{P}_1 = \frac{n}{n^2-1}$

$\mathbf{P}_2 = \frac{1}{2}\Big\{ \quad + \quad - \frac{1}{n+1} \quad \Big\}$

$\mathbf{P}_2 = \frac{1}{2}\Big\{ \quad - \quad - \frac{1}{n-1} \quad \Big\}$

Table 9.3 $SU(n)\ V \otimes A$ Clebsch-Gordan series.

$\mathbf{R}$ projects onto the defining space and satisfies the characteristic equation

$$\mathbf{R}^2 = \frac{n^2-1}{n}\mathbf{R} . \tag{9.62}$$

The corresponding projection operators (3.48) are

$$\mathbf{P}_1 = \frac{n}{n^2-1} \,,$$
$$\mathbf{P}_4 = - \frac{n}{n^2-1} . \tag{9.63}$$

$\mathbf{Q}$ takes a single eigenvalue on the $\mathbf{P}_1$ subspace

$$\mathbf{QR} = = -\frac{1}{n}\mathbf{R} . \tag{9.64}$$

$\mathbf{Q}^2$ is computed by inserting the adjoint rep projection operator (9.57):

$$\mathbf{Q}^2 = = - \frac{1}{n} . \tag{9.65}$$

The projection on the $\mathbf{P}_4$ subspace yields the characteristic equation

$$\mathbf{P}_4(\mathbf{Q}^2 - 1) = 0 \,, \tag{9.66}$$

with the associated projection operators

$$\mathbf{P}_2 = \frac{1}{2}\mathbf{P}_4(1+\mathbf{Q}) \tag{9.67}$$
$$= \frac{1}{2}\left\{ - \frac{n}{n^2-1} \right\}\left\{ + \right\}$$
$$= \frac{1}{2}\left\{ + - \frac{1}{n+1} \right\} ,$$
$$\mathbf{P}_3 = \frac{1}{2}\mathbf{P}_4(1-\mathbf{Q})$$
$$= \frac{1}{2}\left\{ - - \frac{1}{n-1} \right\} . \tag{9.68}$$

The dimensions of the two subspaces are computed by taking traces of their projection operators:

$$d_2 = \mathrm{tr}\,\mathbf{P}_2 = \boxed{\mathrm{P}_2} = \frac{1}{2}\left\{ + - \frac{1}{n+1} \right\}$$
$$= \frac{1}{2}\left(nN + N - N/(n+1)\right) = \frac{1}{2}(n-1)n(n+2) \tag{9.69}$$

and similarly for d_3. This is tabulated in table 9.3.

9.11.1 Algebra of invariants

Mostly for illustration purposes, let us now perform the same calculation by utilizing the algebra of invariants method outlined in section 3.4. A possible basis set, picked from the $V \otimes A \to V \otimes A$ linearly independent tree invariants, consists of

$$(\mathbf{e}, \mathbf{R}, \mathbf{Q}) = \left(\;\;, \;\;, \;\; \right) . \qquad (9.70)$$

The multiplication table (3.42) has been worked out in (9.62), (9.64), and (9.65). For example, the $(t_\alpha)_\beta{}^\gamma$ matrix rep for $\mathbf{Q}\mathbf{t}$ is

$$\sum_{\gamma \in T} (\mathbf{Q})_\beta{}^\gamma \mathbf{t}_\gamma = \mathbf{Q} \begin{pmatrix} \mathbf{e} \\ \mathbf{R} \\ \mathbf{Q} \end{pmatrix} = \begin{pmatrix} 0 & 0 & 1 \\ 0 & -1/n & 0 \\ 1 & -1/n & 0 \end{pmatrix} \begin{pmatrix} \mathbf{e} \\ \mathbf{R} \\ \mathbf{Q} \end{pmatrix} \qquad (9.71)$$

and similarly for $\mathbf{R}$. In this way, we obtain the $[3 \times 3]$ matrix rep of the algebra of invariants

$$\{\mathbf{e}, \mathbf{R}, \mathbf{Q}\} = \left\{ \begin{pmatrix} 1 & 0 & 0 \\ 0 & 1 & 0 \\ 0 & 0 & 1 \end{pmatrix}, \begin{pmatrix} 0 & 1 & 0 \\ 0 & n - \frac{1}{n} & 0 \\ 0 & -1/n & 0 \end{pmatrix}, \begin{pmatrix} 0 & 0 & 1 \\ 0 & -1/n & 0 \\ 1 & -1/n & 0 \end{pmatrix} \right\} . \qquad (9.72)$$

From (9.62) we already know that the eigenvalues of $\mathbf{R}$ are $\{0, 0, n - 1/n\}$. The last eigenvalue yields the projection operator $\mathbf{P}_1 = (n - 1/n)^{-1}$, but the projection operator $\mathbf{P}_4$ yields a 2-dimensional degenerate rep. $\mathbf{Q}$ has three distinct eigenvalues $\{-1/n, 1, -1\}$ and is thus more interesting; the corresponding projection operators fully decompose the $V \otimes A$ space. The $-1/n$ eigenspace projection operator is again $\mathbf{P}_1$, but $\mathbf{P}_4$ is split into two subspaces, verifying (9.68) and (9.67):

$$\mathbf{P}_2 = \frac{(\mathbf{Q} + \mathbf{1})(\mathbf{Q} + \frac{1}{n}\mathbf{1})}{(1+1)(1+1/n)} = \frac{1}{2}\left(\mathbf{1} + \mathbf{Q} - \frac{1}{n+1}\mathbf{R}\right)$$
$$\mathbf{P}_3 = \frac{(\mathbf{Q} - \mathbf{1})(\mathbf{Q} + \frac{1}{n}\mathbf{1})}{(-1-1)(-1+1/n)} = \frac{1}{2}\left(\mathbf{1} - \mathbf{Q} - \frac{1}{n-1}\mathbf{R}\right) . \qquad (9.73)$$

We see that the matrix rep of the algebra of invariants is an alternative tool for implementing the full reduction, perhaps easier to implement as a computation than an out and out birdtracks evaluation.

To summarize, the invariant matrix $\mathbf{R}$ projects out the 1-particle subspace $\mathbf{P}_1$. The particle exchange matrix $\mathbf{Q}$ splits the remainder into the irreducible $V \otimes A$ subspaces $\mathbf{P}_2$ and $\mathbf{P}_3$.

9.12 $SU(n)$ TWO-INDEX ADJOINT TENSORS

Consider the Kronecker product of two adjoint reps. We want to reduce the space of tensors $x_{ij} \in A \otimes A$, with $i = 1, 2, \ldots N$. The first decomposition is the obvious decomposition (9.4) into the symmetric and antisymmetric subspaces,

$$1 = \mathbf{S} + \mathbf{A}$$
$$= \;\; + \;\; . \qquad (9.74)$$

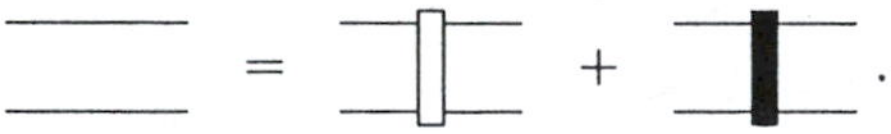

The symmetric part can be split into the trace and the traceless part, as in (9.54):

$$\mathbf{S} = \frac{1}{N}\mathbf{T} + \mathbf{P}_S$$

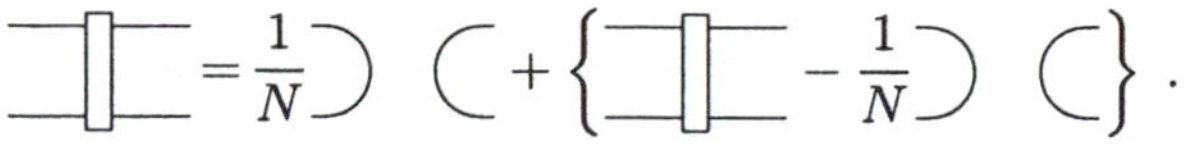

$$= \frac{1}{N} + \left\{ - \frac{1}{N} \right\} . \qquad (9.75)$$

Further decomposition can be effected by studying invariant matrices in the $V^2 \otimes \overline{V}^2$ space. We can visualize the relation between $A \otimes A$ and $V^2 \otimes \overline{V}^2$ by the identity

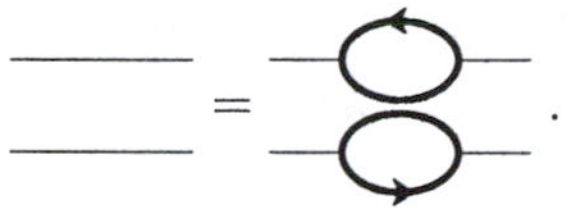

$$= . \qquad (9.76)$$

This suggests the introduction of two invariant matrices:

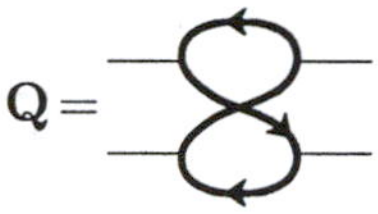

$$\mathbf{Q} = \qquad (9.77)$$

$$\mathbf{R} = = . \qquad (9.78)$$

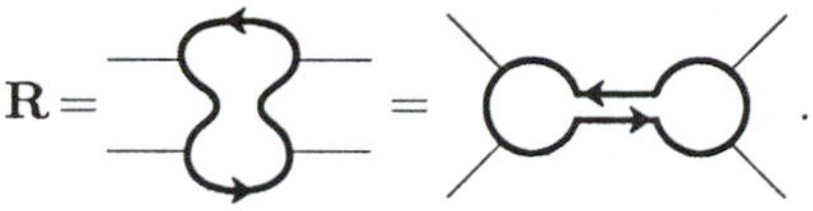

$\mathbf{R}$ can be decomposed by (9.54) into a singlet and the adjoint rep

$$\mathbf{R} = + \tfrac{1}{n}$$
$$= \mathbf{R}' + \tfrac{1}{n}\mathbf{T} . \qquad (9.79)$$

The singlet has already been taken into account in the trace-traceless tensor decomposition (9.75). The $\mathbf{R}'$ projection on the antisymmetric subspace is

$$\mathbf{AR'A} = . \qquad (9.80)$$

By the Lie algebra (4.47),

$$(\mathbf{AR'A})^2 = \frac{1}{16} = \frac{n}{8} = \frac{n}{2}\mathbf{AR'A} , \qquad (9.81)$$

and the associated projection operators,

$$(\mathbf{P}_5)_{ij,kl} = \frac{1}{2n} C_{ijm} C_{mlk} = \frac{1}{2n}$$

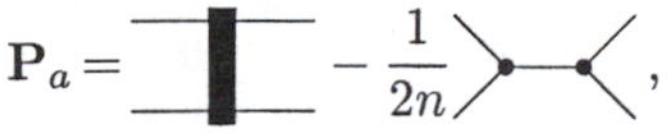

$$\mathbf{P}_a = - \frac{1}{2n} , \qquad (9.82)$$

split the antisymmetric subspace into the adjoint rep and a remainder. On the symmetric subspace (9.75), $\mathbf{R}'$ acts as $\mathbf{P}_S \mathbf{R}' \mathbf{P}_S$. As $\mathbf{R}'\mathbf{T} = 0$, this is the same as $\mathbf{SR'S}$. Consider

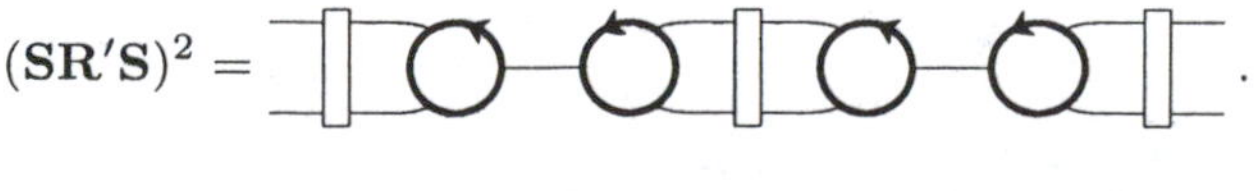

$$(\mathbf{SR'S})^2 = .$$

We compute

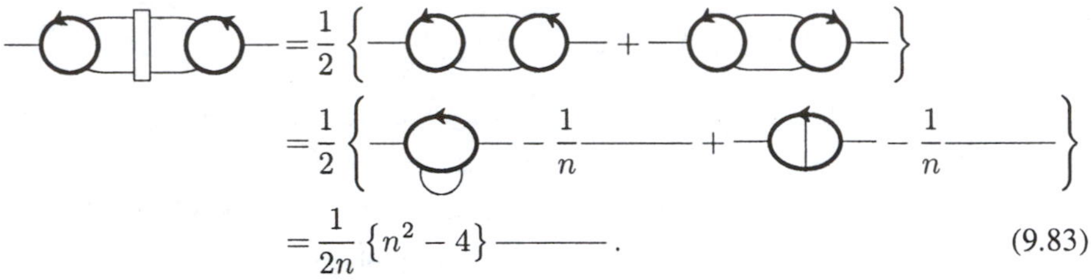

$$= \frac{1}{2}\left\{ \quad + \quad \right\}$$

$$= \frac{1}{2}\left\{ \quad - \frac{1}{n} \quad + \quad - \frac{1}{n} \quad \right\}$$

$$= \frac{1}{2n}\left\{n^2 - 4\right\} \quad . \tag{9.83}$$

Hence, $\mathbf{SR'S}$ satisfies the characteristic equation

$$\left(\mathbf{SR'S} - \frac{n^2-4}{2n}\right)\mathbf{SR'S} = 0\,. \tag{9.84}$$

The associated projection operators split up the traceless symmetric subspace (9.75) into the adjoint rep and a remainder:

$$\mathbf{P}_2 = \frac{2n}{n^2-4}\mathbf{SR'S} = \frac{2n}{n^2-4} \quad , \tag{9.85}$$

$$\mathbf{P}_{2'} = \mathbf{P}_S - \mathbf{P}_2\,. \tag{9.86}$$

The Clebsch-Gordan coefficients for $\mathbf{P}_2$ are known as the Gell-Mann d_{ijk} tensors [137]:

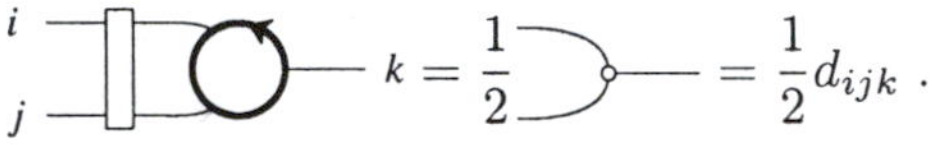

$$k = \frac{1}{2} \quad = \frac{1}{2} d_{ijk}\,. \tag{9.87}$$

For $SU(3)$, $\mathbf{P}_2$ is the projection operator $(\underline{8}\otimes\underline{8})$ symmetric $\to \underline{8}$. In terms of d_{ijk}'s, we have

$$(\mathbf{P}_2)_{ij,k\ell} = \frac{n}{2(n^2-4)} d_{ijm} d_{mk\ell} = \frac{n}{2(n^2-4)} \quad , \tag{9.88}$$

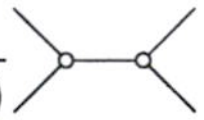

with the normalization

$$d_{ijk} d_{kj\ell} = \quad = \frac{2(n^2-4)}{n}\delta_{i\ell}\,. \tag{9.89}$$

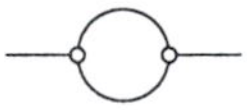

Next we turn to the decomposition of the symmetric subspace induced by matrix $\mathbf{Q}$ (9.77). $\mathbf{Q}$ commutes with $\mathbf{S}$:

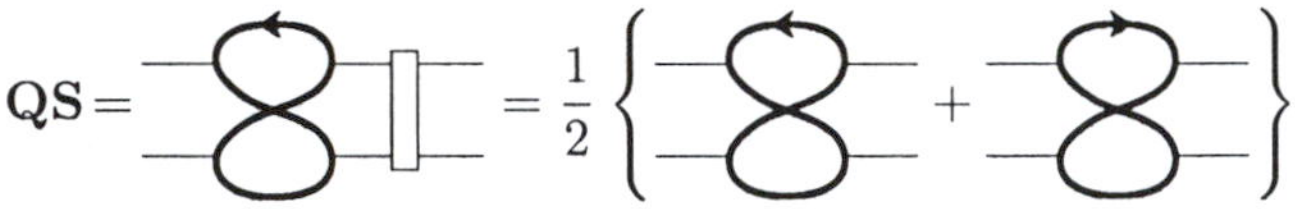

$$\mathbf{QS} = \quad = \frac{1}{2}\left\{ \quad + \quad \right\}$$

$$= \mathbf{SQ} = \mathbf{SQS}\,. \tag{9.90}$$

On the 1-dimensional subspace in (9.75), it takes eigenvalue $-1/n$

$$\mathbf{TQ} = \quad = -\frac{1}{n}\mathbf{T}\,; \tag{9.91}$$

so $\mathbf{Q}$ also commutes with the projection operator $\mathbf{P}_S$ from (9.75),

$$\mathbf{QP}_S = \mathbf{Q}\left(\mathbf{S} - \frac{1}{n^2-1}T\right) = \mathbf{P}_S\mathbf{Q}\,. \tag{9.92}$$

$\mathbf{Q}^2$ is easily evaluated by inserting the adjoint rep projection operators (9.54)

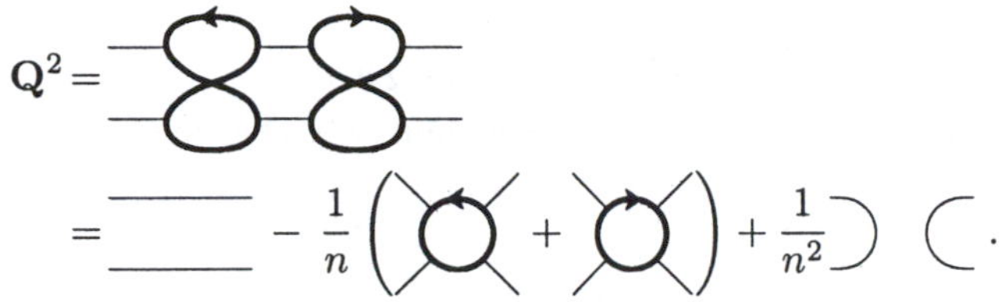

$$\mathbf{Q}^2 = \quad = \quad - \frac{1}{n}\left(\quad + \quad \right) + \frac{1}{n^2} \quad . \tag{9.93}$$

Projecting on the traceless symmetric subspace gives

$$\mathbf{P}_S\left(\mathbf{Q}^2 - 1 + \frac{n^2-4}{n^2}\mathbf{P}_2\right) = 0\,. \tag{9.94}$$

On the $\mathbf{P}_2$ subspace $\mathbf{Q}$ gives

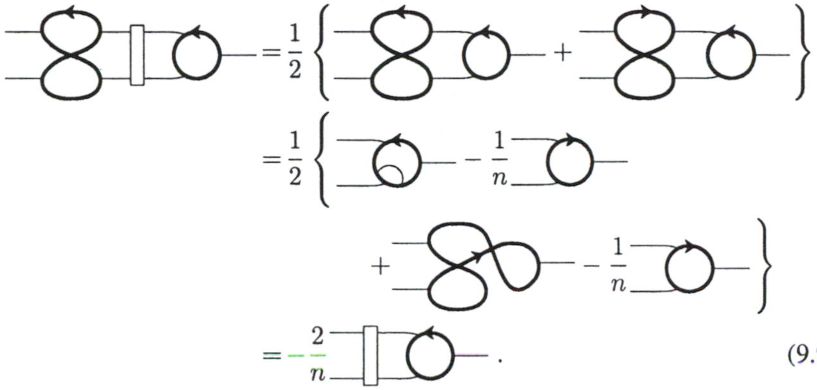

$$= \frac{1}{2}\left\{ \quad + \quad \right\}$$

$$= \frac{1}{2}\left\{ \quad - \frac{1}{n} \quad + \quad - \frac{1}{n} \quad \right\}$$

$$= -\frac{2}{n} \quad . \tag{9.95}$$

Hence, $\mathbf{Q}$ has a single eigenvalue,

$$\mathbf{Q}\mathbf{P}_2 = -\frac{2}{n}\mathbf{P}_2\,, \tag{9.96}$$

and does not decompose the $\mathbf{P}_2$ subspace; this is as it should be, as $\mathbf{P}_2$ is the adjoint rep and is thus irreducible. On $\mathbf{P}_{2'}$ subspace (9.93) yields a characteristic equation

$$\mathbf{P}_{2'}(\mathbf{Q}^2 - 1) = 0\,,$$

with the associated projection operators

$$\mathbf{P}_3 = \frac{1}{2}\mathbf{P}_{2'}(1 - \mathbf{Q}) \tag{9.97}$$

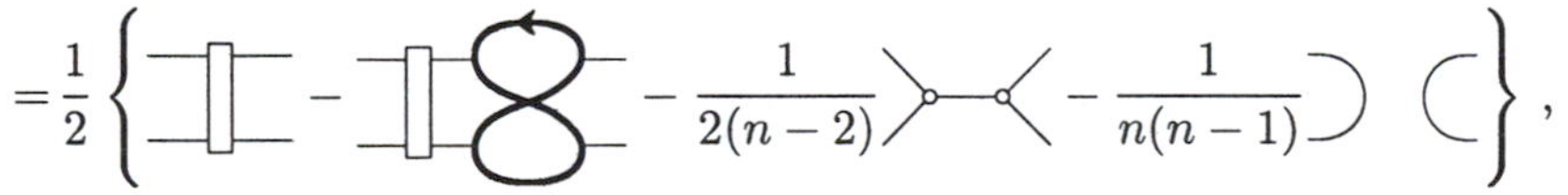

$$= \frac{1}{2}\left\{ \quad - \quad - \frac{1}{2(n-2)} \quad - \frac{1}{n(n-1)} \quad \right\},$$

$$\mathbf{P}_4 = \frac{1}{2}\mathbf{P}_{2'}(1 + \mathbf{Q}) = \frac{1}{2}(\mathbf{P}_S - \mathbf{P}_1)(1 + \mathbf{Q})$$

$$= \frac{1}{2}\left(\mathbf{P}_S - \mathbf{P}_1 + \mathbf{S}\mathbf{Q} - \frac{1}{n^2-1}\mathbf{T}\mathbf{Q} + \frac{2}{n}\mathbf{P}_1\right)$$

$$= \frac{1}{2}\left(\mathbf{S} + \mathbf{S}\mathbf{Q} - \frac{n-2}{n}\mathbf{P}_1 - \frac{1}{n(n+1)}\mathbf{T}\right) \tag{9.98}$$

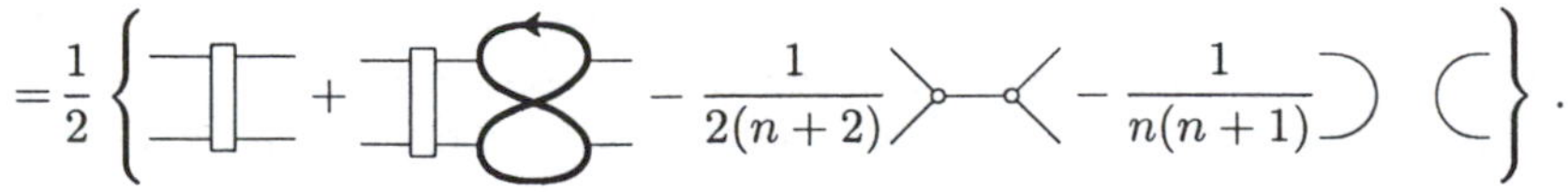

$$= \frac{1}{2}\left\{ \quad + \quad - \frac{1}{2(n+2)} \quad - \frac{1}{n(n+1)} \quad \right\}.$$

This completes the reduction of the symmetric subspace in (9.74). As in (9.90), $\mathbf{Q}$ commutes with $\mathbf{A}$

$$\mathbf{QA} = \mathbf{AQ} = \mathbf{AQA} \,. \tag{9.99}$$

On the antisymmetric subspace, the $\mathbf{Q}^2$ equation (9.93) becomes

$$0 = \mathbf{A}\left(\mathbf{Q}^2 - 1 + \frac{2}{n}\mathbf{R}\right), \qquad \mathbf{A} = \mathbf{A}(\mathbf{Q}^2 - 1 - \mathbf{P}_A) \,. \tag{9.100}$$

The adjoint rep (9.82) should be irreducible. Indeed, it follows from the Lie algebra, that $\mathbf{Q}$ has zero eigenvalue for any simple group:

$$\mathbf{P}_5\mathbf{Q} = \frac{1}{C_A} [\text{birdtrack}] = 0 \,. \tag{9.101}$$

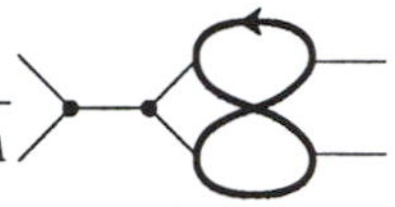

On the remaining antisymmetric subspace $\mathbf{P}_a$ (9.100) yields the characteristic equation

$$\mathbf{P}_a(\mathbf{Q}^2 - 1) = 0 \,, \tag{9.102}$$

with corresponding projection operators

$$\mathbf{P}_6 = \frac{1}{2}\mathbf{P}_a(1+\mathbf{Q}) = \frac{1}{2}\mathbf{A}(1+\mathbf{Q}-\mathbf{P}_A) = \frac{1}{2}\left\{ [\text{birdtrack}] + [\text{birdtrack}] - \frac{1}{C_A} [\text{birdtrack}] \right\}, \tag{9.103}$$

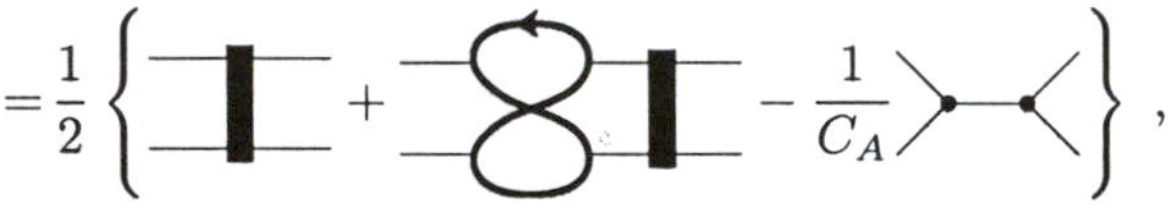

$$\mathbf{P}_7 = \frac{1}{2}\mathbf{P}_a(1-\mathbf{Q}) = \frac{1}{2}\left\{ [\text{birdtrack}] - [\text{birdtrack}] - \frac{1}{C_A} [\text{birdtrack}] \right\}. \tag{9.104}$$

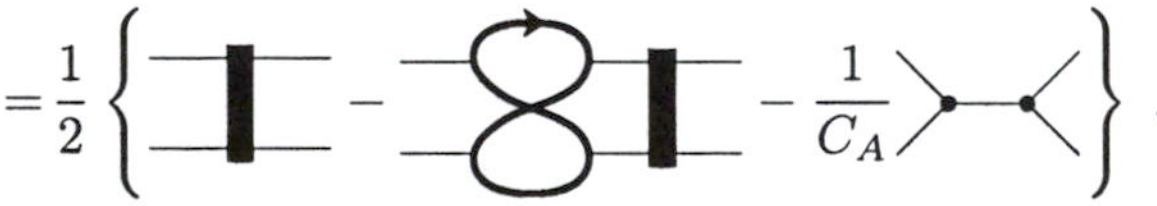

To compute the dimensions of these reps we need

$$\operatorname{tr}\mathbf{AQ} = [\text{birdtrack}] = \frac{1}{2}\left\{ [\text{birdtrack}] - [\text{birdtrack}] \right\} = 0 \,, \tag{9.105}$$

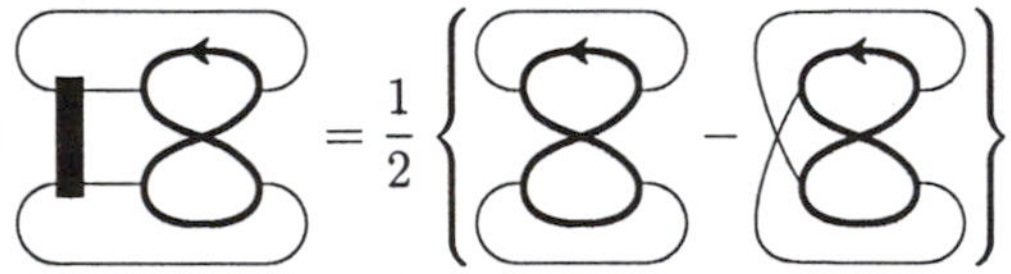

so both reps have the same dimension

$$d_6 = d_7 = \frac{1}{2}(\operatorname{tr}\mathbf{A} - \operatorname{tr}\mathbf{P}_A) = \frac{1}{2}\left\{\frac{(n^2-1)(n^2-2)}{2} - n^2 - 1\right\} = \frac{(n^2-1)(n^2-4)}{4} \,. \tag{9.106}$$

Indeed, the two reps are conjugate reps. The identity

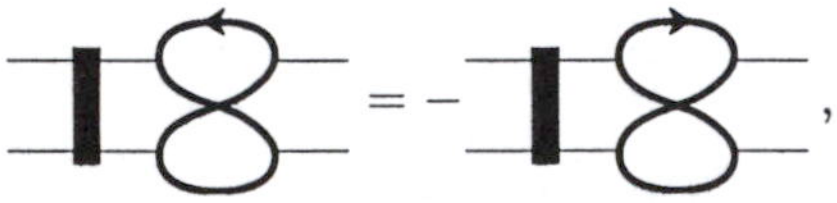

$$[\text{birdtrack}] = -[\text{birdtrack}] \,, \tag{9.107}$$

obtained by interchanging the two left adjoint rep legs, implies that the projection operators (9.103) and (9.104) are related by the reversal of the loop arrow. This is the birdtrack notation for complex conjugation (see section 4.1).

This decomposition of two $SU(n)$ adjoint reps is summarized in table 9.4.

9.13 CASIMIRS FOR THE FULLY SYMMETRIC REPS OF $SU(n)$

In this section we carry out a few explicit birdtrack casimir evaluations.

Consider the fully symmetric Kronecker product of p particle reps. Its Dynkin label (defined on page 106) is $(p, 0, 0 \ldots 0)$, and the corresponding Young tableau is a row of p boxes: 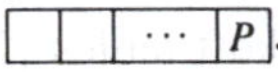. The projection operator is given by (6.4)

$$\mathbf{P}_S = \mathbf{S} =$$ 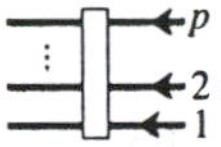 ,

and the generator (4.40) in the symmetric rep is

$$T^i = p$$ 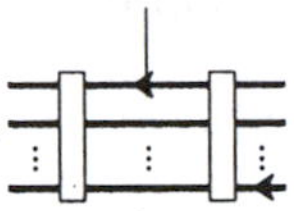 . (9.108)

To compute the casimirs, we introduce matrices:

$$X = x_i T^i = p$$ 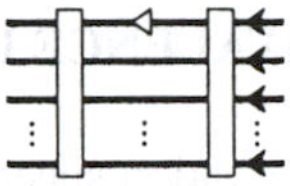

$$X_a^b = x_i (T^i)_a^b = $$  . (9.109)

We next compute the powers of X:

$X^2 = p\Big\{$ 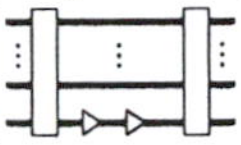 $+ (p-1)$ 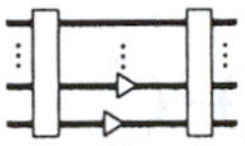 $\Big\}$

$X^3 = p\Big\{$ 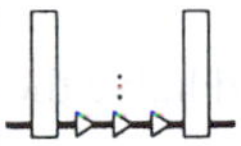 $+ 3(p-1)$ 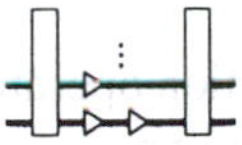 $+ (p-1)(p-2)$ 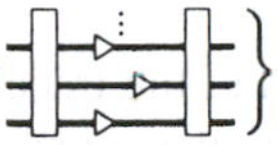 $\Big\}$

$X^4 = p\Big\{$ 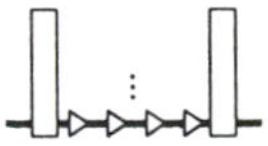 $+ 4(p-1)$ 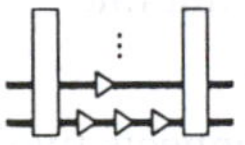 $+ 3(p-1)$ 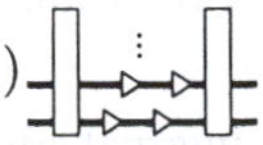

$+6(p-1)(p-2)$ 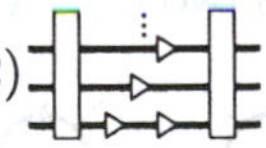 $+ (p-1)(p-2)(p-3)$ 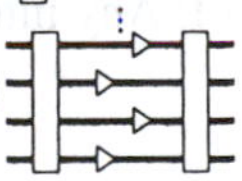 $\Big\}$

$\vdots$ (9.110)

The tr X^k are then

$$\mathrm{tr}\, X^0 = d_s \binom{n+p-1}{p} \qquad \text{(see (6.13))} \tag{9.111}$$

$$\mathrm{tr}\, X = 0 \qquad \text{(semisimplicity)} \tag{9.112}$$

$$\mathrm{tr}\, X^2 = d_s \frac{p(p+n)}{n(n+1)} \mathrm{tr}\, x^2 \tag{9.113}$$

$$\begin{aligned} \mathrm{tr}\, X^3 &= \frac{d_s}{n} p \left(1 + 3\frac{p-1}{n+1} + 2\frac{(p-1)(p-2)}{(n+1)(n+2)}\right) \mathrm{tr}\, x^3 \\ &= \frac{(n+p)!(n+2p)}{(n+2)!(p-1)!} \mathrm{tr}\, x^3 = d_s \frac{p(n+p)(n+2p)}{n(n+1)(n+2)} \mathrm{tr}\, x^3 \end{aligned} \tag{9.114}$$

$$\begin{aligned} \mathrm{tr}\, X^4 = d\frac{p}{n} \Bigg\{ &\left(1 + 7\frac{p-1}{n+1} + 12\frac{p-1}{n+1}\frac{p-2}{n+2} + 6\frac{p-1}{n+1}\frac{p-2}{n+2}\frac{p-3}{n+3}\right) \mathrm{tr}\, x^4 \\ &+ \frac{p-1}{n+1}\left(3 + 6\frac{p-2}{n+2} + 3\frac{p-2}{n+2}\frac{p-3}{n+3}\right) \left(\mathrm{tr}\, x^2\right)^2 \Bigg\} . \end{aligned} \tag{9.115}$$

The quadratic Dynkin index is given by the ratio of tr X^2 and $\mathrm{tr}_A X^2$ for the adjoint rep (7.30):

$$\ell_2 = \frac{\mathrm{tr}\, X^2}{\mathrm{tr}_A\, X^2} = \frac{d_s p(p+n)}{2n^2(n+1)} \,. \tag{9.116}$$

To take a random example from the Patera-Sankoff tables [273], the $SU(6)$ rep dimension and Dynkin index

rep	dim	ℓ_2
(0,0,0,0,0,14)	11628	6460

(9.117)

check with the above expressions.

9.14 $SU(n)$, $U(n)$ EQUIVALENCE IN ADJOINT REP

The following simple observation speeds up evaluation of pure adjoint rep group-theoretic weights $(3n\text{-}j)$'s for $SU(n)$: The adjoint rep weights for $U(n)$ and $SU(n)$ are identical. This means that we can use the $U(n)$ adjoint projection operator

$$U(n):\quad [\text{diagram}] \tag{9.118}$$

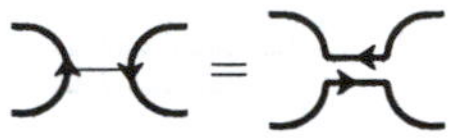

instead of the traceless $SU(n)$ projection operator (9.54), and halve the number of terms in the expansion of each adjoint line.

Proof: Any internal adjoint line connects two C_{ijk}'s:

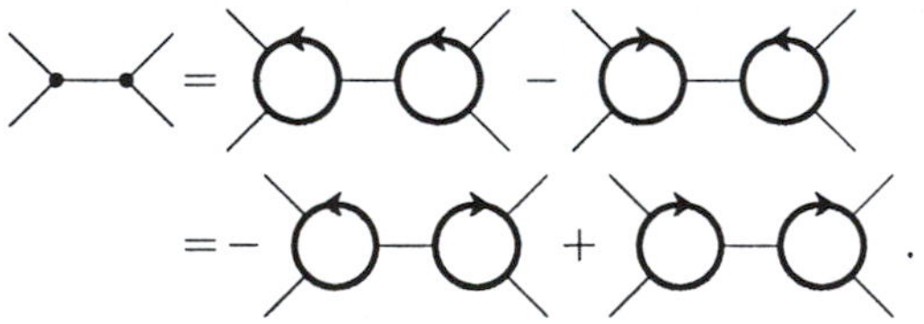

[diagram]

The trace part of (9.54) cancels on each line; hence, it does not contribute to the pure adjoint rep diagrams. As an example, we reevaluate the adjoint quadratic casimir for $SU(n)$:

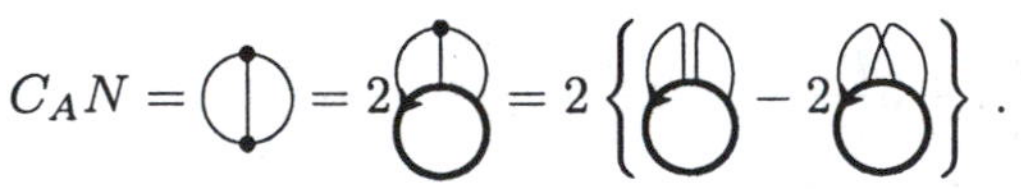

$$C_A N = [\text{diagram}] = 2[\text{diagram}] = 2\left\{[\text{diagram}] - 2[\text{diagram}]\right\} .$$

Now substitute the $U(n)$ adjoint projection operator (9.118):

$$C_A N = 2\left\{[\text{diagram}] - 2[\text{diagram}]\right\} = 2n(n^2-1) \,,$$

in agreement with the first exercise of section 2.2.

9.15 SOURCES

Sections 9.3–9.9 of this chapter are based on Elvang *et al.* [113]. The introduction to the Young tableaux folows ref. [113], which, in turn, is based on Lichtenberg [214] and Hamermesh [153]. The rules for reduction of direct products follow Lichtenberg [214], stated here as in ref. [112]. The construction of the Young projection operators directly from the Young tableaux is described in van der Waerden [334], who ascribes the idea to von Neumann.

R. Penrose's papers are the first (known to the authors) to cast the Young projection operators into a diagrammatic form. Here we use Penrose diagrammatic notation for symmetrization operators [280], Levi-Civita tensors [282], and "strand networks" [281]. For several specific, few-particle examples, diagrammatic Young projection operators were constructed by Canning [41], Mandula [227], and Stedman [318]. A diagrammatic construction of the $U(n)$ Young projection operators for *any* Young tableau was outlined in the unpublished ref. [186], without proofs; the proofs of appendix B that the Young projection operators so constructed are unique were given in ref. [112].

			Symmetric								Antisymmetric				
	$\underline{V}_A \otimes \underline{V}_A$	$=$	V_1	$\oplus$	V_2	$\oplus$	V_3	$\oplus$	V_4	$\oplus$	V_5	$\oplus$	V_6	$\oplus$	V_7
Dimensions	$(n^2-1)^2$	$=$	1	$+$	(n^2-1)	$+$	$\frac{n^2(n-3)(n+1)}{4}$	$+$	$\frac{n^2(n+3)(n-1)}{4}$	$+$	(n^2-1)	$+$	$\frac{(n^2-1)(n^2-4)}{4}$	$+$	$\frac{(n^2-1)(n^2-4)}{4}$
Dynkin indices	$2(n^2-1)$	$=$	0	$+$	1	$+$	$\frac{n(n-3)}{2}$	$+$	$\frac{n(n+3)}{2}$	$+$	1	$+$	$\frac{n^2-4}{2}$	$+$	$\frac{n^2-4}{2}$
$SU(3)$ example:															
Dimensions	8^2	$=$	1	$+$	8	$+$	0	$+$	27	$+$	8	$+$	10	$+$	$\overline{10}$
Indices	$2 \cdot 8$	$=$	0	$+$	1	$+$	0	$+$	9	$+$	1	$+$	5/2	$+$	5/2
$SU(4)$ example:															
	$(101) \otimes (101)$	$=$	(000)	$\oplus$	(101)	$\oplus$	(020)	$\oplus$	(202)	$\oplus$	(101)	$\oplus$	(012)	$\oplus$	(210)
Dimensions	15^2	$=$	1	$+$	15	$+$	20	$+$	84	$+$	15	$+$	45	$+$	$\overline{45}$
Indices	$2 \cdot 15$	$=$	0	$+$	1	$+$	2	$+$	14	$+$	1	$+$	6	$+$	6

Projection operators

$\mathbf{P}_1 = \frac{1}{n^2-1}$

$\mathbf{P}_2 = \frac{n}{2(n^2-4)}$

$\mathbf{P}_3 = \frac{1}{2}\Big\{ \quad - \quad - \frac{1}{2(n-2)} \quad - \frac{1}{n(n-1)} \quad \Big\}$

$\mathbf{P}_4 = \frac{1}{2}\Big\{ \quad + \quad - \frac{1}{2(n+2)} \quad - \frac{1}{n(n+1)} \quad \Big\}$

$\mathbf{P}_5 = \frac{1}{2n}$

$\mathbf{P}_6 = \frac{1}{2}\Big\{ \quad + \quad - \frac{1}{2n} \quad \Big\}$

$\mathbf{P}_7 = \frac{1}{2}\Big\{ \quad - \quad - \frac{1}{2n} \quad \Big\}$

Table 9.4 $SU(n)$, $n \geq 3$ Clebsch-Gordan series for $A \otimes A$.

Chapter Ten

Orthogonal groups

Orthogonal group $SO(n)$ is the group of transformations that leaves invariant a symmetric quadratic form $(q,q) = g_{\mu\nu}q^{\mu}q^{\nu}$:

$$g_{\mu\nu} = g_{\nu\mu} = \mu \longleftarrow\!\circ\!\longrightarrow \nu \qquad \mu, \nu = 1, 2, \ldots, n\,. \tag{10.1}$$

If (q,q) is an invariant, so is its complex conjugate $(q,q)^* = g^{\mu\nu}q_{\mu}q_{\nu}$, and

$$g^{\mu\nu} = g^{\nu\mu} = \mu \longrightarrow\!\circ\!\longleftarrow \nu \tag{10.2}$$

is also an invariant tensor. The matrix $A_{\mu}^{\nu} = g_{\mu\sigma}g^{\sigma\nu}$ must be proportional to unity, as otherwise its characteristic equation would decompose the defining n-dimensional rep. A convenient normalization is

$$g_{\mu\sigma}g^{\sigma\nu} = \delta_{\mu}^{\nu}$$

$$\longleftarrow\!\circ\!\longrightarrow\!\circ\!\longleftarrow \;=\; \longleftarrow\,. \tag{10.3}$$

As the indices can be raised and lowered at will, nothing is gained by keeping the arrows. Our convention will be to perform all contractions with metric tensors with upper indices and omit the arrows and the open dots:

$$g^{\mu\nu} \equiv \mu \;\text{———}\; \nu\,. \tag{10.4}$$

All other tensors will have lower indices. For example, Lie group generators $(T_i)_{\mu}{}^{\nu}$ from (4.31) will be replaced by

$$(T_i)_{\mu}{}^{\nu} = \;\cdots \;\to\; (T_i)_{\mu\nu} = \;\cdots\,.$$

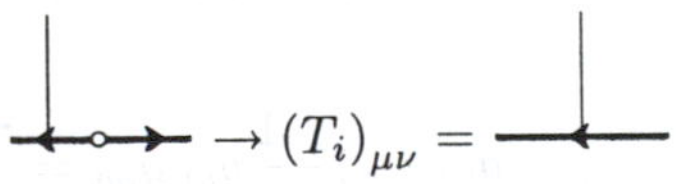

The invariance condition (4.36) for the metric tensor

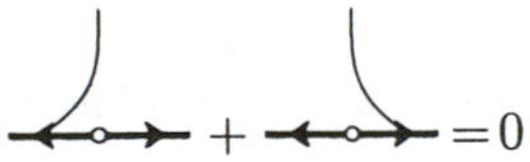

$$\cdots + \cdots = 0$$

$$(T_i)_{\mu}{}^{\sigma}g_{\sigma\nu} + (T_i)_{\nu}{}^{\sigma}g_{\mu\sigma} = 0 \tag{10.5}$$

becomes, in this convention, a statement that the $SO(n)$ generators are antisymmetric:

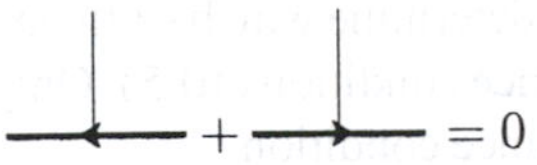

$$\cdots + \cdots = 0$$

$$(T_i)_{\mu\nu} = -\,(T_i)_{\nu\mu}\,. \tag{10.6}$$

Our analysis of the reps of $SO(n)$ will depend only on the existence of a symmetric metric tensor and its invertability, and not on its eigenvalues. The resulting Clebsch-Gordan series applies both to the compact $SO(n)$ and noncompact orthogonal groups, such as the Minkowski group $SO(1,3)$. In this chapter, we outline the construction of $SO(n)$ tensor reps. Spinor reps will be taken up in chapter 11.

10.1 TWO-INDEX TENSORS

In section 9.1 we have decomposed the $SU(n)$ 2-index tensors into symmetric and antisymmetric parts. For $SO(n)$, the rule is to lower all indices on all tensors, and the symmetric state projection operator (9.2) is replaced by

$$S_{\mu\nu,\rho\sigma} = g_{\rho\rho'} g_{\sigma\sigma'} S_{\mu\nu,}{}^{\rho'\sigma'}$$
$$= \frac{1}{2}\left(g_{\mu\sigma}g_{\nu\rho} + g_{\mu\rho}g_{\nu\sigma}\right)$$

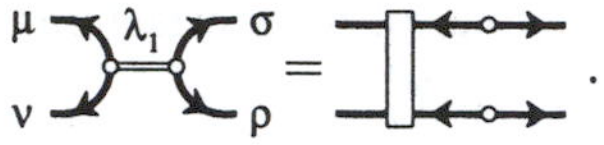

From now on, we drop all arrows and $g^{\mu\nu}$'s and write (9.4) as

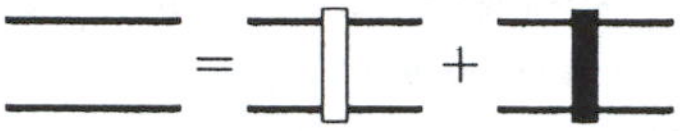

$$g_{\mu\sigma}g_{\nu\rho} = \frac{1}{2}(g_{\mu\sigma}g_{\nu\rho} + g_{\mu\rho}g_{\nu\sigma}) + \frac{1}{2}(g_{\mu\sigma}g_{\nu\rho} - g_{\mu\rho}g_{\nu\sigma}) \,. \qquad (10.7)$$

The new invariant, specific to $SO(n)$, is the index contraction:

$$\mathbf{T}_{\mu\nu,\rho\sigma} = g_{\mu\nu}g_{\rho\sigma}\,, \qquad \mathbf{T} = \,. \qquad (10.8)$$

The characteristic equation for the trace invariant

$$\mathbf{T}^2 = = n\mathbf{T} \qquad (10.9)$$

yields the trace and the traceless part projection operators (9.53), (9.54). As $\mathbf{T}$ is symmetric, $S\mathbf{T} = \mathbf{T}$, only the symmetric subspace is resolved by this invariant. The final decomposition of $SO(n)$ 2-index tensors is
traceless symmetric:

$$(P_2)_{\mu\nu,\rho\sigma} = \frac{1}{2}\left(g_{\mu\sigma}g_{\nu\rho} + g_{\mu\rho}g_{\nu\sigma}\right) - \frac{1}{n} g_{\mu\nu}g_{\rho\sigma} = \quad - \frac{1}{n} \,, \qquad (10.10)$$

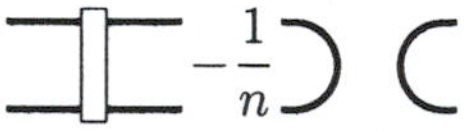

$$\text{singlet:} \quad (P_1)_{\mu\nu,\rho\sigma} = \frac{1}{n} g_{\mu\nu}g_{\rho\sigma} = \frac{1}{n} \,, \qquad (10.11)$$

$$\text{antisymmetric:} \quad (P_3)_{\mu\nu,\rho\sigma} = \frac{1}{2}\left(g_{\mu\sigma}g_{\nu\rho} - g_{\mu\rho}g_{\nu\sigma}\right) = \,. \qquad (10.12)$$

The adjoint rep (9.57) of $SU(n)$ is decomposed into the traceless symmetric and the antisymmetric parts. To determine which of them is the new adjoint rep, we substitute them into the invariance condition (10.5). Only the antisymmetric projection operator satisfies the invariance condition

$$+ = 0 \,,$$

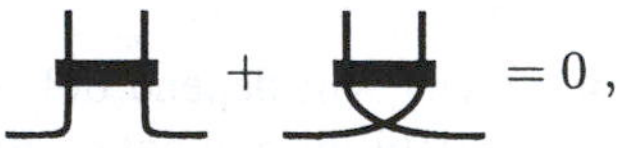

so the adjoint rep projection operator for $SO(n)$ is

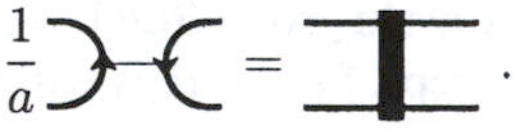

$$\frac{1}{a} = \,. \qquad (10.13)$$

Young tableaux	□×□	=	•	+	 (two boxes in a column)	+	□□
Dynkin labels	(10...) × (10...)	=	(00...)	+	(010...)	+	(20...)
Dimensions	n^2	=	1	+	$\frac{n(n-1)}{2}$	+	$\frac{(n+2)(n-1)}{2}$
Dynkin indices	$2n\frac{1}{n-2}$	=	0	+	1	+	$\frac{n+2}{n-2}$
Projectors	 [diagram]	=	$\frac{1}{n}$ [diagram]	+	[diagram]	+	{[diagram] $-\frac{1}{n}$ [diagram]}

Table 10.1 $SO(n)$ Clebsch-Gordan series for $V \otimes V$.

The dimension of $SO(n)$ is given by the trace of the adjoint projection operator:

$$N = \operatorname{tr} \mathbf{P}_A = [\text{diagram}] = \frac{n(n-1)}{2} . \tag{10.14}$$

Dimensions of the other reps and the Dynkin indices (see section 7.5) are listed in table 10.1.

10.2 MIXED ADJOINT ⊗ DEFINING REP TENSORS

The mixed adjoint-defining rep tensors are decomposed in the same way as for $SU(n)$. The intermediate defining rep state matrix $\mathbf{R}$ (9.60) satisfies the characteristic equation

$$\mathbf{R}^2 = [\text{diagram}] = \frac{n-1}{2}\mathbf{R} . \tag{10.15}$$

The corresponding projection operators are

$$\mathbf{P}_1 = \frac{2}{n-1} [\text{diagram}] ,$$
$$\mathbf{P}_2 = [\text{diagram}] - \frac{2}{n-1} [\text{diagram}] . \tag{10.16}$$

The eigenvalue of $\mathbf{Q}$ from (9.61) on the defining subspace can be computed by inserting the adjoint projection operator (10.13):

$$\mathbf{QR} = [\text{diagram}] = \frac{1}{2}\mathbf{R} . \tag{10.17}$$

$\mathbf{Q}^2$ is also computed by inserting (10.13):

$$\mathbf{Q}^2 = [\text{diagram}] = \frac{1}{2}\left\{ [\text{diagram}] - [\text{diagram}] \right\} = \frac{1}{2}(1 - \mathbf{Q}) . \tag{10.18}$$

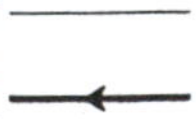

The eigenvalues are $\{-1, \frac{1}{2}\}$, and the associated projection operators (3.48) are

$$\mathbf{P}_2 = \mathbf{P}_4 \frac{2}{3}(1+\mathbf{Q}) = \frac{2}{3}\left(1 - \frac{2}{n-1}\mathbf{R}\right)(1+\mathbf{Q}) = \frac{2}{3}\left(1 + \mathbf{Q} - \frac{3}{n-1}\mathbf{R}\right)$$

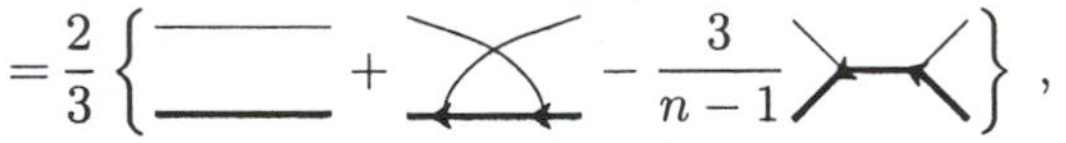

$$= \frac{2}{3}\left\{ \quad + \quad - \frac{3}{n-1} \quad \right\}, \tag{10.19}$$

$$\mathbf{P}_3 = \mathbf{P}_4 \frac{1}{3}(1-2\mathbf{Q}) = \frac{1}{3}\left\{ \quad - 2 \quad \right\}. \tag{10.20}$$

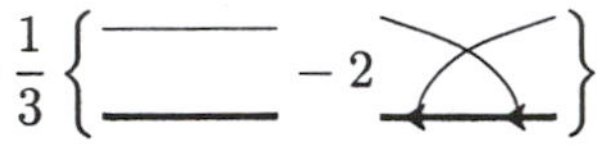

This decomposition is summarized in table 10.2. The same decomposition can be obtained by viewing the $SO(n)$ defining-adjoint tensors as $\boxminus \otimes \square$ products, and starting with the $SU(n)$ decomposition along the lines of section 9.2.

10.3 TWO-INDEX ADJOINT TENSORS

The reduction of the 2-index adjoint rep tensors proceeds as for $SU(n)$. The annihilation matrix $\mathbf{R}$ (9.78) induces decomposition of (10.11) through (10.12) into three tensor spaces

$$\mathbf{R} = \tag{10.21}$$

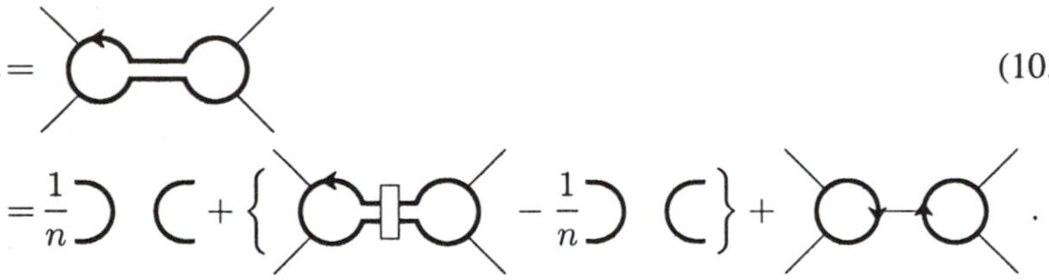

$$= \frac{1}{n} \quad + \left\{ \quad - \frac{1}{n} \quad \right\} + \quad .$$

On the antisymmetric subspace, the last term projects out the adjoint rep:

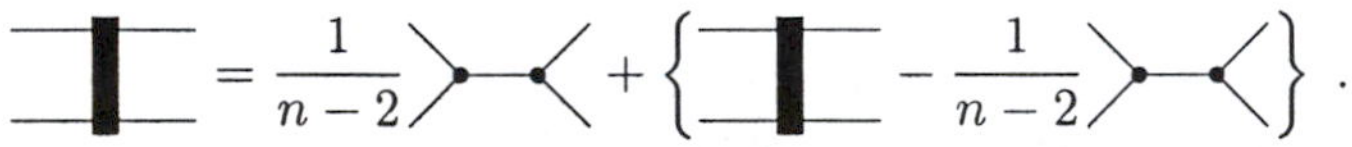

$$= \frac{1}{n-2} \quad + \left\{ \quad - \frac{1}{n-2} \quad \right\}. \tag{10.22}$$

The last term in (10.21) does not affect the symmetric subspace

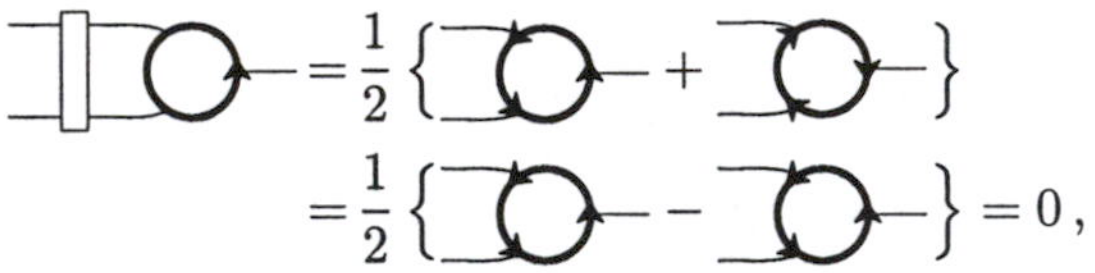

$$= \frac{1}{2}\left\{ \quad + \quad \right\}$$

$$= \frac{1}{2}\left\{ \quad - \quad \right\} = 0, \tag{10.23}$$

because of the antisymmetry of the $SO(n)$ generators ($d_{ijk} = 0$ for orthogonal groups). The second term in (10.21),

$$\mathbf{RS} = \quad - \frac{1}{n} \quad , \tag{10.24}$$

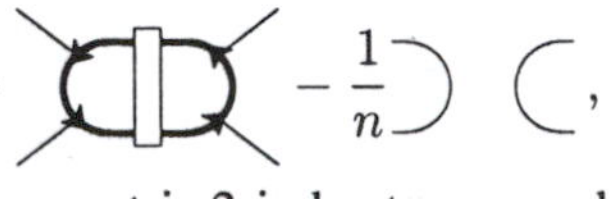

projects out the intermediate symmetric 2-index tensors subspace. To normalize it, we compute $(\mathbf{RS})^2$:

$$(\mathbf{RS})^2 = \quad - \frac{2}{n} \quad + \frac{n-1}{2n} \quad$$

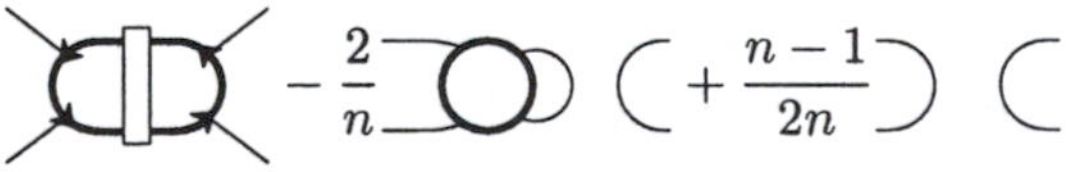

$$= \frac{n-2}{4}\mathbf{RS}. \tag{10.25}$$

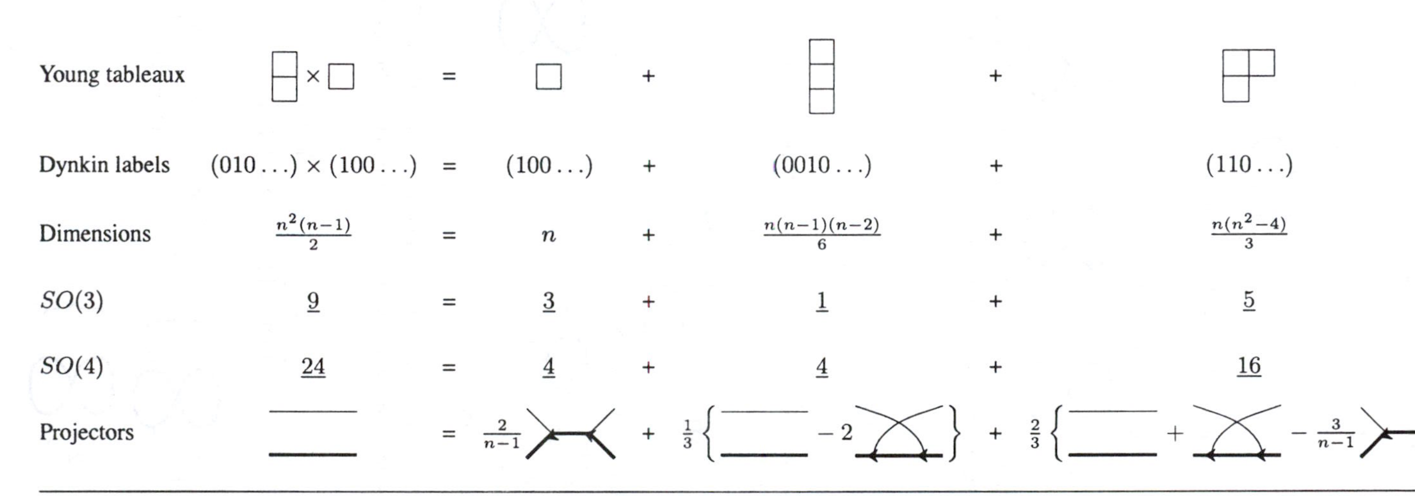

Young tableaux	×	=		+		+	
Dynkin labels	(010 . . .) × (100 . . .)	=	(100 . . .)	+	(0010 . . .)	+	(110 . . .)
Dimensions	$\frac{n^2(n-1)}{2}$	=	n	+	$\frac{n(n-1)(n-2)}{6}$	+	$\frac{n(n^2-4)}{3}$
$SO(3)$	$\underline{9}$	=	$\underline{3}$	+	$\underline{1}$	+	$\underline{5}$
$SO(4)$	$\underline{24}$	=	$\underline{4}$	+	$\underline{4}$	+	$\underline{16}$
Projectors		=	$\frac{2}{n-1}$	+	$\frac{1}{3}\{\ -2\ \}$	+	$\frac{2}{3}\{\ +\ -\frac{3}{n-1}\ \}$

Table 10.2 $SO(n)$ $A \otimes V$ Clebsch-Gordan series.

RS decomposes the symmetric 2-index adjoint subspace into

$$= \frac{2}{n(n-1)} + \left\{ - - \frac{2}{n(n-1)} \right\}$$

$$\mathbf{P}_2 = = \frac{4}{n-2} \left\{ - \frac{1}{n} \right\} . \qquad (10.26)$$

Because of the antisymmetry of the $SO(n)$ generators, the index interchange matrix (9.77) is symmetric,

$$S\mathbf{Q} = S\mathbf{Q}^* = \mathbf{Q}$$

$$= = , \qquad (10.27)$$

so it cannot induce a decomposition of the antisymmetric subspace in (10.22). Here $\mathbf{Q}^*$ indicates the diagram for $\mathbf{Q}$ with the arrow reversed. On the singlet subspace it has eigenvalue $\frac{1}{2}$:

$$\mathbf{Q}T = = \frac{1}{2}T . \qquad (10.28)$$

On the symmetric 2-index defining rep tensors subspace, its eigenvalue is also $\frac{1}{2}$, as the evaluation by the substitution of adjoint projection operators by (10.13) yields

$$\mathbf{QR} = = \frac{1}{2}S\mathbf{R} . \qquad (10.29)$$

Q^2 is evaluated in the same manner:

$$\mathbf{Q}^2 = = \frac{1}{2}\left\{ - \right\}$$

$$= \frac{1}{2}S(1 - \mathbf{Q}) . \qquad (10.30)$$

Thus, $\mathbf{Q}$ satisfies the same characteristic equation as in (10.18). The corresponding projection operators decompose the symmetric subspace (the third term in (10.26)) into

$$\mathbf{P}_3 = \left\{ - - \frac{2}{n(n-1)} \right\} \frac{2}{3} \left\{ + \right\}$$

$$= \frac{2}{3} \left\{ + \right\} - - \frac{2}{n(n-1)} , \qquad (10.31)$$

$$\mathbf{P}_4 = \frac{1}{3} \left\{ - 2 \right\} . \qquad (10.32)$$

			Symmetric								Antisymmetric		
	A ⊗ A	=	V_1	⊕	V_2	⊕	V_3	⊕	V_4	⊕	V_5	⊕	V_6
Young tableaux	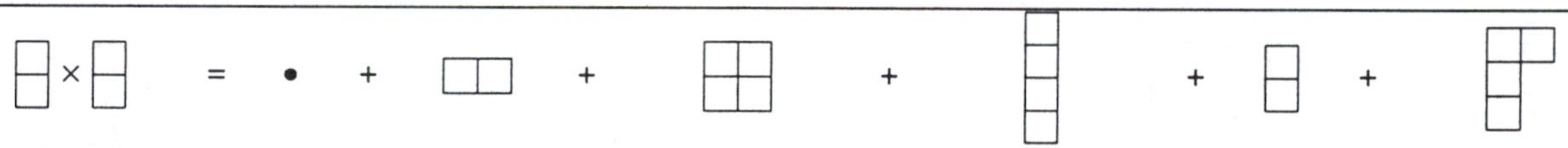	=	•	+		+		+		+		+	
Dynkin labels	(010...)×(010...)	=	(00...)	+	(20...)	+	(02...)	+	(00010...)	+	(010...)	+	(1010...)
Dimensions	$\frac{n^2(n-1)^2}{4}$	=	1	+	$\frac{(n-1)(n+2)}{2}$	+	$\frac{(n-3)n(n+1)(n+2)}{12}$	+	$\frac{n(n-1)(n-2)(n-3)}{24}$	+	$\frac{n(n-1)}{2}$	+	$\frac{n(n+2)(n-1)(n-3)}{8}$
SO(3)	9	=	1	+	5	+	0	+	0	+	3	+	0
SO(4)	36	=	1	+	9	+	10	+	1	+	6	+	9
SO(5)=Sp(4)	100	=	1	+	14	+	35	+	5	+	10	+	35
SO(6)=SU(4)	225	=	1	+	20	+	84	+	15	+	15	+	(45+45)
SO(7)	441	=	1	+	27	+	168	+	35	+	21	+	189
SO(8)	784	=	1	+	35	+	300	+	70	+	28	+	350
SO(9)	1296	=	1	+	44	+	495	+	126	+	36	+	594
SO(10)	2025	=	1	+	54	+	770	+	210	+	45	+	945

Projection operators

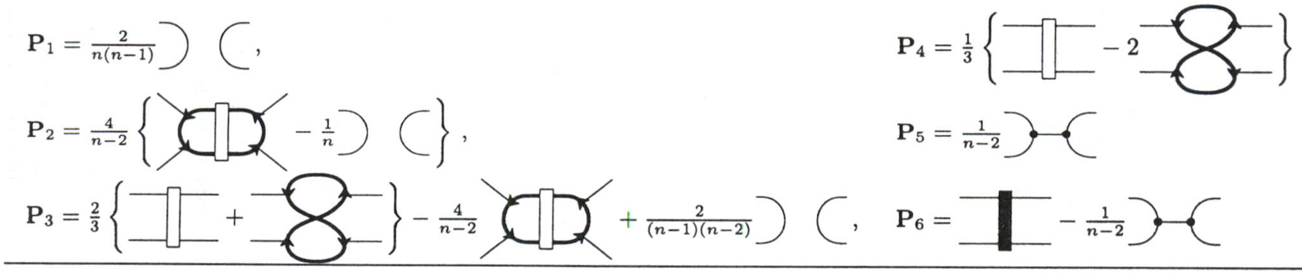

$\mathbf{P}_1 = \frac{2}{n(n-1)}$,

$\mathbf{P}_2 = \frac{4}{n-2}\{ \quad - \frac{1}{n} \quad \}$,

$\mathbf{P}_3 = \frac{2}{3}\{ \quad + \quad \} - \frac{4}{n-2} \quad + \frac{2}{(n-1)(n-2)}$,

$\mathbf{P}_4 = \frac{1}{3}\{ \quad - 2 \quad \}$

$\mathbf{P}_5 = \frac{1}{n-2}$

$\mathbf{P}_6 = \quad - \frac{1}{n-2}$

Table 10.3 $SO(n)$, $n \geq 3$ Clebsch-Gordan series for $A \otimes A$.

This Clebsch-Gordan series is summarized in table 10.3.

The reduction of 2-index adjoint tensors, outlined above, is patterned after the reduction for $SU(n)$. Another, fully equivalent approach, is to consider the $SO(n)$ 2-index adjoint tensors as ⊟ ⊗ ⊟ products and start from the decomposition of section 9.5. This will be partially carried out in section 10.5.

10.4 THREE-INDEX TENSORS

In the reduction of the 2-index tensors in section 10.1, the new $SO(n)$ invariant was the index contraction (10.8). In general, for a multi-index tensor, the $SU(n) \to SO(n)$ reduction is due to the additional index contraction invariants. Consider the fully symmetric 3-index $SU(n)$ state in table 9.1. The new $SO(n)$ invariant matrix on this space is

$$\mathbf{R} = \quad . \tag{10.33}$$

This is a projection onto the defining rep. The normalization follows from

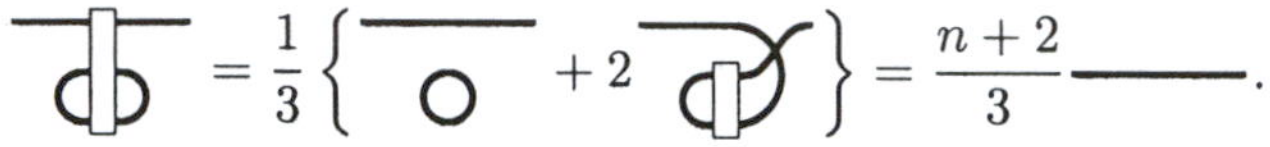

$$= \frac{1}{3}\left\{ \quad + 2 \quad \right\} = \frac{n+2}{3} \quad . \tag{10.34}$$

The □□□ rep of $SU(n)$ thus splits into

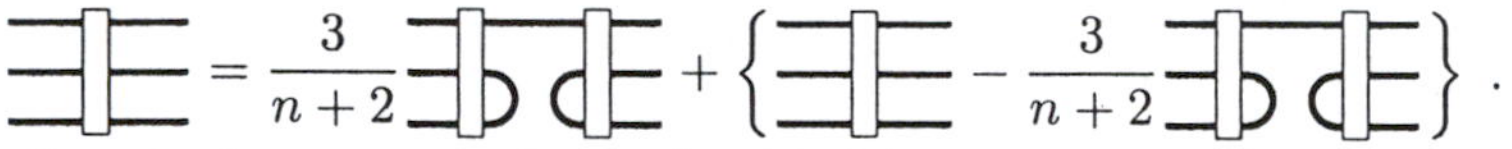

$$= \frac{3}{n+2} \quad + \left\{ \quad - \frac{3}{n+2} \quad \right\} . \tag{10.35}$$

On the mixed symmetry subspace in table 9.1, one can try various index contraction matrices $\mathbf{R}_i$. However, their projections $\mathbf{P}_2\mathbf{R}_i\mathbf{P}_2$ are all proportional to

$$\quad . \tag{10.36}$$

The normalization is fixed by

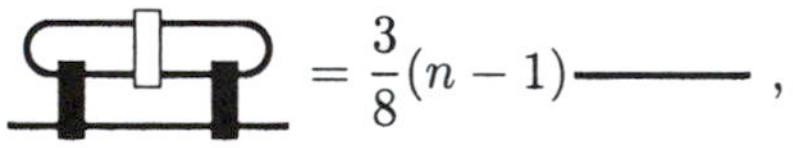

$$= \frac{3}{8}(n-1) \quad , \tag{10.37}$$

and the mixed symmetry rep of $SU(n)$ in (9.12) splits as

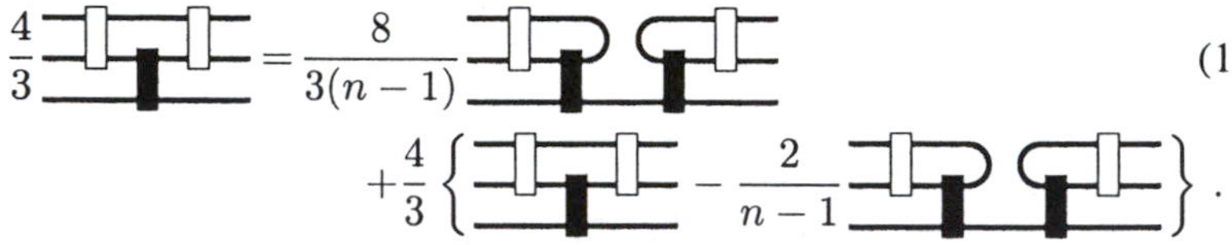

$$\frac{4}{3} \quad = \frac{8}{3(n-1)} \quad + \frac{4}{3}\left\{ \quad - \frac{2}{n-1} \quad \right\} . \tag{10.38}$$

The other mixed symmetry rep in table 9.1 splits in analogous fashion. The fully antisymmetric space is not affected by contractions, as

$$= 0 \tag{10.39}$$

by the symmetry of $g_{\mu\nu}$. Besides, as ⊟ is the adjoint rep, we have already performed the ⊟ ⊗ □ decomposition in the preceding section. The full Clebsch-Gordan series for the $SO(n)$ 3-index tensors is given in table 10.4.

	$V\otimes V\otimes V$		V_1	$\oplus$	V_2	$\oplus$	V_3	$\oplus$	V_4	$\oplus$	V_5	$\oplus$	V_6	$\oplus$	V_7
											$A\otimes V$				
Young tableaux	1 × 2 × 3	=	□□□	+	□	+	1 2 / 3	+	□	+	1 3 / 2	+	□	+	□ / □ / □
Dynkin labels			(30...)	+	(10...)	+	(110...)	+	(10...)	+	(110...)	+	(10...)	+	(0010...)
Dimensions	n^3	=	$\frac{(n-1)n(n+4)}{6}$	+	n	+	$\frac{n(n^2-4)}{3}$	+	n	+	$\frac{n(n^2-4)}{3}$	+	n	+	$\frac{n(n-1)(n-2)}{6}$
SO(3)	27	=	7	+	3	+	5	+	3	+	5	+	3	+	1
SO(4)	64	=	16	+	4	+	16	+	4	+	16	+	4	+	4

Projection operators

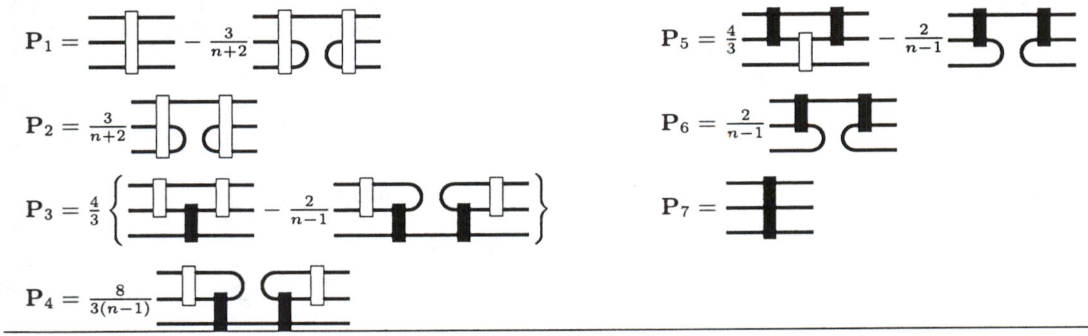

$\mathbf{P}_1 = \; - \frac{3}{n+2}$

$\mathbf{P}_2 = \frac{3}{n+2}$

$\mathbf{P}_3 = \frac{4}{3}\left\{ \; - \frac{2}{n-1} \; \right\}$

$\mathbf{P}_4 = \frac{8}{3(n-1)}$

$\mathbf{P}_5 = \frac{4}{3} \; - \frac{2}{n-1}$

$\mathbf{P}_6 = \frac{2}{n-1}$

$\mathbf{P}_7 =$

Table 10.4 $SO(n)$ Clebsch-Gordan series for $V\otimes V\otimes V$.

10.5 GRAVITY TENSORS

In a different application of birdtracks, we now change the language and construct the "irreducible rank-four gravity curvature tensors." The birdtrack notation for Young projection operators had originally been invented by Penrose [280] in this context. The Riemann-Christoffel curvature tensor has the following symmetries [336]:

$$\begin{aligned} R_{\alpha\beta\gamma\delta} &= -R_{\beta\alpha\gamma\delta} \\ R_{\alpha\beta\gamma\delta} &= R_{\gamma\delta\alpha\beta} \\ R_{\alpha\beta\gamma\delta} + R_{\beta\gamma\alpha\beta} + R_{\gamma\alpha\beta\delta} &= 0 \,. \end{aligned} \tag{10.40}$$

Introducing birdtrack notation for the Riemann tensor

$$R_{\alpha\beta\gamma\delta} = \text{[birdtrack: R box with legs } \alpha,\beta,\gamma,\delta] \,, \tag{10.41}$$

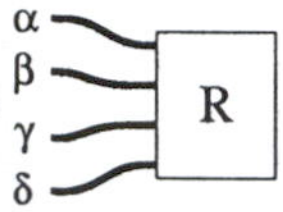

we can state the above symmetries as

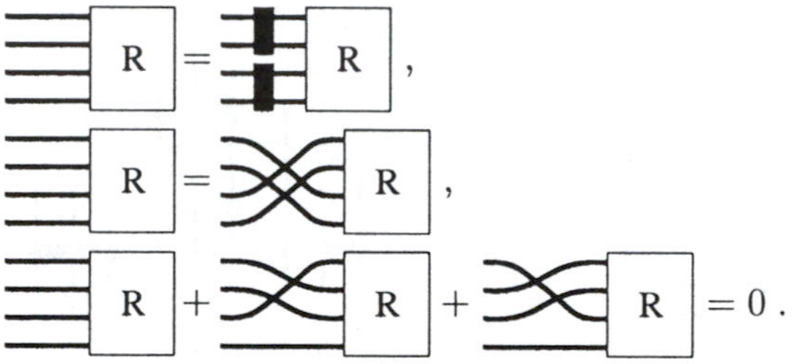

$$\text{[birdtrack]} \tag{10.42}$$

$$\text{[birdtrack]} \tag{10.43}$$

$$\text{[birdtrack]} = 0 \,. \tag{10.44}$$

The first condition says that R lies in ⊟ ⊗ ⊟ subspace. We have decomposed this subspace in table 9.2. The second condition says that R lies in ⊟ ↔ ⊟ interchange-symmetric subspace, which splits into ⊞ and ⊟ subspaces:

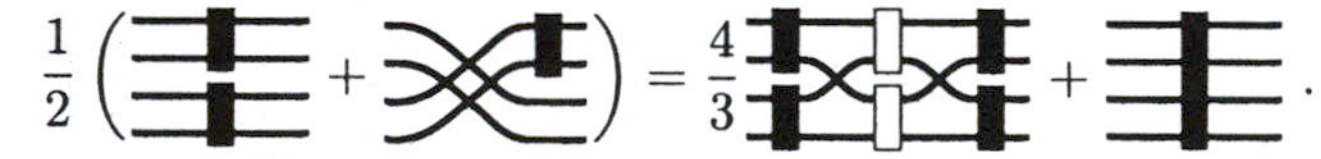

$$\frac{1}{2}\left(\text{[birdtrack]} + \text{[birdtrack]}\right) = \frac{4}{3}\text{[birdtrack]} + \text{[birdtrack]} \,. \tag{10.45}$$

The third condition says that R has no components in the ⊟ space:

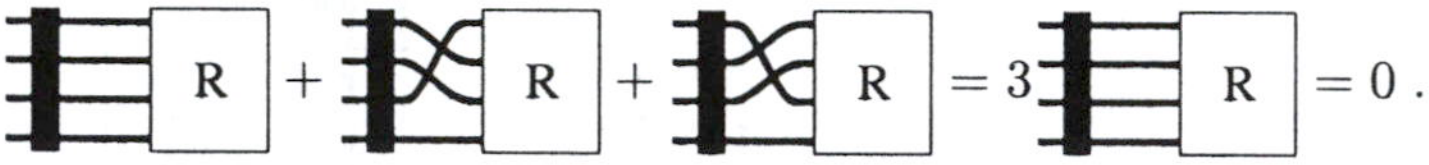

$$\text{[birdtrack]} + \text{[birdtrack]} + \text{[birdtrack]} = 3\,\text{[birdtrack]} = 0 \,. \tag{10.46}$$

Hence, the Riemann tensor is a pure ⊞ tensor, whose symmetries are summarized by the ⊞ rep projection operator [280]:

$$(P_R)_{\alpha\beta\gamma\delta,}{}^{\delta'\gamma'\beta'\alpha'} = \frac{4}{3}\,\text{[birdtrack with legs } \alpha,\beta,\gamma,\delta \text{ and } \alpha',\beta',\gamma',\delta'] \tag{10.47}$$

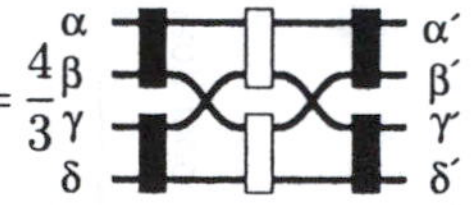

$$(P_R R)_{\alpha\beta\gamma\delta} = (P_R)_{\alpha\beta\gamma\delta,}{}^{\delta'\gamma'\beta'\alpha'} R_{\alpha'\beta'\gamma'\delta'} = R_{\alpha\beta\gamma\delta}$$

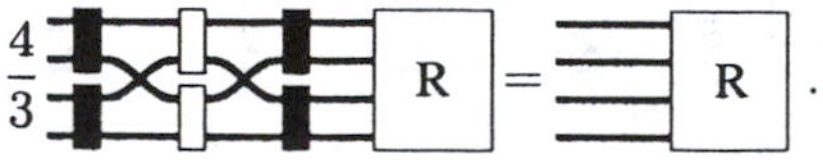

$$\frac{4}{3}\,\text{[birdtrack]} = \text{[birdtrack]} \,. \tag{10.48}$$

This compact statement of the Riemann tensor symmetries yields immediately the number of independent components of $\mathbf{R}_{\alpha\beta\gamma\delta}$, *i.e.*, the dimension of the reps in table 9.2:

$$d_R = \mathrm{tr}\,\mathbf{P}_R = \frac{n^2(n^2-1)}{12} \,. \tag{10.49}$$

The Riemann tensor has the symmetries of the rep of $SU(n)$. However, gravity is also characterized by the symmetric tensor $g_{\alpha\beta}$, that induces local $SO(n)$ invariance (more precisely $SO(1, n-1)$, but compactness is not important here). The extra invariants built from $g_{\alpha\beta}$'s decompose $SU(n)$ reps into sums of $SO(n)$ reps.

The $SU(n)$ subspace, corresponding to , is decomposed by the $SO(n)$ intermediate 2-index state contraction matrix

$$\mathbf{Q} = \,. \tag{10.50}$$

The intermediate 2-index subspace splits into three irreducible reps by (10.11)-(10.12):

$$\mathbf{Q} = \frac{1}{n} + \left\{ - \frac{1}{n} \right\} + $$

$$= \mathbf{Q}_0 + \mathbf{Q}_S + \mathbf{Q}_A \,. \tag{10.51}$$

The Riemann tensor is symmetric under the interchange of index pairs, so the antisymmetric 2-index state does not contribute

$$\mathbf{P_R Q}_A = 0 \,. \tag{10.52}$$

The normalization of the remaining two projectors is fixed by computation of $\mathbf{Q}_S^2, \mathbf{Q}_0^2$:

$$\mathbf{P}_0 = \frac{2}{n(n-1)} \,, \tag{10.53}$$

$$\mathbf{P}_S = \frac{4}{n-2}\left\{ - \frac{1}{n} \right\} \,. \tag{10.54}$$

This completes the $SO(n)$ reduction of the $SU(n)$ rep (10.48):

$SU(n)$	$\to$	$SO(n)$					
	$\to$		+		+	$\circ$	
$\mathbf{P}_R$	=	$\mathbf{P}_W$	+	$\mathbf{P}_S$	+	$\mathbf{P}_0$	(10.55)
$\frac{n^2(n^2-1)}{12}$	=	$\frac{(n+2)(n+1)n(n-3)}{12}$	+	$\frac{(n+2)(n-1)}{2}$	+	1	

Here the projector for the traceless tensor is given by $\mathbf{P}_W = \mathbf{P}_R - \mathbf{P}_S - \mathbf{P}_0$:

$$\mathbf{P}_W = \frac{4}{3} - \frac{4}{n-2} + \frac{2}{(n-1)(n-2)} \,. \tag{10.56}$$

The above three projectors project out the standard relativity tensors:

Curvature scalar:

$$R = - [\text{birdtrack}] = R^{\mu}{}_{\nu\mu}{}^{\nu} \tag{10.57}$$

Traceless Ricci tensor:

$$R_{\mu\nu} - \frac{1}{n} g_{\mu\nu} R = - [\text{birdtrack}] + \frac{1}{n} [\text{birdtrack}] \tag{10.58}$$

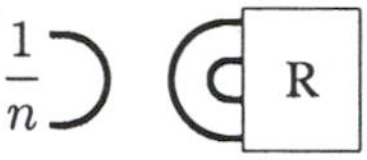

Weyl tensor:

$$\begin{aligned} C_{\lambda\mu\nu\kappa} &= (\mathbf{P}_W R)_{\lambda\mu\nu\kappa} \\ &= [\text{birdtrack}] - \frac{4}{n-2} [\text{birdtrack}] + \frac{2}{(n-1)(n-2)} [\text{birdtrack}] \\ &= R_{\lambda\mu\nu\kappa} + \frac{1}{n-2} (g_{\mu\nu} R_{\lambda\kappa} - g_{\lambda\nu} R_{\mu\kappa} - g_{\mu\kappa} R_{\lambda\nu} + g_{\lambda\kappa} R_{\mu\nu}) \\ &\quad - \frac{1}{(n-1)(n-2)} (g_{\lambda\kappa} g_{\mu\nu} - g_{\lambda\nu} g_{\mu\kappa}) R \,. \end{aligned} \tag{10.59}$$

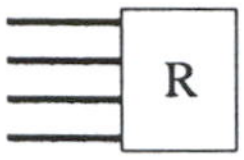

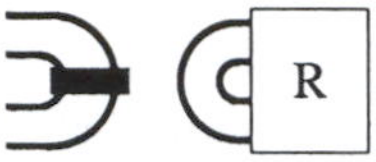

The numbers of independent components of these tensors are given by the dimensions of corresponding subspaces in (10.55). The Ricci tensor contributes first in three dimensions, and the Weyl tensor first in four, so we have

$$\begin{aligned} n = 2: \quad R_{\lambda\mu\nu\kappa} &= (P_0 R)_{\lambda\mu\nu\kappa} = \tfrac{1}{2}(g_{\lambda\nu} g_{\mu\kappa} - g_{\lambda\kappa} g_{\mu\nu}) R \\ n = 3: \quad &= g_{\lambda\nu} R_{\mu\kappa} - g_{\mu\nu} R_{\lambda\kappa} + g_{\mu\kappa} R_{\lambda\nu} - g_{\lambda\kappa} R_{\mu\nu} \\ &\quad - \tfrac{1}{2}(g_{\lambda\nu} g_{\mu\kappa} - g_{\lambda\kappa} g_{\mu\nu}) R \,. \end{aligned} \tag{10.60}$$

The last example of this section is an application of birdtracks to general relativity index manipulations. The object is to find the characteristic equation for the Riemann tensor in *four dimensions*. We contract (6.24) with two Riemann tensors:

$$0 = [\text{birdtrack}] \,. \tag{10.61}$$

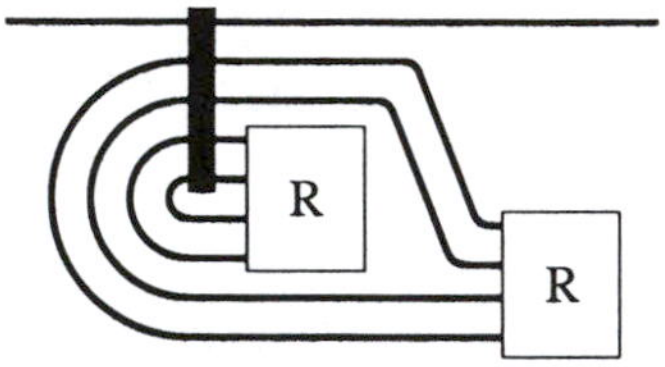

Expanding with (6.19) we obtain the characteristic equation

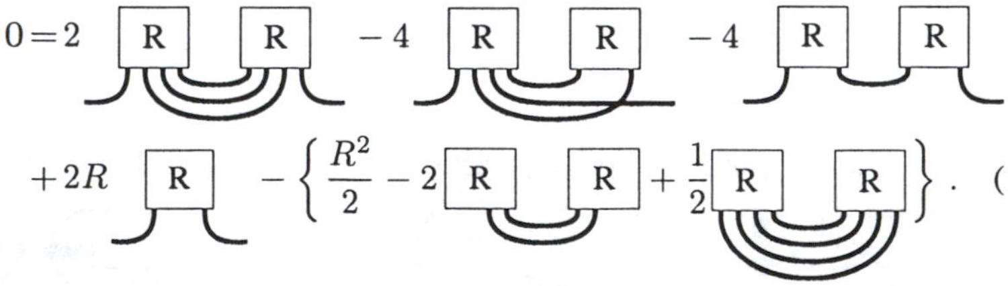

$$\begin{aligned} 0 = 2 [\text{birdtrack}] - 4 [\text{birdtrack}] - 4 [\text{birdtrack}] \\ + 2R [\text{birdtrack}] - \left\{ \frac{R^2}{2} - 2 [\text{birdtrack}] + \frac{1}{2} [\text{birdtrack}] \right\} . \end{aligned} \tag{10.62}$$

For example, this identity has been used by Adler *et al.*, eq. (E2) in ref. [5].

10.6 $SO(n)$ DYNKIN LABELS

In general, one has to distinguish between the odd- and the even-dimensional orthogonal groups, as well as their spinor and nonspinor reps. In this chapter, we study only the tensor reps; spinor reps will be taken up in chapter 11.

For $SO(2r+1)$ reps there are r Dynkin labels $(a_1 a_2 \ldots a_{r-1} Z)$. If Z is odd, the rep is spinor; if Z is even, it is tensor. For the tensor reps, the corresponding Young tableau in the Fischler notation [122] is given by

$$(a_1 a_2 \ldots a_{r-1} Z) \rightarrow (a_1 a_2 \ldots a_{r-1} \frac{Z}{2} 00 \ldots) \,. \tag{10.63}$$

For example, for $SO(7)$ rep (102) we have

$$(102) \rightarrow (1010 \ldots) = \yng(2,1,1) \,. \tag{10.64}$$

For orthogonal groups, the Levi-Civita tensor can be used to convert a long column of k boxes into a short column of $(2r+1-k)$ boxes. The highest column that cannot be shortened by this procedure has r boxes, where r is the rank of $SO(2r+1)$.

For $SO(2r)$ reps, the last two Dynkin labels are spinor roots $(a_1 a_2 \ldots a_{r-2} Y Z)$. Tensor reps have $Y + Z =$ even. However, as spinors are complex, tensor reps can also be complex, conjugate reps being related by

$$(a_1 a_2 \ldots Y Z) = (a_1 a_2 \ldots Z Y)^* \,. \tag{10.65}$$

For $Z \geq Y$, $Z + Y$ even, the corresponding Young tableau is given by

$$(a_1 a_2 \ldots a_{r-2} Y Z) \rightarrow (a_1 a_2 \ldots a_{r-2} \frac{Z-Y}{2} 00 \ldots) \,. \tag{10.66}$$

The Levi-Civita tensor can be used to convert long columns into short columns. For columns of r boxes, the Levi-Civita tensor splits $O(2r)$ reps into conjugate pairs of $SO(2r)$ reps.

We find the formula of King [191] and Murtaza and Rashid [251] the most convenient among various expressions for the dimensions of $SO(n)$ tensor reps given in the literature. If the Young tableau λ is represented as in section 9.3, the list of the row lengths $[\lambda_1, \lambda_2, \ldots \lambda_\kappa]$, then the dimension of the corresponding $SO(n)$ rep is given by

$$d_\lambda = \frac{d_S}{p!} \prod_{i=1}^{k} \frac{(\lambda_i + n - k - i - 1)!}{(n - 2i)!} \prod_{j=1}^{k} (\lambda_i + \lambda_j + n - i - j) \,. \tag{10.67}$$

Here p is the total number of boxes, and d_S is the dimension of the symmetric group rep computed in (9.16). For $SO(2r)$ and $\kappa = r$, this rep is reducible and splits into a conjugate pair of reps. For example,

$$\begin{aligned}
d_{\yng(2,1)} &= \frac{1}{\young(31,1)} \cdot (n+2)n(n-2) = \frac{n(n^2-4)}{3} \\
d_{\yng(2,1,1)} &= \frac{(n+2)n(n-1)(n-3)}{8} \\
d_{\yng(2,2)} &= \frac{(n+2)(n+1)n(n-3)}{12} \,,
\end{aligned} \tag{10.68}$$

in agreement with (10.55). Even though the Dynkin labels distinguish $SO(2r+1)$ from $SO(2r)$ reps, this distinction is significant only for the spinor reps. The tensor reps of $SO(n)$ have the same Young tableaux for the even and the odd n's.

Chapter Eleven

Spinors

P. Cvitanović and A. D. Kennedy

In chapter 10 we have discussed the tensor reps of orthogonal groups. However, the spinor reps of $SO(n)$ also play a fundamental role in physics, both as reps of space-time symmetries (Pauli spin matrices, Dirac gamma matrices, fermions in D-dimensional supergravities), and as reps of internal symmetries ($SO(10)$ grand unified theory, for example). In calculations of radiative corrections, the QED spin traces can easily run up to traces of products of some twelve gamma matrices [195], and efficient evaluation algorithms are of great practical importance. A most straightforward algorithm would evaluate such a trace in some $11!! = 11 \cdot 9 \cdot 7 \cdot 5 \cdot 3 \simeq 10,000$ steps. Even computers shirk such tedium. A good algorithm, such as the ones we shall describe here, will do the job in some $6^2 \simeq 100$ steps.

Spinors came to Cartan [43] as an unexpected fruit of his labors on the complete classification of reps of the simple Lie groups. Dirac [95] rediscovered them while looking for a linear version of the relativistic Klein-Gordon equation. He introduced matrices γ_μ, which were required to satisfy

$$(p_0\gamma_0 + p_1\gamma_1 + \ldots)^2 = (p_0^2 - p_1^2 - p_2^2 - \ldots) \,. \tag{11.1}$$

For $n = 4$ he constructed γ's as $[4 \times 4]$ complex matrices. For $SO(2r)$ and $SO(2r+1)$ γ-matrices were constructed explicitly as $[2^r \times 2^r]$ complex matrices by Weyl and Brauer [344].

In the early days, such matrices were taken as a literal truth, and Klein and Nishina [196] are reputed to have computed their celebrated Quantum Electrodynamics crosssection by multiplying γ-matrices by hand. Every morning, day after day, they would multiply away explicit $[4\times4]$ γ_μ matrices and sum over μ's. In the afternoon, they would meet in the cafeteria of the Niels Bohr Institute to compare their results.

Nevertheless, all information that is actually needed for spin traces evaluation is contained in the Dirac algebraic condition (11.1), and today the Klein-Nishina trace over Dirac γ's is a textbook exercise, reducible by several applications of the Clifford algebra condition on γ-matrices:

$$\{\gamma_\mu, \gamma_\nu\} = \gamma_\mu\gamma_\nu + \gamma_\nu\gamma_\mu = 2g_{\mu\nu}\,\mathbf{1} \,. \tag{11.2}$$

Iterative application of this condition immediately yields a spin traces evaluation algorithm in which the only residue of γ-matrices is the normalization factor $\mathrm{tr}\,\mathbf{1}$. However, this simple algorithm is inefficient in the sense that it requires a combinatorially large number of evaluation steps. The most efficient algorithm on the market (for any $SO(n)$) appears to be the one given by Kennedy [185, 81]. In

Kennedy's algorithm, one views the spin trace to be evaluated as a $3n$-j coefficient. Fierz [120] identities are used to express this $3n$-j coefficient in terms of 6-j coefficients (see section 11.3). Gamma matrices are $[2^{n/2} \times 2^{n/2}]$ in even dimensions, $[2^{(n-1)/2} \times 2^{(n-1)/2}]$ in odd dimensions, and at first sight it is not obvious that a smooth analytic continuation in dimension should be possible for spin traces. The reason why the Kennedy algorithm succeeds is that spinors are really not there at all. Their only role is to restrict the $SO(n)$ Clebsch-Gordan series to fully antisymmetric reps. The corresponding 3-j and 6-j coefficients are relatively simple combinatoric numbers, with analytic continuations in terms of gamma functions. The case of four spacetime dimensions is special because of the reducibility of $SO(4)$ to $SU(2) \otimes SU(2)$. Farrar and Neri [115], who as of April 18, 1983, have computed in excess of 58,149 Feynman diagrams, have used this structure to develop a very efficient method for evaluating $SO(4)$ spinor expressions. An older technique, described here in section 11.8, is the Kahane [178] algorithm, which implements diagrammatically the Chisholm [55] identities. REDUCE, an algebra manipulation program written by Hearn [159], uses the Kahane algorithm. Thörnblad [323] has used $SO(4) \subset SO(5)$ embedding to speedup evaluation of traces for massive fermions.

This chapter is based on ref. [81].

11.1 SPINOGRAPHY

Kennedy [185] introduced diagrammatic notation for γ-matrices

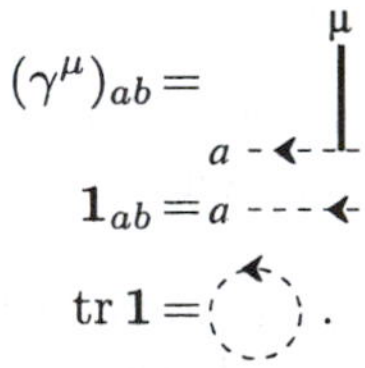

$$(\gamma^\mu)_{ab} = \quad \text{[diagram: solid line } \mu \text{ on dashed line } a \leftarrow b], \qquad a, b = 1, 2, \ldots, 2^{n/2} \text{ or } 2^{(n-1)/2}$$

$$\mathbf{1}_{ab} = a \text{ - - -} \leftarrow \text{- - - } b, \qquad \mu = 1, 2, \ldots, n$$

$$\mathrm{tr}\,\mathbf{1} = \text{[dashed loop]}\,. \tag{11.3}$$

In this context, birdtracks go under the name "spinography." For notational simplicity, we take all γ-indices to be lower indices and omit arrows on the n-dimensional rep lines. The n-dimensional rep is drawn by a solid directed line to conform to the birdtrack notation of chapter 4. For QED and QCD spin traces, one might prefer the conventional Feynman diagram notation,

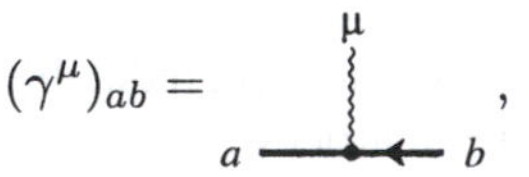

$$(\gamma^\mu)_{ab} = \quad \text{[diagram: wavy line } \mu \text{ on solid line } a \leftarrow b]\,,$$

where the photons/gluons are in the n-dimensional rep of $SO(3,1)$, and electrons are spinors. We eschew such notation here, as it would conflict with $SO(n)$ birdtracks of chapter 10. The Clifford algebra anticommutator condition (11.2) is given by

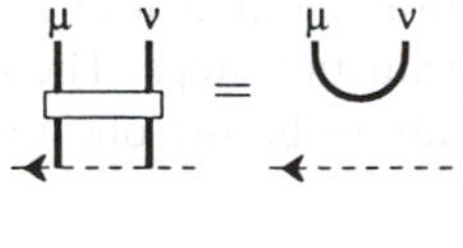

$$\text{[diagram: } \mu\ \nu \text{ symmetrized on dashed line]} = \text{[diagram: } \mu\ \nu \text{ joined, dashed line]}\,. \tag{11.4}$$

For antisymmetrized products of γ-matrices, this leads to the relation

$$1\,2\,3 \ldots p \;=\; 1\,2 \ldots p \;+ (p-1)\; 1\,2 \ldots p \quad (11.5)$$

(we leave the proof as an exercise). Hence, any product of γ-matrices can be expressed as a sum over antisymmetrized products of γ-matrices. For example, substitute the Young projection operators from figure 9.1 into the products of two and three γ-matrices and use the Clifford algebra (11.4):

$$= \quad + \quad (11.6)$$

$$= \quad + $$

$$= \quad + \left\{ \quad - \quad + \quad \right\}, \textit{etc.}. \quad (11.7)$$

Only the fully antisymmetrized products of γ's are immune to reduction by (11.4). Hence, the antisymmetric tensors

$$
\begin{aligned}
\Gamma^{(0)} &= \mathbf{1} = \; = \boxed{0} \\
\Gamma^{(1)}_{\mu} &= \gamma_\mu = \; \mu \; = \boxed{1} \\
\Gamma^{(2)}_{\mu\nu} &= \tfrac{1}{2}[\gamma_\mu, \gamma_\nu] = \; \mu\,\nu \; = \boxed{2} \\
\Gamma^{(3)}_{\mu\nu\sigma} &= \gamma_{[\mu}\gamma_\nu\gamma_{\sigma]} = \; \mu\,\nu\,\sigma \; = \boxed{3} \\
\Gamma^{(a)}_{\mu_1\nu_2\ldots\mu_a} &= \gamma_{[\mu_1}\gamma_{\mu_2}\cdots\gamma_{\mu_a]} = \; \mu_1 \cdots \mu_a \; = \boxed{a}
\end{aligned}
\quad (11.8)
$$

provide a complete basis for expanding products of γ-matrices. Applying the anticommutator (11.4) to a string of γ's, we can move the first γ all the way to the right and obtain

$$= 2 \quad - $$

$$= 2 \quad - 2 \quad + (-1)^2 \quad = \ldots \quad (11.9)$$

$$\frac{1}{2}\left(1\,2\,3 \ldots p \; + (-1)^p \; 1\,2\,3 \ldots p \right) =$$

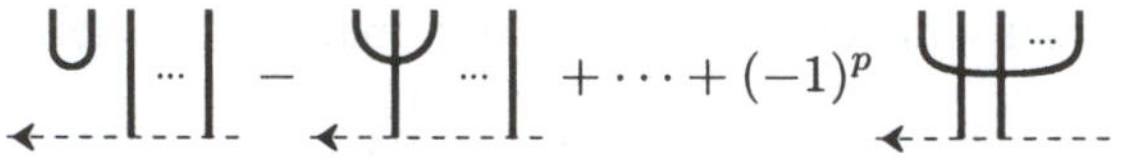

$$+\cdots+(-1)^p$$

$$\frac{1}{2}(\gamma^{\mu_1}\gamma^{\mu_2}\ldots\gamma^{\mu_p}\pm\gamma^{\mu_2}\ldots\gamma^{\mu_p}\gamma^{\mu_1})= \\ g^{\mu_1\mu_2}\gamma^{\mu_3}\ldots\gamma^{\mu_p}-g^{\mu_1\mu_3}\gamma^{\mu_2}\gamma^{\mu_p}+\ldots \qquad (11.10)$$

This identity has three immediate consequences:

(i) Traces of odd numbers of γ's vanish for n even.

(ii) Traces of even numbers of γ's can be evaluated recursively.

(iii) The result does not depend on the direction of the spinor line.

According to (11.10), any γ-matrix product can be expressed as a sum of terms involving $g_{\mu\nu}$'s and the antisymmetric basis tensors $\Gamma^{(a)}$, so in order to prove (i) we need only to consider traces of $\Gamma^{(a)}$ for a odd. This may be done as follows:

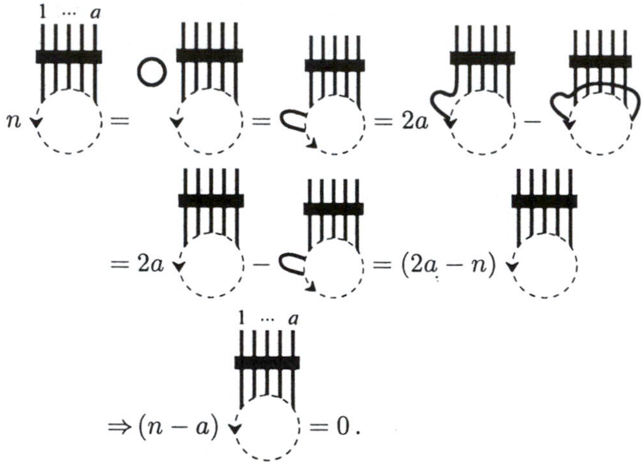

$1 \cdots a$

$$n \;=\; = \; = 2a \; - \; = 2a \; - \; = (2a-n)$$

$1 \cdots a$

$$\Rightarrow (n-a) \; = 0\,. \qquad (11.11)$$

In the third step we have used (11.10) and the fact that a is odd. Hence, $\operatorname{tr}\Gamma^{(a)}$ vanishes for all odd a if n is even. If n is odd, $\operatorname{tr}\Gamma^{(n)}$ does not vanish because by (6.28),

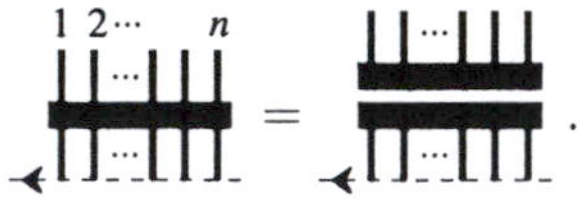

$1\ 2 \cdots\ n$

$$= \,. \qquad (11.12)$$

The n-dimensional analogue of the γ_5,

$$\varepsilon^{\mu\nu\ldots\sigma}\gamma_\mu\gamma_\nu\ldots\gamma_\sigma\,, \qquad (11.13)$$

commutes with all γ-matrices, and, by Schur's lemma, it must be a multiple of the unit matrix, so it cannot be traceless. This proves (i). (11.10) relates traces of length p to traces of length $p-2$, so (ii) gives

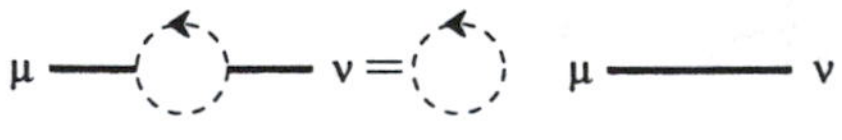

$$\mu \;\; \nu = \;\; \mu \;\; \nu$$

$$\operatorname{tr}\gamma_\mu\gamma_\nu = (\operatorname{tr}\mathbf{1})\, g_{\mu\nu}\,, \tag{11.14}$$

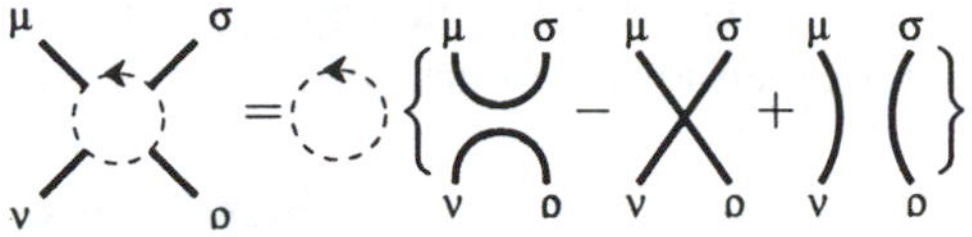

$$\operatorname{tr}\gamma_\mu\gamma_\nu\gamma_\rho\gamma_\sigma = \operatorname{tr}\mathbf{1}\,\{g_{\mu\nu}g_{\rho\sigma} - g_{\mu\rho}g_{\nu\sigma} + g_{\mu\nu}g_{\nu\rho}\}\,, \tag{11.15}$$

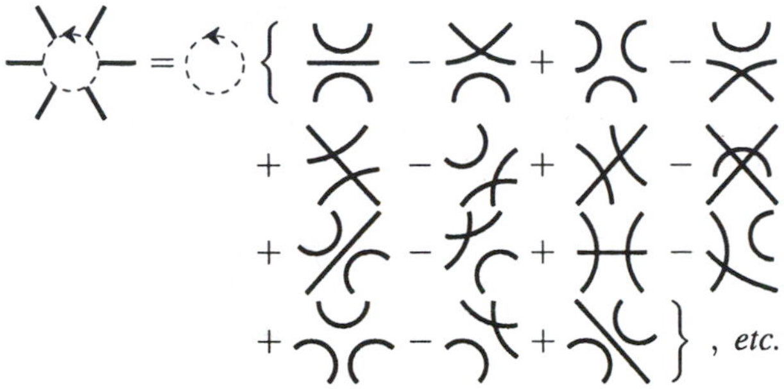

(11.16)

The result is always the $(2p-1)!!$ ways of pairing $2p$ indices with p Kronecker deltas. It is evident that nothing depends on the direction of spinor lines, as spinors are remembered only by an overall normalization factor $\operatorname{tr}\mathbf{1}$. The above identities are in principle a solution of the spinor traces evaluation problem. In practice they are intractable, as they yield a factorially growing number of terms in intermediate steps of trace evaluation.

11.2 FIERZING AROUND

The algorithm (11.16) is too cumbersome for evaluation of traces of more than four or six γ-matrices. A more efficient algorithm is obtained by going to the Γ basis (11.8). Evaluation of traces of two and three Γ's is a simple combinatoric exercise using the expansion (11.16). Any term in which a pair of $g_{\mu\nu}$ indices gets antisymmetrized vanishes:

$$= 0\,. \tag{11.17}$$

That implies that Γ's are *orthogonal*:

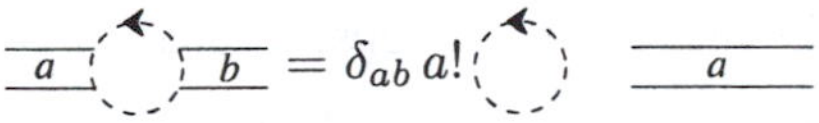

$$= \delta_{ab}\, a! \,. \tag{11.18}$$

Here $a!$ is the number of terms in the expansion (11.16) that survive antisymmetrization (11.18). A trace of three Γ's is obtained in the same fashion:

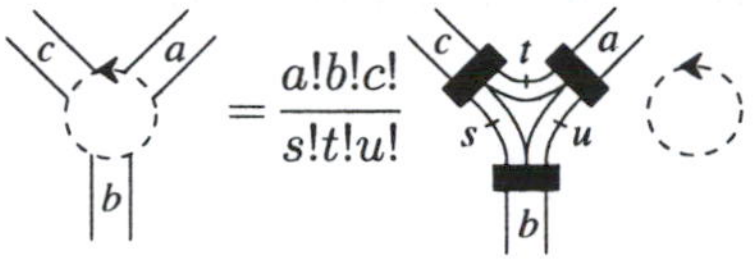

$$= \frac{a!b!c!}{s!t!u!}$$

$$s = \frac{1}{2}(b+c-a)\,, \qquad t = \frac{1}{2}(c+a-b)\,, \qquad u = \frac{1}{2}(a+b-c)\,.$$

As the Γ's provide a complete basis, we can express a product of two Γ matrices as a sum over Γ's, with the extra indices carried by $g_{\mu\nu}$'s. From symmetry alone we know that terms in this expansion are of the form

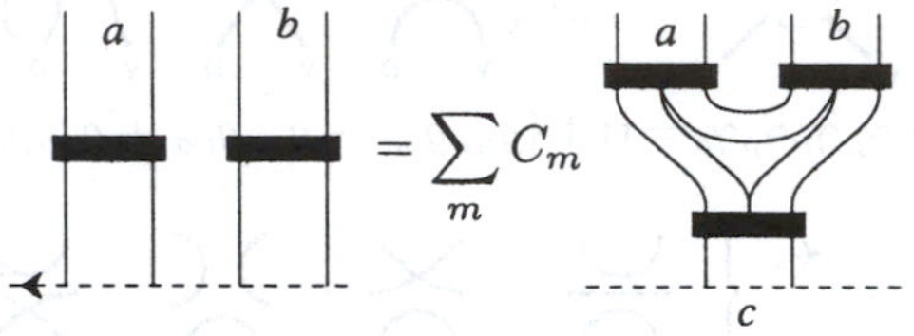

$$= \sum_m C_m \quad . \qquad (11.19)$$

The coefficients C_m can be computed by tracing both sides with Γ^c and using the orthogonality relation (11.18):

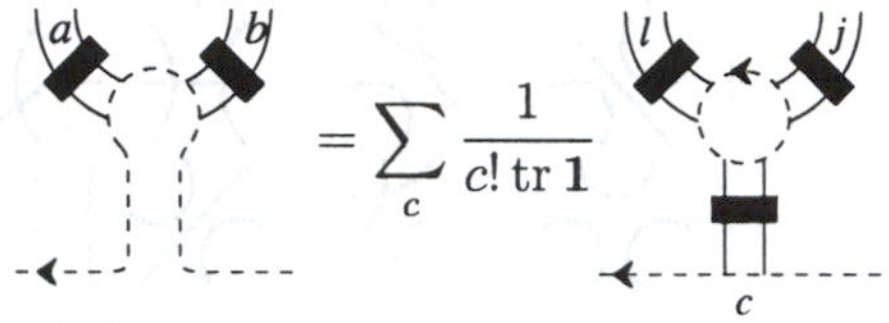

$$= \sum_c \frac{1}{c!\,\mathrm{tr}\,\mathbf{1}} \quad . \qquad (11.20)$$

We do not have to consider traces of four or more Γ's, as they can all be reduced to three-Γ traces by the above relation.

Let us now streamline the birdtracks. The orthogonality of Γ's (11.18) enables us to introduce projection operators

$$(P_a)_{cd,ef} = \frac{1}{a!\ \mathrm{tr}\,\mathbf{1}} \left(\gamma_{[\mu_1}\gamma_{\mu_2}\cdots\gamma_{\mu_a]}\right)_{ab} \left(\gamma^{\mu_a}\ldots\gamma^{\mu_2}\gamma^{\mu_1}\right)_{cd}$$

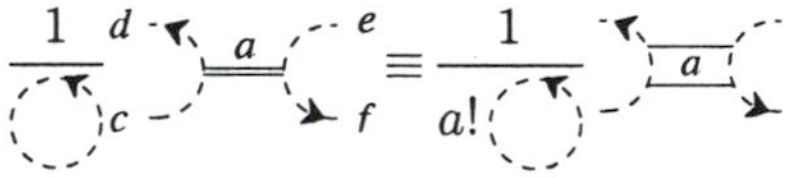

$$\frac{1}{\bigcirc} \equiv \frac{1}{a!\bigcirc} \quad . \qquad (11.21)$$

The factor of tr $\mathbf{1}$ on the left-hand side is a convenient (but inessential) normalization convention. It is analogous to the normalization factor a in (4.29):

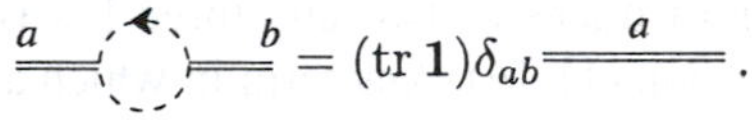

$$= (\mathrm{tr}\,\mathbf{1})\delta_{ab} \quad . \qquad (11.22)$$

With this normalization, each spinor loop will carry factor $(\mathrm{tr}\,\mathbf{1})^{-1}$, and the final results will have no tr $\mathbf{1}$ factors. $a, b, \ldots$ are rep labels, not indices, and the repeated index summation convention does not apply. Only the fully antisymmetric $SO(n)$ reps occur, so a single integer (corresponding to the number of boxes in the single Young tableau column) is sufficient to characterize a rep.

For the trivial and the single γ-matrix reps, we shall omit the labels,

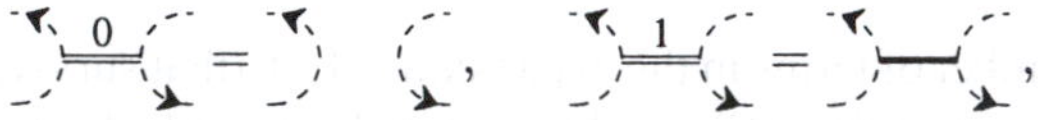

$$= \quad , \qquad = \quad , \qquad (11.23)$$

in keeping with the original definitions (11.3). The 3-Γ trace (11.19) defines a 3-vertex

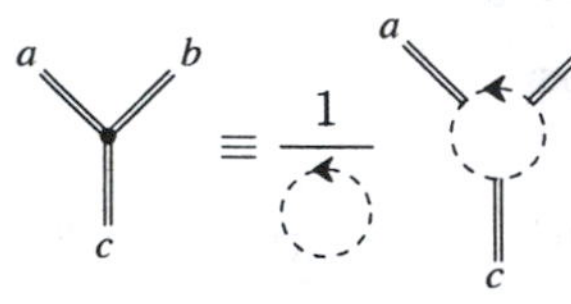

$$\equiv \frac{1}{\bigcirc} \qquad (11.24)$$

that is nonzero only if $a+b+c$ is even, and if a, b, and c satisfy the triangle inequalities $|a-b| \leq c \leq |a+b|$. We apologize for using a, b, c both for the $SO(n)$ antisymmetric representations labels, and for spinor indices in (11.3), but the Latin alphabet has only so many letters. It is important to note that in this definition the spinor loop runs anticlockwise, as this vertex can change sign under interchange of two legs. For example, by (11.19),

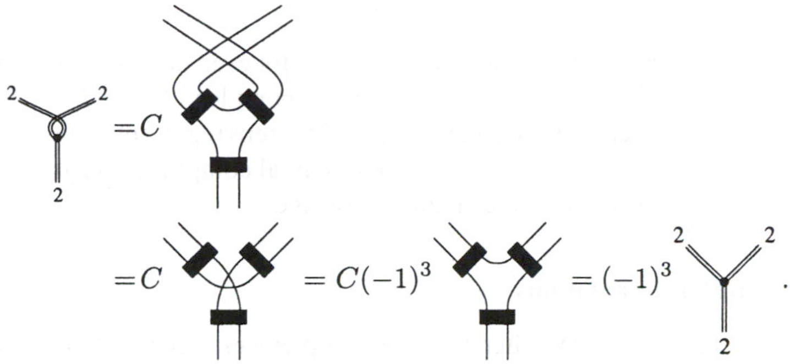

$$= C \quad = C \quad = C(-1)^3 \quad = (-1)^3 \quad . \quad (11.25)$$

This vertex couples three adjoint representations (10.13) of $SO(n)$, and the sign rule is the usual rule (4.46) for the antisymmetry of C_{ijk} constants. The general sign rule follows from (11.19):

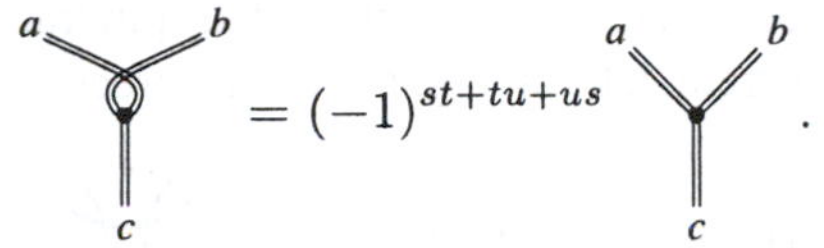

$$= (-1)^{st+tu+us} \quad . \qquad (11.26)$$

The projection operators $\mathbf{P}_a$ (11.21) satisfy the completeness relation (5.8):

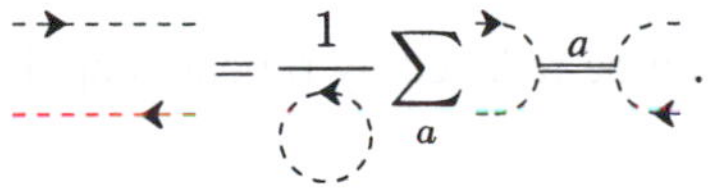

$$= \frac{1}{} \sum_a \quad . \qquad (11.27)$$

This follows from the completeness of Γ's, used in deriving (11.20). We have already drawn the left-hand side of (11.20) in such a way that the completeness relation (11.27) is evident:

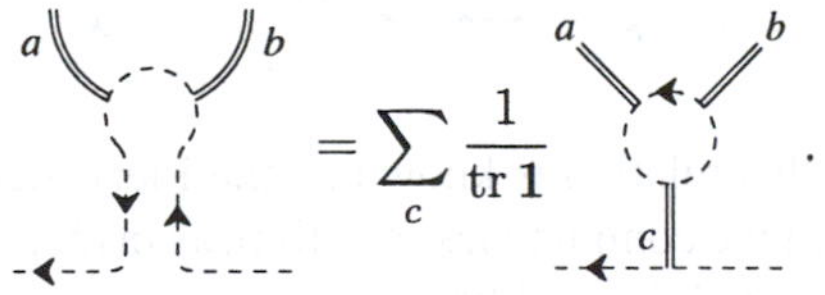

$$= \sum_c \frac{1}{\mathrm{tr}\,\mathbf{1}} \quad .$$

In terms of the vertex (11.24) we get

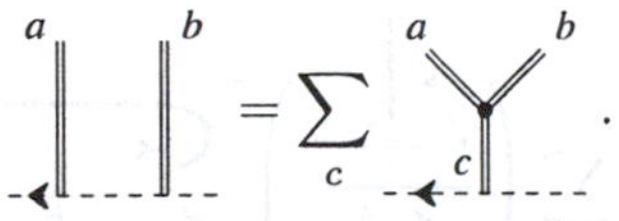

$$= \sum_c \quad . \qquad (11.28)$$

In this way we can systematically replace a string of γ-matrices by trees of 3-vertices.

Before moving on, let us check the completeness of $\mathbf{P}_a$. $\mathbf{P}_a$ projects spinor $\otimes$ antispinor $\rightarrow$ antisymmetric a-index tensor rep of $SO(n)$. Its dimension was computed in (6.21):

$$d_a = \mathrm{tr}\,\mathbf{P}_a = \frac{1}{\mathrm{tr}\,\mathbf{1}} \quad = \quad = \binom{n}{a} \,. \qquad (11.29)$$

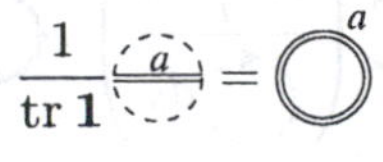

d_a is automatically equal to zero for $n < a$; this guarantees the correctness of treating (11.28) as an arbitrarily large sum, even though for a given n it terminates at $a = n$. Tracing both sides of the completeness relation (11.27), we obtain a dimension sum rule:

$$(\mathrm{tr}\,\mathbf{1})^2 = \sum_a d_a = \sum_{a=0}^{n} \binom{n}{a} = (1+1)^n = 2^n\,. \qquad (11.30)$$

This confirms the results of Weyl and Brauer [344]: for even dimensions the number of components is 2^n, so Γ's can be represented by complex $[2^{n/2} \times 2^{n/2}]$ matrices. For odd dimensions there are two inequivalent spinor reps represented by $[2^{\,(n-1)/2} \times 2^{(n-1)/2}]$ matrices (see section 11.7). This inessential complication has no bearing on the evaluation algorithm we are about to describe.

11.2.1 Exemplary evaluations

What have we accomplished? Iterating the completeness relation (11.28) we can make γ-matrices disappear altogether, and spin trace evaluation reduces to combinatorics of 3-vertices defined by the right-hand side of (11.19). This can be done, but is it any quicker than the simple algorithm (11.16)? The answer is yes: high efficiency can be achieved by viewing a complicated spin trace as a $3n$-j coefficient of section 5.2. To be concrete, take an eight γ-matrix trace as an example:

$$\mathrm{tr}(\gamma_\mu\gamma_\nu\gamma_\alpha\gamma_\beta\gamma^\nu\gamma^\mu\gamma^\beta\gamma^\alpha) =$$

(11.31)

Such a $3n$-j coefficient can be reduced by repeated application of the recoupling relation (5.13)

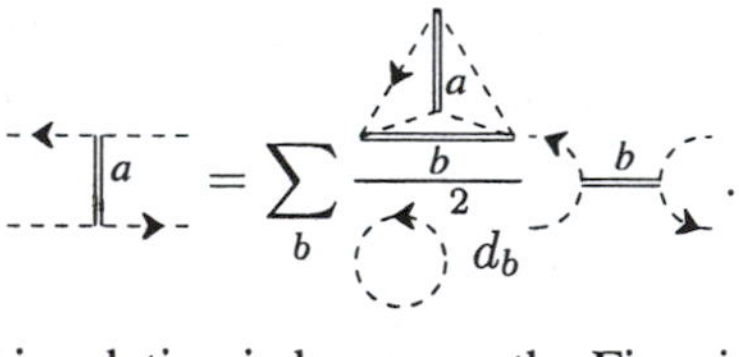

(11.32)

In the present context this relation is known as the Fierz identity [120]. It follows from two applications of the completeness relation, as in (5.13). Now we can redraw the 12-j coefficient from (11.31) and fierz on

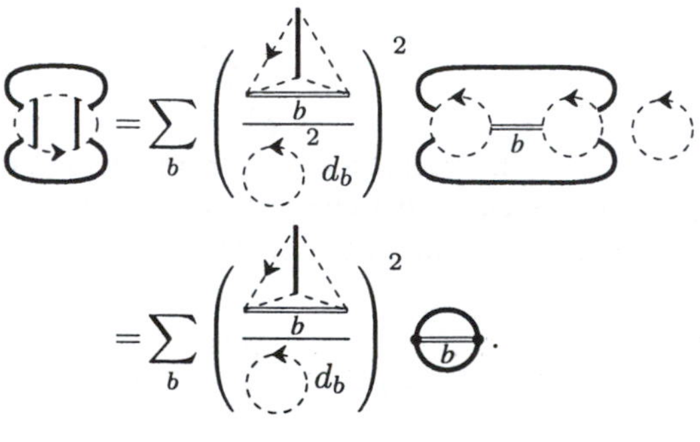

(11.33)

Another example is the reduction of a vertex diagram, a special case of the Wigner-Eckart theorem (5.24):

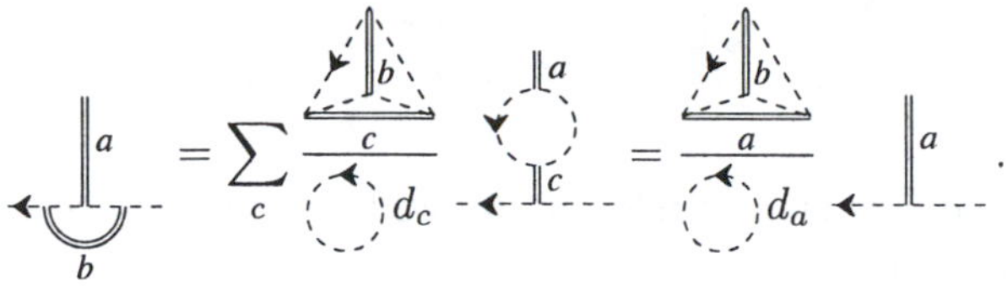

$$(11.34)$$

As the final example we reduce a trace of ten matrices:

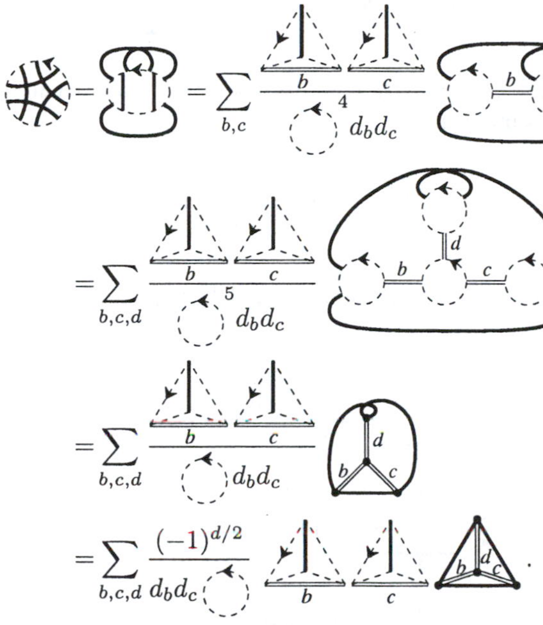

$$(11.35)$$

In this way, any spin trace can be reduced to a sum over 6-j and 3-j coefficients. Our next task is to evaluate these.

11.3 FIERZ COEFFICIENTS

The 3-j coefficient in (11.33) can be evaluated by substituting (11.19) and doing "some" combinatorics

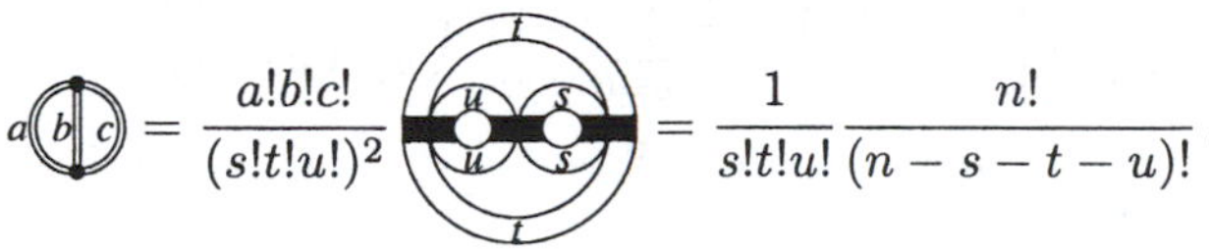

$$= \frac{a!b!c!}{(s!t!u!)^2} \quad = \frac{1}{s!t!u!} \frac{n!}{(n-s-t-u)!} \,. \tag{11.36}$$

s, t, u are defined in (11.19). Note that $a+b+c = 2(s+t+u)$, and $a+b+c$ is even, otherwise the traces in the above formula vanish.

The 6-j coefficients in the Fierz identity (11.32) are not independent of the above 3-j coefficients. Redrawing a 6-j coefficient slightly, we can apply the completeness

relation (11.28) to obtain

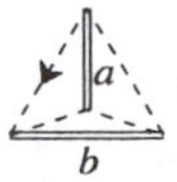 $=$  $= \frac{1}{}$  $\sum_c$ 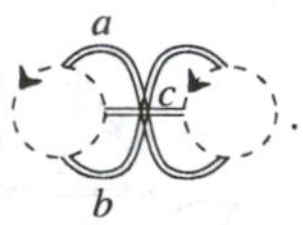 .

Interchanging j and k by the sign rule (11.26), we express the 6-j coefficient as a sum over 3-j coefficients:

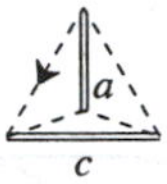 $=$  $\sum_c (-1)^{st+tu+us}$  . (11.37)

Using relations $t = a - u$, $s = b - u$, $a + t + u = a + b - u$, we can replace [48] the sum over c by the sum over u:

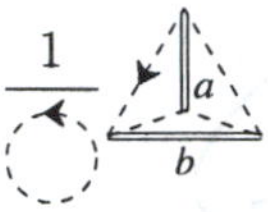 $= (-1)^{ab}\binom{b}{n}\sum_u (-1)^u \binom{b}{u}\binom{n-b}{a-u}$. (11.38)

u ranges from 0 to a or b, whichever is smaller, and the 6-j's for low values of a are particularly simple

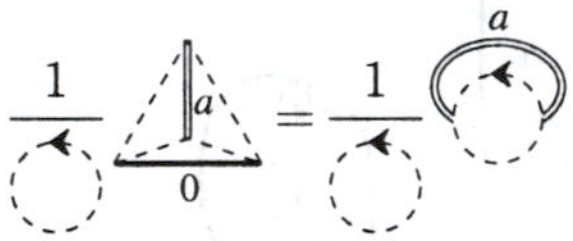 $= d_a$, (11.39)

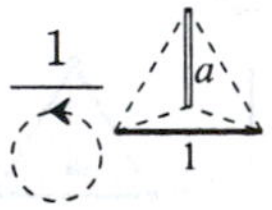 $= (-1)^a(n-2a)d_a$, (11.40)

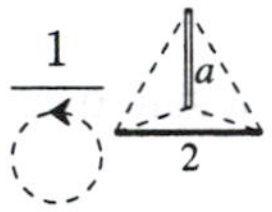 $= \frac{(n-2a)^2 - n}{2} d_a$. (11.41)

⋮

Kennedy [185] has tabulated Fierz coefficients [120, 278, 278] F_{bc}, $b, c \leq 6$. They are related to 6-j's by

$F_{bc} = \frac{b!}{c!}\frac{1}{d_a}$ 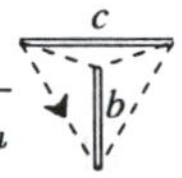 $= (-1)^{bc}\frac{b!}{c!}\sum_{a=0}^{b}(-1)^u\binom{a}{u}\binom{n-a}{b-u}$. (11.42)

11.4 6-j COEFFICIENTS

To evaluate (11.35) we need 6-j coefficients for six antisymmetric tensor reps of $SO(n)$. Substitutions (11.24), (11.21), and (11.19) lead to a strand-network [280]

expression for a 6-j coefficient,

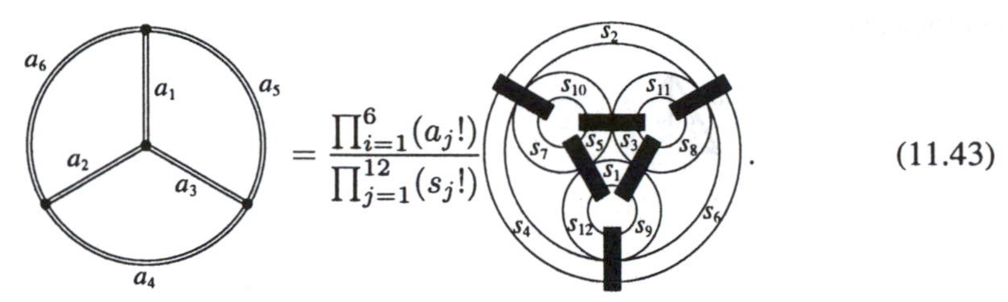

$$= \frac{\prod_{i=1}^{6}(a_j!)}{\prod_{j=1}^{12}(s_j!)} \tag{11.43}$$

Pick out a line in a strand, and follow its possible routes through the strand network. Seven types of terms give nonvanishing contributions: four "mini tours"

(11.44)

and three "grand tours"

(11.45)

Let the numbers of lines in different tours be $t_1, t_2, t_3, t_4, t_5, t_6$ and t_7. A nonvanishing contribution to the 6-j coefficient (11.43) corresponds to a partition of twelve strands, $s_1, s_2, \ldots, s_{12}$ into seven tours $t_1, t_2, \ldots, t_7$

$$M(t_1) = \tag{11.46}$$

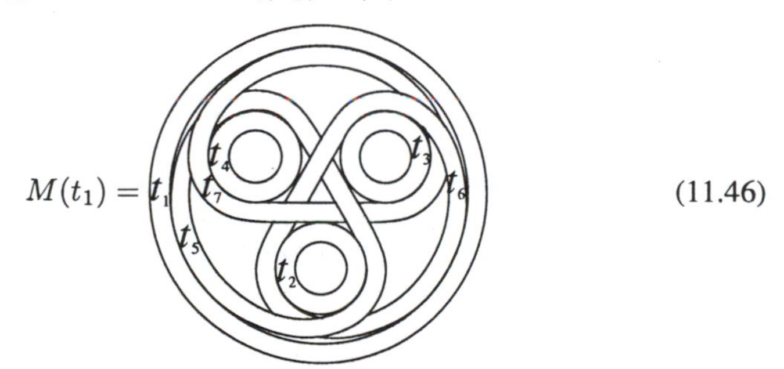

Comparing with (11.43), we see that each s_i is a sum of two t_i's: $s_1 = t_2 + t_7$, $s_2 = t_1 + t_7$, *etc.* It is sufficient to specify one t_1; this fixes all t_i's. Now one stares at the above figure and writes down

$$M(t_1) = \binom{n}{t} \frac{t!}{\prod_{i=1}^{7} t_i!} \frac{\prod_{i=1}^{12} s_i!}{\prod_{j=1}^{7} a_j!}, \qquad t = t_1 + t_2 + \ldots + t_7 \tag{11.47}$$

(a well-known theorem states that combinatorial factors cannot be explained [162]). The $\binom{n}{t}$ factor counts the number of ways of coloring $t_1 + t_2 + \ldots + t_7$ lines with n different colors. The second factor counts the number of distinct partitions of t lines into seven strands $t_1, t_2 \ldots, t_7$. The last factor again comes from the projector operator normalizations and the number of ways of coloring each strand and cancels against the corresponding factor in (11.43). Summing over the allowed partitions (for example, taking $0 \le t_1 \le s_2$), we finally obtain an expression for the 6-j coefficients:

$$= \sum_t \binom{n}{t} \frac{t!}{t_1!t_2!t_3!t_4!t_5!t_6!t_7!}$$

$$\begin{aligned}
t_1 &= -\frac{a_1+a_2+a_3}{2}+t & t_5 &= \frac{a_1+a_3+a_4+a_6}{2}-t \\
t_2 &= -\frac{a_1+a_5+a_6}{2}+t & t_6 &= \frac{a_1+a_2+a_4+a_5}{2}-t \\
t_3 &= -\frac{a_2+a_4+a_6}{2}+t & t_7 &= \frac{a_2+a_3+a_5+a_6}{2}-t \\
t_4 &= -\frac{a_3+a_4+a_5}{2}+t\,. & & \qquad (11.48)
\end{aligned}$$

The summation in (11.48) is over all values of t, such that all the t_i are nonnegative integers. The 3-j (11.36) is a special case of the 6-j (11.48). The 3-j's and 6-j's evaluated here, for all reps antisymmetric, should suffice in most applications.

The above examples show how Kennedy's method produces the n-dimensional spinor reductions needed for the dimensional regularization [161]. Its efficiency pays off only for longer spin traces. Each γ-pair contraction produces one 6-j symbol, and the completeness relation sums do not exceed the number of pair contractions, so for $2p$ γ-matrices the evaluation does not exceed p^2 steps. This is far superior to the initial algorithm (11.16).

Finally, a comment directed at the reader wary of analytically continuing in n while relying on completeness sums (de Wit and 't Hooft [94, 303] anomalies). Trouble could arise if, as we continued to low n, the $k > n$ terms in the completeness sum (11.27) gave nonvanishing contributions. We have explicitly noted that the dimension, 3-j and 6-j coefficients do vanish for any rep if $k > n$. The only danger arises from the Fierz coefficients (11.32): a ratio of 6-j and d can be finite for $j > n$. However, one is saved by the projection operator in the Fierz identity (11.32). This projection operator will eventually end up in some 6-j or 3-j coefficient without d in the denominator (as in (11.33)), and the whole term will vanish for $k > j$.

11.5 EXEMPLARY EVALUATIONS, CONTINUED

Now that we have explicit formulas for all 3-j and 6-j coefficients, we can complete the evaluation of examples commenced in section 11.2.1. The eight γ-matrix trace (11.33) is given by

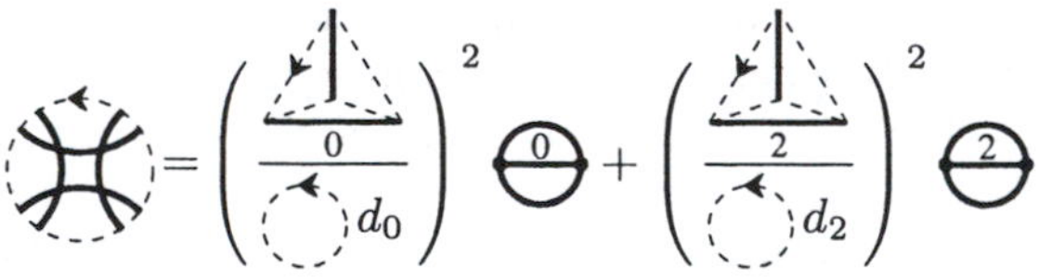

$$= n + n(n-1)(n-4)^2\,, \qquad (11.49)$$

and the ten γ-matrix trace (11.35) by

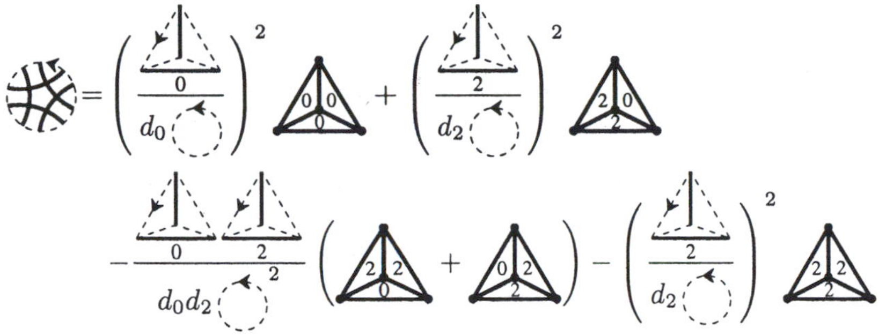

$$= n^3 + n(n-1)(n-4)^2 - 2n^2(n-1)(n-4)$$
$$-n(n-1)(n-2)(n-4)^2$$
$$= n^3 - n(n-1)(n-4)(n^2 - 5n + 12)\,. \qquad (11.50)$$

11.6 INVARIANCE OF γ-MATRICES

The above discussion of spinors did not follow the systematic approach of section 3.4 that we employ everywhere else in this monograph: start with a list of primitive invariants, find the characteristic equations they satisfy, construct projection operators, and identify the invariance group. In the present case, the primitive invariants are $g_{\mu\nu}$, δ_{ab} and $(\gamma_\mu)_{ab}$. We could retroactively construct the characteristic equation for $Q_{ab,cd} = (\gamma_\mu)_{ad}(\gamma_\mu)_{cb}$ from the Fierz identity (11.32), but the job is already done and the n eigenvalues are given by (11.38)–(11.41). The only thing that we still need to do is check that $SO(n)$, the invariance group of $g_{\mu\nu}$, is also the invariance group of $(\gamma_\mu)_{ab}$.

The $SO(n)$ Lie algebra is generated by the antisymmetric projection operator (8.7), or $\Gamma^{(2)}$ in the γ-matrix notation (11.8). The invariance condition (4.36) for γ-matrices is

$$- \quad - \quad = 0\,. \qquad (11.51)$$

To check whether $\Gamma^{(2)}$ respects the invariance condition, we evaluate the first and the term by means of the completeness relation (11.28):

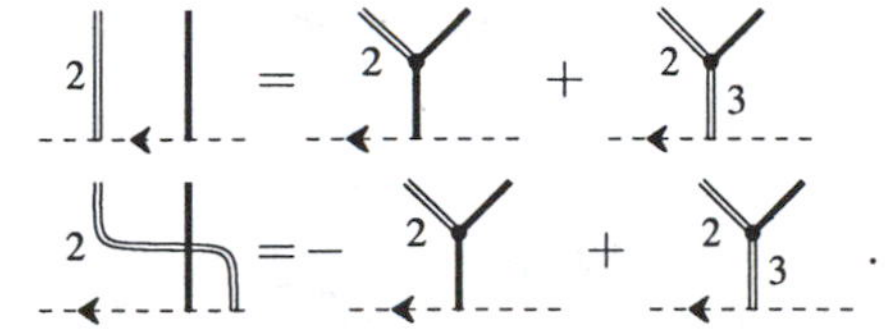

$$= \quad + $$
$$= - \quad + \quad .$$

The minus sign comes from the sign rule (11.26). Subtracting, we obtain

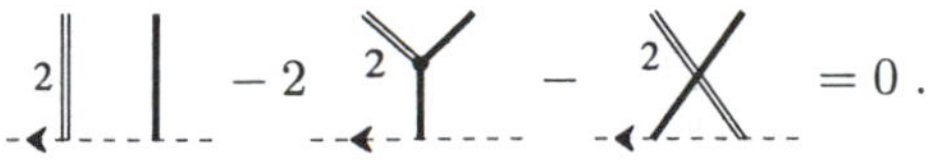

$$-2 \quad - \quad = 0\,.$$

This already has the form of the invariance condition (11.51), modulo normalization convention. To fix the normalization, we go back to definitions (11.8), (11.24), (11.19):

$$= 0 \, . \qquad (11.52)$$

The invariance condition (11.51) now fixes the relative normalizations of generators in the n-dimensional and spinor rep. If we take (8.7) for the n-dimensional rep

$$(T_{\mu\nu})_{\rho\sigma} = \qquad (11.53)$$

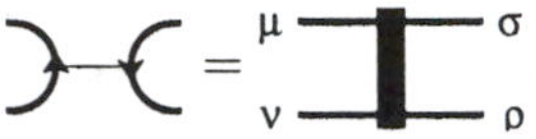

then the normalization of the generators in the spinor rep is

$$(T_{\mu\nu})_{ab} = \frac{1}{4}$$

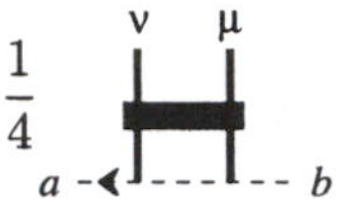

$$= \frac{1}{8}[\gamma_\nu, \gamma_\mu] \, . \qquad (11.54)$$

The γ-matrix invariance condition (11.51) written out in the tensor notation is

$$[T_{\mu\nu}, \gamma_\sigma] = \frac{1}{2}(g_{\mu\sigma}\gamma_\nu - g_{\nu\sigma}\gamma_\mu) \, . \qquad (11.55)$$

If you prefer generators $(T_i)_{ab}$ indexed by the adjoint rep index $i = 1, 2, \ldots, N$, then you can use spinor rep generators defined as

$$(T_i)_{ab} = \qquad (11.56)$$

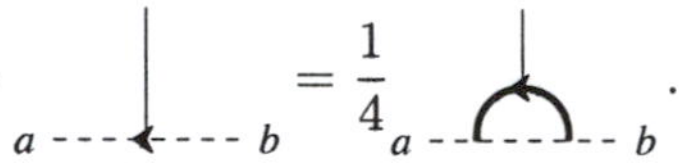

Now we can compute various casimirs for spinor reps. For example, the Dynkin index (section 7.5) for the lowest-dimensional spinor rep is given by

$$\ell =$$

$$= \frac{\mathrm{tr}\,\mathbf{1}}{8(n-2)} = \frac{2^{[\frac{n}{2}]-3}}{n-2} \, . \qquad (11.57)$$

From the invariance of γ_μ follows invariance of all $\Gamma^{(k)}$. In particular, the invariance condition for $\Gamma^{(2)}$ is the usual Lie algebra condition (4.47) with the structure constants given by (11.25).

11.7 HANDEDNESS

Among the bases (11.8), $\Gamma^{(n)}_{\mu_1\mu_2\ldots\mu_n}$ is special; it projects onto a 1-dimensional space, and the antisymmetrization can be replaced by a pair of Levi-Civita tensors (6.28):

$$\Gamma^{(n)} = \qquad (11.58)$$

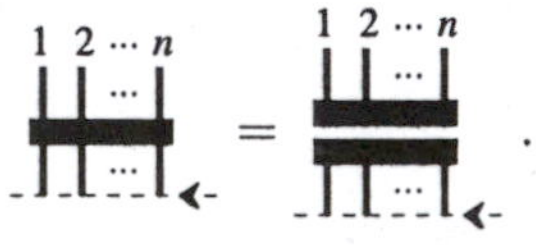

The corresponding clebsches are the generalized "γ_5" matrices,

$$\gamma^* \equiv \frac{1}{\sqrt{n!}}$$

$$= i^{n(n-1)/2}\gamma_1\gamma_2\ldots\gamma_n \, . \qquad (11.59)$$

The phase factor is, as explained in section 4.8, only a nuisance that cancels away in physical calculations. γ^* satisfies a trivial characteristic equation (use (6.28) and (11.18) to evaluate this),

$$(\gamma^*)^2 = \frac{1}{n!} \quad = \frac{1}{n!} \quad = \mathbf{1} \,, \tag{11.60}$$

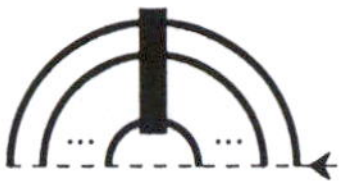

which yields projection operators (4.18):

$$\mathbf{P}_+ = \frac{1}{2}(\mathbf{1} + \gamma^*) \,, \qquad \mathbf{P}_- = \frac{1}{2}(\mathbf{1} - \gamma^*) \,. \tag{11.61}$$

The reducibility of Dirac spinors does not affect the correctness of the Kennedy spin traces algorithm. However, this reduction of Dirac spinors is of physical interest, so we briefly describe the irreducible spinor reps. Let us denote the two projectors diagrammatically by

$$\mathbf{1} = \mathbf{P}_+ + \mathbf{P}_- \tag{11.62}$$

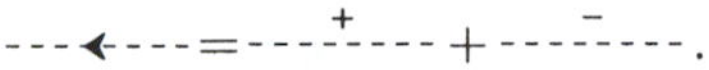

In even dimensions $\gamma_\mu \gamma^* = -\gamma^* \gamma_\mu$, while in odd dimensions $\gamma_\mu \gamma^* = \gamma^* \gamma_\mu$, so

$$n \text{ even:} \quad \gamma_\mu \mathbf{P}_+ = \mathbf{P}_- \gamma_\mu \,, \tag{11.63}$$

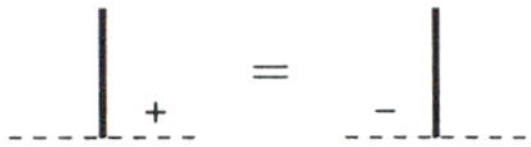

$$n \text{ odd:} \quad \gamma_\mu \mathbf{P}_+ = \mathbf{P}_+ \gamma_\mu \,. \tag{11.64}$$

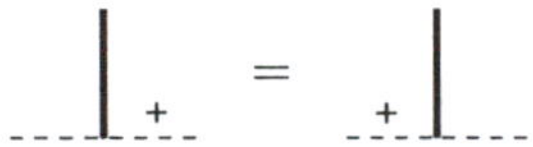

Hence, in the odd dimensions Dirac γ_μ matrices decompose into a pair of conjugate $[2^{(n-1)/2} \times 2^{(n-1)/2}]$ reps:

$$n \text{ odd:} \qquad \gamma_\mu = \mathbf{P}_+ \gamma_\mu \mathbf{P}_+ + \mathbf{P}_- \gamma_\mu \mathbf{P}_- \,, \tag{11.65}$$

and the irreducible spinor reps are of dimension $2^{(n-1)/2}$.

11.8 KAHANE ALGORITHM

For the case of four dimensions, there is a fast algorithm for trace evaluation, due to Kahane [178].

Consider a γ-matrix contraction,

$$\gamma^a \gamma_b \gamma_c \ldots \gamma_d \gamma_a = \,, \tag{11.66}$$

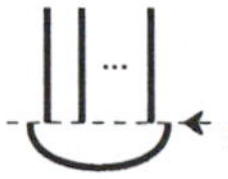

and use the completeness relation (11.27) and the "vertex" formula (11.34):

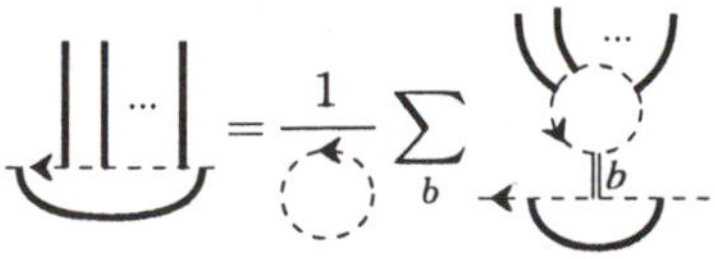

$$= \frac{1}{} \sum_b$$

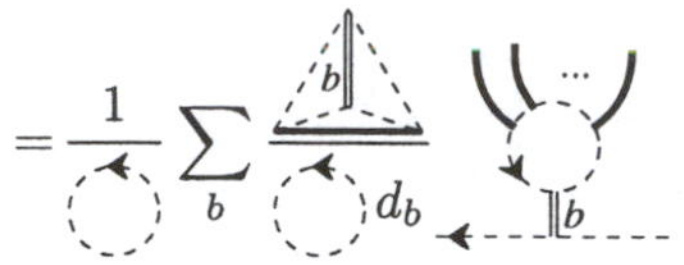

$$= \frac{1}{\circlearrowleft} \sum_b \frac{\triangle_b}{\circlearrowleft\, d_b} \quad \cdots \tag{11.67}$$

For $n = 4$, this sum ranges over $k = 0, 1, 2, 3, 4$. A spinor trace is nonvanishing only for even numbers of γ's, (11.16), so we distinguish the even and the odd cases when substituting the Fierz coefficients (11.40):

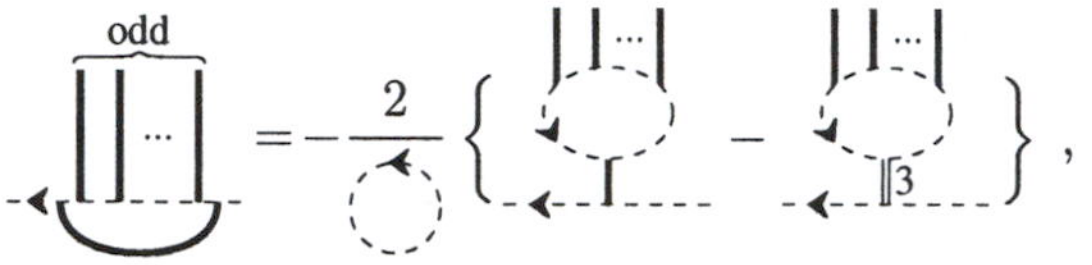

$$\text{odd} \quad = -\frac{2}{\circlearrowleft}\left\{ \quad - \quad \right\}, \tag{11.68}$$

$$\text{even} \quad = \frac{4}{\circlearrowleft}\left\{ \quad - \quad \right\}. \tag{11.69}$$

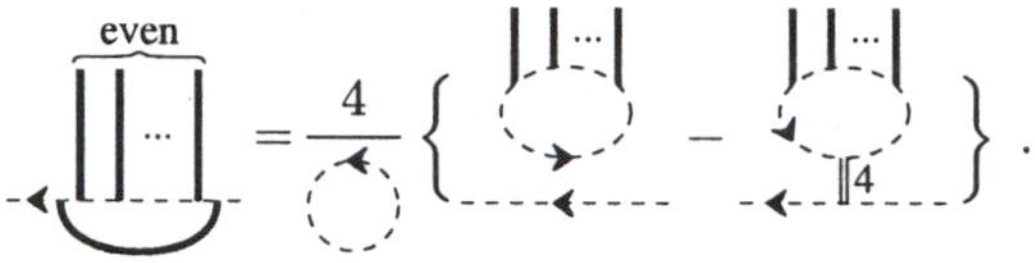

The sign of the second term in (11.68) can be reversed by transposing the three γ's (remember, the arrows on the spinor lines keep track of signs, *cf.* (11.24) and (11.26)):

$$= - \quad = - \quad . \tag{11.70}$$

But now the term in the brackets in (11.68) is just the completeness sum (11.27), and the summation can be dropped:

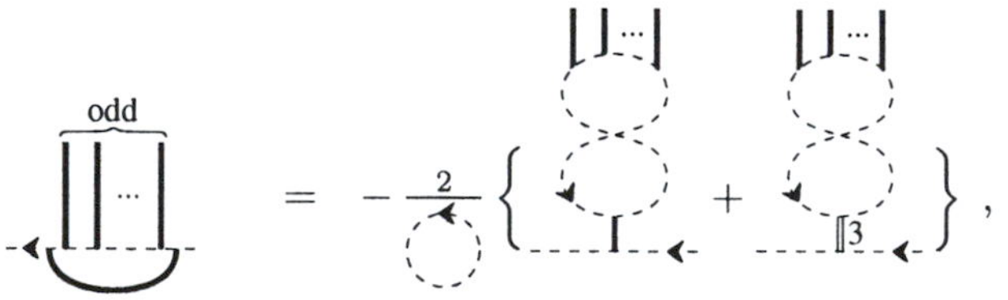

$$\text{odd} \quad = -\frac{2}{\circlearrowleft}\left\{ \quad + \quad \right\},$$

Rule 1:
$$= -2 \tag{11.71}$$

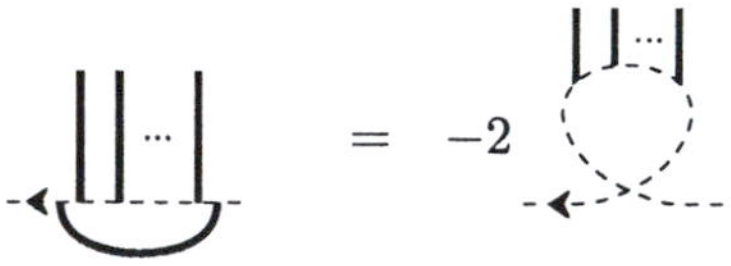

$$\gamma^a \gamma_b \gamma_c \cdots \gamma_d \gamma_a = -2\, \gamma_d \cdots \gamma_c \gamma_b$$

The same trick does not work for (11.69), because there the completeness sum has three terms:

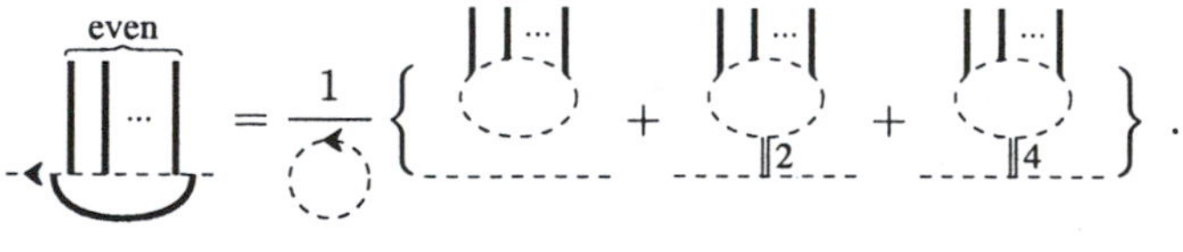

$$\text{even} \quad = \frac{1}{\circlearrowleft}\left\{ \quad + \quad + \quad \right\}. \tag{11.72}$$

However, as $\gamma_{[a}\gamma_{b]} = -\gamma_{[b}\gamma_{a]}$

$$= - \quad , \tag{11.73}$$

the sum of $\gamma_a\gamma_b \ldots \gamma_d$ and its transpose $\gamma_d \ldots \gamma_b\gamma_a$ has a two-term completeness sum:

$$\text{even} \quad + \quad = \frac{2}{\;} \left\{ \quad + \quad 4 \right\} . \tag{11.74}$$

Finally, we can change the sign of the second term in (11.69) by using $\{\gamma_5, \gamma_a\} = 0$;

Rule 2: $\text{even} = 2\left\{ \quad + \quad \right\}$

$$\gamma^e\gamma_a\gamma_b \ldots \gamma_c\gamma_d\gamma_e = 2\left\{\gamma_d\gamma_a\gamma_b \ldots \gamma_c + \gamma_c \ldots \gamma_b\gamma_a\gamma_d\right\} . \tag{11.75}$$

This rule and rule (11.71) enable us to remove γ-contractions ("internal photon lines") one by one, at most doubling the number of terms at each step. These rules are special to $n = 4$ and have no n-dimensional generalization.

Chapter Twelve

Symplectic groups

Symplectic group $Sp(n)$ is the group of all transformations that leave invariant a skew symmetric $(p, q) = f_{ab} p^a q^b$:

$$f_{ab} = - f_{ba} \qquad a, b = 1, 2, \ldots n$$

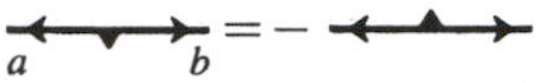

$$n \text{ even} . \tag{12.1}$$

The birdtrack notation is motivated by the need to distinguish the first and the second index: it is a special case of the birdtracks for antisymmetric tensors of even rank (6.57). If (p, q) is an invariant, so is its complex conjugate $(p, q)^* = f^{ba} p_a q_b$, and

$$f^{ab} = - f^{ba} \tag{12.2}$$

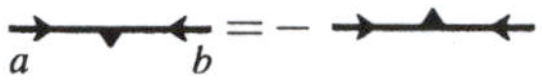

is also an invariant tensor. The matrix $A_a^{\ b} = f_{ac} f^{cb}$ must be proportional to unity, as otherwise its characteristic equation would decompose the defining n-dimensional rep. A convenient normalization is

$$f_{ac} f^{cb} = - \delta_a^b \tag{12.3}$$

Indices can be raised and lowered at will, so the arrows on lines can be dropped. However, omitting symplectic invariants (the black triangles) is not recommended, as without them it is hard to keep track of signs. Our convention will be to perform all contractions with f^{ab} and omit the arrows but not the symplectic invariants:

$$f^{ab} = \quad . \tag{12.4}$$

All other tensors will have lower indices. The Lie group generators $(T_i)_a^{\ b}$ will be replaced by

$$(T_i)_{ab} = (T_i)_a^{\ c} f_{cb} = \quad . \tag{12.5}$$

The invariance condition (4.36) for the symplectic invariant tensor is

$$+ \quad = 0$$

$$(T_i)_{ac} f_{cb} + f_{ac} (T_i)_{cb} = 0 . \tag{12.6}$$

A skew-symmetric matrix f_{ab} has the inverse in (12.3) only if $\det f \neq 0$. That is possible only in even dimensions [121, 144], so $Sp(n)$ can be realized only for even n.

In this chapter we shall outline the construction of $Sp(n)$ tensor reps. They are obtained by contracting the irreducible tensors of $SU(n)$ with the symplectic invariant f^{ab} and decomposing them into traces and traceless parts. The representation theory for $Sp(n)$ is analogous in step-by-step fashion to the representation theory for $SO(n)$. This arises because the two groups are related by supersymmetry, and in chapter 13 we shall exploit this connection by showing that all group-theoretic weights for the two groups are related by analytic continuation into negative dimensions.

12.1 TWO-INDEX TENSORS

The decomposition goes the same way as for $SO(n)$, section 10.1. The matrix (10.8), given by

$T =$  , (12.7)

satisfies the same characteristic equation (10.9) as for $SO(n)$. Now T is antisymmetric, $AT = T$, and only the antisymmetric subspace gets decomposed. $Sp(n)$ 2-index tensors decompose as

singlet: $(P_1)_{ab,cd} = \frac{1}{n} f_{ab} f_{cd} = \frac{1}{n}$ 

antisymmetric: $(P_2)_{ab,cd} = \frac{1}{2}(f_{ad} f_{bc} - f_{ac} f_{bd}) - \frac{1}{n} f_{ab} f_{cd}$

$=$ 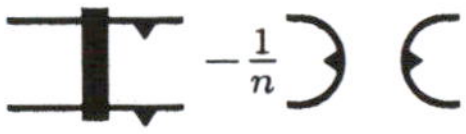 (12.8)

symmetric: $(P_3)_{ab,cd} = \frac{1}{2}(f_{ad} f_{bc} + f_{ac} f_{bd}) =$ 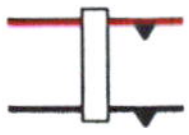 .

The $SU(n)$ adjoint rep (10.14) is now split into traceless symmetric and antisymmetric parts. The adjoint rep of $Sp(n)$ is given by the symmetric subspace, as only $\mathbf{P}_3$ satisfies the invariance condition (12.6):

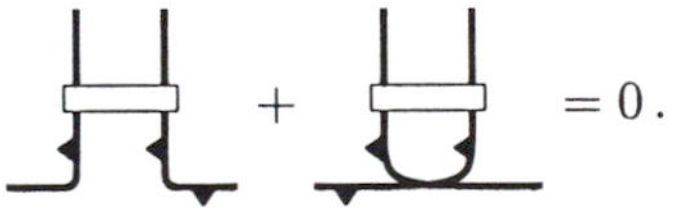

Hence, the adjoint rep projection operator for $Sp(n)$ is given by

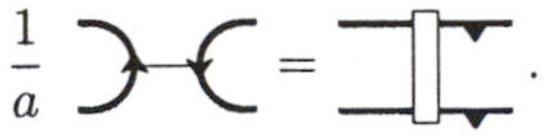 . (12.9)

The dimension of $Sp(n)$ is

$N = \mathrm{tr}\,\mathbf{P}_A =$ 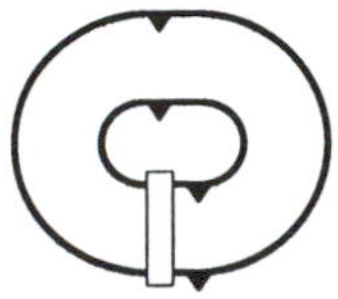 $= \frac{n(n+1)}{2}$. (12.10)

Young tableaux	 □⊗□	=	•	+	□□	+	□ over □ (two-box column)
Dynkin labels	(10...) × (10...)	=	(00...)	+	(010...)	+	(20...)
Dimensions	n^2	=	1	+	$\frac{n(n+1)}{2}$	+	$\frac{(n-2)(n+1)}{2}$
Dynkin indices	$\frac{2n}{n+2}$	=	0	+	1	+	$\frac{n-2}{n+2}$
Projectors	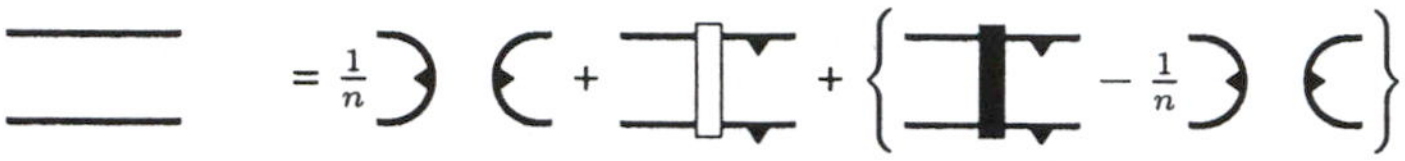	=	$\frac{1}{n}$	+		+	$\left\{ \; - \frac{1}{n} \; \right\}$

Table 12.1 $Sp(n)$ Clebsch-Gordan series for $V \otimes V$.

Remember that all contractions are carried out by f^{ab} — hence the symplectic invariants in the trace expression. Dimensions of the other reps and the Dynkin indices (see section 7.5) are listed in table 12.1.

We could continue as for the $SO(n)$ case, with $A \otimes V$, $V \otimes V \otimes V, \cdots$ decompositions, but that would turn out to be a step-by-step repetition of chapter 10. As we shall show next, reps of $SO(n)$ and $Sp(n)$ are related by a "negative dimensional" duality, so there is no need to work out the $Sp(n)$ reps separately.

Chapter Thirteen

Negative dimensions

P. Cvitanović and A. D. Kennedy

A cursory examination of the expressions for the dimensions and the Dynkin indices listed in tables 7.3 and 7.5, and in the tables of chapter 9, chapter 10, and chapter 12, reveals intriguing symmetries under substitution $n \to -n$. This kind of symmetry is best illustrated by the reps of $SU(n)$; if λ stands for a Young tableau with p boxes, and $\overline{\lambda}$ for the transposed tableau obtained by flipping λ across the diagonal (*i.e.*, exchanging symmetrizations and antisymmetrizations), then the dimensions of the corresponding $SU(n)$ reps are related by

$$SU(n): \qquad d_\lambda(n) = (-1)^p d_{\overline{\lambda}}(-n) \,. \tag{13.1}$$

This is evident from the standard recipe for computing the $SU(n)$ rep dimensions (section 9.3), as well as from the expressions listed in the tables of chapter 9. In all cases, exchanging symmetrizations and antisymmetrizations amounts to replacing n by $-n$.

Here we shall prove the following:

Negative Dimensionality Theorem 1: For any $SU(n)$ invariant scalar exchanging symmetrizations and antisymmetrizations is equivalent to replacing n by $-n$:

$$SU(n) = \overline{SU}(-n) \,. \tag{13.2}$$

Negative Dimensionality Theorem 2: For any $SO(n)$ invariant scalar there exists the corresponding $Sp(n)$ invariant scalar (and vice versa), obtained by exchanging symmetrizations and antisymmetrizations, replacing the $SO(n)$ symmetric bilinear invariant g_{ab} by the $Sp(n)$ antisymmetric bilinear invariant f_{ab}, and replacing n by $-n$:

$$SO(n) = \overline{Sp}(-n) \,, \qquad Sp(n) = \overline{SO}(-n) \,. \tag{13.3}$$

The bars on $\overline{SU}$, $\overline{Sp}$, $\overline{SO}$ indicate interchange of symmetrizations and antisymmetrizations. In chapter 14 we shall extend the relation (13.3) to spinorial representations of $SO(n)$.

Such relations are frequently noted in literature: Parisi and Sourlas [269] have suggested that a Grassmann vector space of dimension n can be interpreted as an ordinary vector space of dimension $-n$. Penrose [280] has introduced the term "negative dimensions" in his construction of $SU(2) \simeq Sp(2)$ reps as $SO(-2)$. King [191] has proved that the dimension of any irreducible rep of $Sp(n)$ is equal to that of $SO(n)$ with symmetrizations exchanged with antisymmetrizations (the transposed Young tableau), and n replaced by $-n$. Mkrtchyan [245] has observed this relation for the QCD loop equations. With the advent of supersymmetries, $n \to -n$ relations have become commonplace, as they are built into the structure of groups such as the orthosymplectic group $OSp(b, f)$.

Various examples of $n \to -n$ relations cited in the literature are all special cases of the theorems that we now prove. The birdtrack proof is simpler than the published proofs for the special cases. Some highly nontrivial examples of $n \to -n$ symmetries for the exceptional groups [78] will be discussed in chapter 18 and chapter 20, where we show that the negative-dimensional cousins of $SO(4)$ are $E_7(56), D_6(32), \cdots$, and that for $SU(3)$ the $n \to -n$ symmetry leads to $E_6(27), \cdots$.

13.1 $SU(n) = \overline{SU}(-n)$

As we have argued in section 5.2, all physical consequences of a symmetry (rep dimensions, level splittings, *etc.*) can be expressed in terms of invariant scalars. The primitive invariant tensors of $SU(n)$ are the Kronecker tensor δ^a_b and the Levi-Civita tensor $\epsilon_{a_1 \cdots a_n}$. All other invariants of $SU(n)$ are built from these two objects. A scalar ($3n$-j coefficient, vacuum bubble) is a tensor object with all indices contracted, which in birdtrack notation corresponds to a diagram with no external legs. Thus, in scalars, Levi-Civita tensors can appear only in pairs (the lines must end somewhere), and by (6.28) the Levi-Civita tensors combine to antisymmetrizers. Consequently $SU(n)$ invariant scalars are all built only from symmetrizers and antisymmetrizers. Expanding all symmetry operators in an $SU(n)$ vacuum bubble gives a sum of entangled loops. Each loop is worth n, so each term in the sum is a power of n, and therefore an $SU(n)$ invariant scalar is a polynomial in n.

The idea of the proof is illustrated by the following typical computation: evaluate, for example, the $SU(n)$ 9-j coefficient for recoupling of three antisymmetric rank-2 reps:

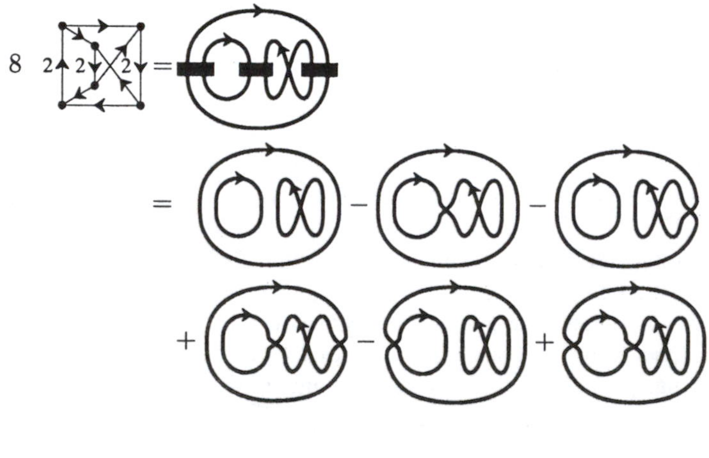

$$+ \quad - $$

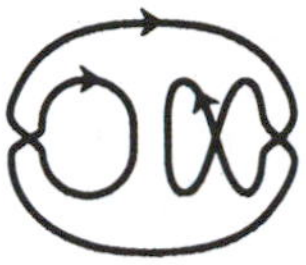

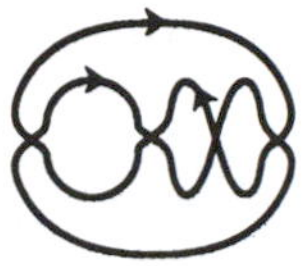

$$= n^3 - n^2 - n^2 + n - n^2 + n + n - n^2$$
$$= n(n-1)(n-3) \,. \tag{13.4}$$

Notice that in the expansion of the symmetry operators the graphs with an odd number of crossings give an even power of n, and vice versa. If we change the three symmetrizers into antisymmetrizers, the terms that change the sign are exactly those with an even number of crossings. The crossing in the original graph that had nothing to do with any symmetry operator, appears in every term of the expansion, and thus does not affect our conclusion; an exchange of symmetrizations and antisymmetrizations amounts to substitution $n \to -n$. The overall sign is only a matter of convention; it depends on how we define the vertices in the $3n$-j's.

The proof for the general $SU(n)$ case is even simpler than the above example: Consider the graph corresponding to an arbitrary $SU(n)$ scalar, and expand all its symmetry operators as in (13.4). The expansion can be arranged (in any of many possible ways) as a sum of pairs of form

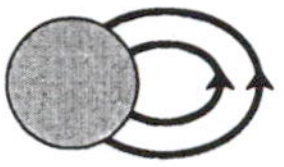

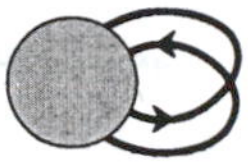

$$\ldots + \quad \pm \quad + \ldots \,, \tag{13.5}$$

with a plus sign if the crossing arises from a symmetrization, and a minus sign if it arises from an antisymmetrization. The gray blobs symbolize the tangle of lines common to the two terms. Each graph consists only of closed loops, *i.e.*, a definite power of n, and thus uncrossing two lines can have one of two consequences. If the two crossed line segments come from the same loop, then uncrossing splits this into two loops, whereas if they come from two loops, it joins them into one loop. The power of n is changed by the uncrossing:

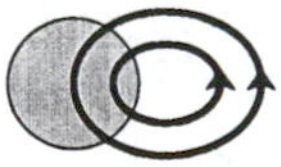

$$= n \quad . \tag{13.6}$$

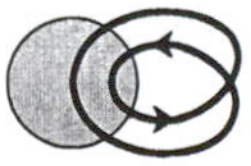

Hence, the pairs in the expansion (13.5) always differ by $n^{\pm 1}$, and exchanging symmetrizations and antisymmetrizations has the same effect as substituting $n \to -n$ (up to an irrelevant overall sign). This completes the proof of (13.2).

Some examples of $n \to -n$ relations for $SU(n)$ reps:

1. Dimensions of the fully symmetric reps (6.13) and the fully antisymmetric reps (6.21) are related by the Beta-function analytic continuation formula

$$\frac{n!}{(n-p)!} = (-1)^p \frac{(-n+p-1)!}{(-n-1)!} \,. \tag{13.7}$$

2. The reps (9.13) and (9.14) correspond to the 2-index symmetric, antisymmetric tensors, respectively. Therefore, their dimensions in figure 9.1 are related by $n \to -n$.

3. The reps (9.79) and (9.80) (see also table 7.5) are related by $n \to -n$ for the same reason.

4. section 9.9.

13.2 $SO(n) = \overline{Sp}(-n)$

In addition to δ^a_b and $\varepsilon_{ab...d}$, $SO(n)$ preserves a symmetric bilinear invariant g_{ab}, for which we have introduced open circle birdtrack notation in (10.1). Such open circles can occur in $SO(n)$ $3n$-j graphs, flipping the line directions. The Levi-Civita tensor still cannot occur, as directed lines, starting on an ε tensor, would have to end on a g tensor, that gives zero by symmetry. $Sp(n)$ differs from $SO(n)$ by having a skew-symmetric f_{ab}, for which we have introduced birdtrack notation in (12.1). In $Sp(n)$ we can convert a Levi-Civita tensor with upper indices into one with lower indices by contracting with n f's, with the appropriate power of det f appearing. We can therefore eliminate pairs of Levi-Civita tensors. A single Levi-Civita tensor can still appear in an $Sp(n)$ $3n$-j graph, but as

$$[\text{birdtrack}] = \mathrm{Pf}(f)\,, \qquad (13.8)$$

where Pf(f) is the Pfaffian, and $\mathrm{Pf}(f)^2 = \det f$ (that is left as an exercise for the reader). Therefore a Levi-Civita can always be replaced by an antisymmetrization

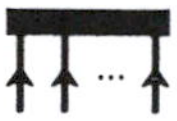

$$[\text{birdtrack}] = (\det f)^{-\frac{1}{2}} [\text{birdtrack}]\,. \qquad (13.9)$$

For any $SO(n)$ scalar there exists a corresponding $Sp(n)$ scalar, obtained by exchanging the symmetrizations and antisymmetrizations *and* the g_{ab}'s and f_{ab}'s in the corresponding graphs. The proof that the two scalars are transformed into each other by replacing n by $-n$, is the same as for $SU(n)$, except that the two line segments at a crossing could come from a new kind of loop, containing g_{ab}'s or f_{ab}'s. In that case, equation (13.6) is replaced by

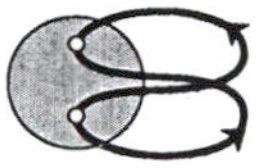

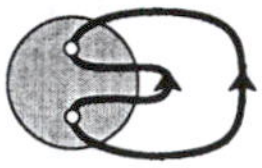

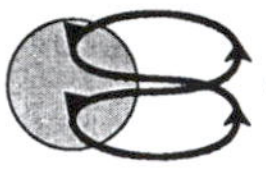

$$[\text{birdtrack}] = [\text{birdtrack}] \quad \Leftrightarrow \quad [\text{birdtrack}] = -[\text{birdtrack}]\,. \qquad (13.10)$$

While now uncrossing the lines does not change the number of loops, changing g_{ab}'s to f_{ab}'s does provide the necessary minus sign. This completes the proof of (13.3) for the tensor reps of $SO(n)$ and $Sp(n)$.

Some examples of $SO(n) = \overline{Sp}(-n)$ relations:

1. The $SO(n)$ antisymmetric adjoint rep (10.13) corresponds to the $Sp(n)$ symmetric adjoint rep (12.9).

2. Compare table 12.1 and table 10.1. See table 7.3, table 7.4, and table 7.2.

3. Penrose [280] binors: $SU(2) = Sp(2) = \overline{SO}(-2)$.

In order to extend the proof to the spinor reps, we will first have to invent the $Sp(n)$ analog of spinor reps. We turn to this task in the next chapter.

Chapter Fourteen

Spinors' symplectic sisters

P. Cvitanović and A. D. Kennedy

Dirac discovered spinors in his search for a vectorial quantity that could be interpreted as a "square root" of the Minkowski 4-momentum squared,

$$(p_1\gamma_1 + p_2\gamma_2 + p_3\gamma_3 + p_4\gamma_4)^2 = -p_1^2 - p_2^2 - p_3^2 + p_4^2.$$

What happens if one extends a Minkowski 4-momentum (p_1, p_2, p_3, p_4) into fermionic, Grassmann dimensions $(p_{-n}, p_{-n+1}, \ldots, p_{-2}, p_{-1}, p_1, p_2, \ldots, p_{n-1}, p_n)$? The Grassmann sector p_μ anticommute and the gamma-matrix relatives in the Grassmann dimensions have to satisfy the Heisenberg algebra commutation relation,

$$[\gamma_\mu, \gamma_\nu] = f_{\mu\nu}\mathbf{1}\,,$$

instead of the Clifford algebra anticommutator condition (11.2), with the bilinear invariant $f_{\mu\nu} = -f_{\nu\mu}$ skew-symmetric in the Grassmann dimensions.

In chapter 12, we showed that the symplectic group $Sp(n)$ is the invariance group of a skew-symmetric bilinear symplectic invariant $f_{\mu\nu}$. In section 14.1, we investigate the consequences of taking γ matrices to be Grassmann valued; we are led to a new family of objects, which we have named *spinsters* [81]. In the literature such reps are called *metaplectic* [335, 309, 192, 322, 300, 102, 193, 222]. Spinsters play a role for symplectic groups analogous to that played by spinors for orthogonal groups. With the aid of spinsters we are able to compute, for example, all the 3-j and 6-j coefficients for symmetric reps of $Sp(n)$. We find that these coefficients are identical with those obtained for $SO(n)$ if we interchange the roles of symmetrization and antisymmetrization and simultaneously replace the dimension n by $-n$. In section 14.2, we make use of the fact that $Sp(2) \simeq SU(2)$ to show that the formulas for $SU(2)$ 3-j and 6-j coefficients are special cases of general expressions for these quantities we derived earlier.

This chapter is based on ref. [81]. For a discussion of the role negative-dimensional groups play in quantum physics, see ref. [102].

14.1 SPINSTERS

The Clifford algebra (11.2) Dirac matrix elements $(\gamma_\mu)_{ab}$ are commuting numbers. In this section we shall investigate consequences of taking γ_μ to be Grassmann valued,

$$(\gamma_\mu)_{ab}(\gamma_\nu)_{cd} = -(\gamma_\nu)_{cd}(\gamma_\mu)_{ab}\,. \tag{14.1}$$

The Grassmann extension of the Clifford algebra (11.2) is

$$\frac{1}{2}[\gamma_\mu, \gamma_\nu] = f_{\mu\nu}\mathbf{1}\,, \qquad \mu, \nu = 1, 2, \ldots, n, \quad n \text{ even}\,. \tag{14.2}$$

The anticommutator gets replaced by a commutator, and the $SO(n)$ symmetric invariant tensor $g_{\mu\nu}$ by the $Sp(n)$ symplectic invariant $f_{\mu\nu}$. Just as the Dirac gamma-matrices lead to spinor reps of $SO(n)$, the Grassmann valued γ_μ give rise to $Sp(n)$ reps, which we shall call *spinsters*. Following the $Sp(n)$ diagrammatic notation for the symplectic invariant (12.1), we represent the defining commutation relation (14.2) by

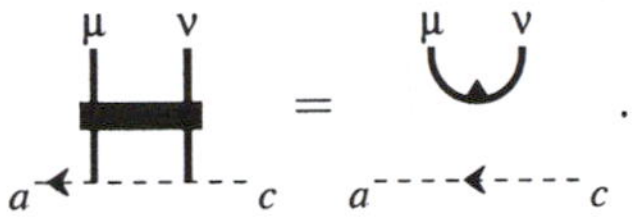

(14.3)

For the symmetrized products of γ matrices, the above commutation relations lead to

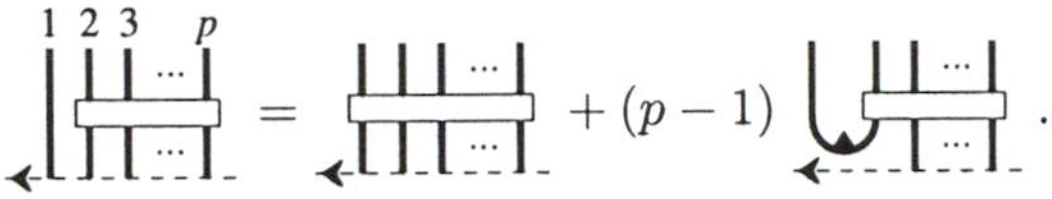

(14.4)

As in chapter 11, this gives rise to a complete basis for expanding products of γ-matrices. Γ's are now the symmetrized products of γ matrices:

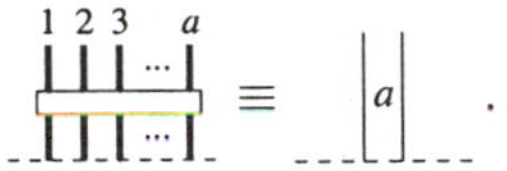

(14.5)

Note that while for spinors the $\Gamma^{(k)}$ vanish by antisymmetry for $k > n$, for spinsters the $\Gamma^{(k)}$'s are nonvanishing for any k, and the number of spinster basis tensors is infinite. However, the reduction of a product of k-γ-matrices involves only a finite number of $\Gamma^{(l)}$, $0 \leqslant l \leqslant k$. As the components $(\gamma_\mu)_{ab}$ are Grassmann valued, spinster traces of even numbers of γ's are anticyclic:

$$\mathrm{tr}\,\gamma_\mu\gamma_\nu = (\gamma_\mu)_{ab}(\gamma_\nu)_{ba} = -\,\mathrm{tr}\,\gamma_\nu\gamma_\mu$$

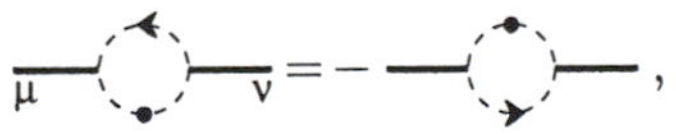

$$\mathrm{tr}\,\gamma_\mu\gamma_\nu\gamma_\rho\gamma_\sigma = -\,\mathrm{tr}\,\gamma_\nu\gamma_\rho\gamma_\sigma\gamma_\mu \tag{14.6}$$

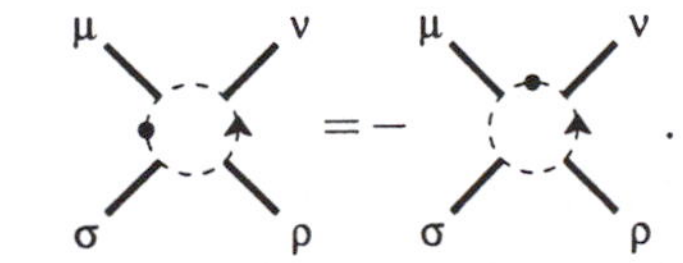

In the diagrammatic notation we indicate the beginning of a spinster trace by a dot. The dot keeps track of the signs in the same way as the symplectic invariant (12.3) for $f_{\mu\nu}$. Indeed, tracing (14.3) we have

$$\mathrm{tr}\,\gamma_\mu\gamma_\nu = f_{\mu\nu}\ \mathrm{tr}\,\mathbf{1} \tag{14.7}$$

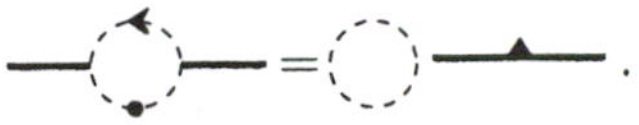

Moving a dot through a γ matrix gives a factor -1, as in (14.6).

Spinster traces can be evaluated recursively, as in (11.7). For a trace of an even number of γ's we have

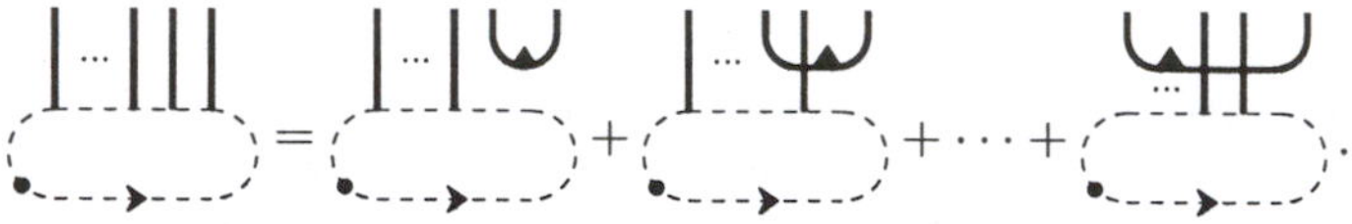

(14.8)

The trace of an odd number of γ's vanishes [81]. Iteration of equation (14.8) expresses a spinster trace as a sum of the $(p-1)!! = (p-1)(p-3)\ldots 5.3.1$ ways of connecting the external legs with $f_{\mu\nu}$. The overall sign is fixed uniquely by the position of the dot on the spinster trace:

(14.9)

and so on (see (11.15)).

Evaluation of traces of several Γ's is again a simple combinatoric exercise. Any term in which a pair of $f_{\mu\nu}$ indices are symmetrized vanishes, which implies that any $\Gamma^{(k)}$ with $k > 0$ is traceless. The Γ's are orthogonal:

$$= a!\,\delta_{ab} \quad . \tag{14.10}$$

The symmetrized product of a $f_{\mu\nu}$'s denoted by

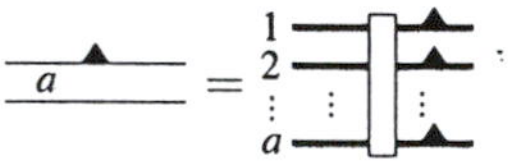

(14.11)

is either symmetric or skew-symmetric:

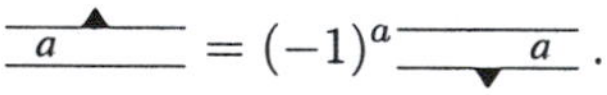

$$= (-1)^a \quad . \tag{14.12}$$

A spinster trace of three symmetric $Sp(n)$ reps defines a 3-vertex:

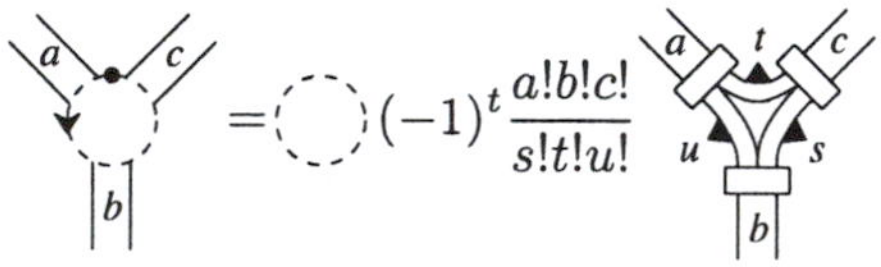

$$= (-1)^t \frac{a!b!c!}{s!t!u!}$$

$$=0 \quad \text{for } a+b+c = \text{odd},$$

$$s = \frac{1}{2}(b+c-a), \quad t=\frac{1}{2}(c+a-b), \quad u = \frac{1}{2}(a+b-c). \tag{14.13}$$

As in (11.20), Γ's provide a complete basis for expanding products of arbitrary numbers of γ matrices:

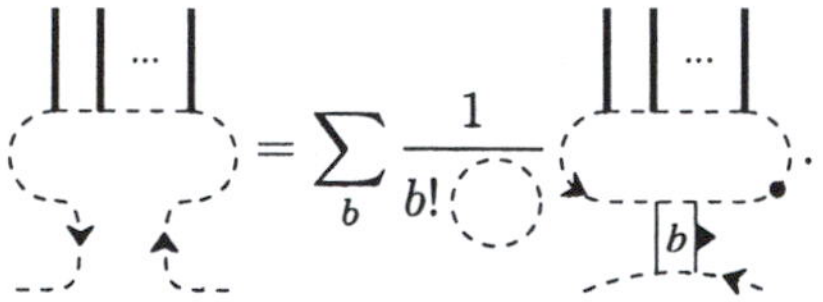

$$= \sum_b \frac{1}{b!} \quad . \tag{14.14}$$

The coupling coefficients in (14.14) are computed as spinster traces using the orthogonality relation (14.10). As only traces of even numbers of γ's are nonvanishing, spinster traces are even Grassmann elements; they thus commute with any other Γ, and all the signs in the above completeness relation are unambiguous.

The orthogonality of Γ's enables us to introduce projection operators and 3-vertices:

$$\frac{1}{\bigcirc} \quad = \frac{1}{a!\bigcirc} \quad , \qquad (14.15)$$

$$= \frac{(-1)^t}{\bigcirc} \quad . \qquad (14.16)$$

The sign factor $(-1)^t$ gives a symmetric definition of the 3-vertex (see (3.11)). It is important to note that the spinster loop runs clockwise in this definition. Because of (3.41), the 3-vertex has a nontrivial symmetry under interchange of two legs:

$$= (-1)^{s+t+u} \quad . \qquad (14.17)$$

Note that this is different from (11.26); one of the few instances of spinsters and spinors differing in a way that cannot be immediately understood as an $n \to -n$ continuation.

The completeness relation (14.14) can be written as

$$= \sum_b \frac{1}{\bigcirc} \quad . \qquad (14.18)$$

The recoupling relation is derived as in the spinor case (11.32):

$$= \sum_b \frac{b}{2 \; d_b} \quad . \qquad (14.19)$$

Here d_b is the dimension of the fully symmetrized b-index tensor rep of $Sp(n)$:

$$d_b = \quad = \quad = \binom{n+b-1}{b} = (-1)^b \binom{-n}{b} . \qquad (14.20)$$

The spinster recoupling coefficients in (14.19) are analogues of the spinor Fierz coefficients in (11.32). Completeness can be used to evaluate spinster traces in the same way as in examples (11.34) to (11.35).

The next step is the evaluation of 3-j's, 6-j's, and spinster recoupling coefficients. The spinster recoupling coefficients can be expressed in terms of 3-j's just as in (11.37):

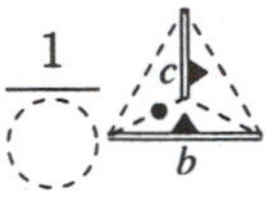

$$\frac{1}{\bigcirc} \; = \sum (-1)^{\frac{a+b+c}{2}} \qquad (14.21)$$

The evaluation of 3-j and 6-j coefficients is again a matter of simple combinatorics:

$$= (-1)^{s+t+u} \binom{n+s+t+u-1}{s+t+u} \frac{(s+t+u)!}{s!t!u!}, \qquad (14.22)$$

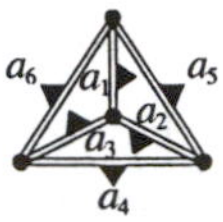

$$= \sum_t \binom{n+t-1}{t} \frac{(-1)^t t!}{t_1!t_2!t_3!t_4!t_5!t_6!t_7!}, \qquad (14.23)$$

with the t_i defined in (11.48).

We close this section with a comment on the dimensionality of spinster reps. Tracing both sides of the spinor completeness relation (11.27), we determine the dimensionality of spinor reps from the sum rule (11.30):

$$(\mathrm{tr}\,\mathbf{1})^2 = \sum_{a=0}^{n} \binom{n}{a} = 2^n .$$

Hence, Dirac matrices (in even dimensions) are $[2^{n/2} \times 2^{n/2}]$, and the range of spinor indices in (11.3) is $a, b = 1, 2, \ldots, 2^{n/2}$.

For spinsters, tracing the completeness relation (14.18) yields (the string of γ matrices was indicated only to keep track of signs for odd b's):

$$= \sum_b \frac{1}{\bigcirc} \; = \sum_b d_b \qquad (14.24)$$

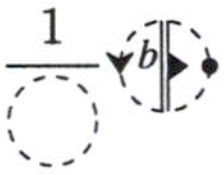

$$(\mathrm{tr}\,\mathbf{1})^2 = \sum_{c=o}^{\infty} \binom{n+b-1}{b} .$$

The spinster trace is infinite. This is the reason why spinster traces are not to be found in the list of the finite-dimensional irreducible reps of $Sp(n)$. One way of making the traces meaningful is to note that in any spinster trace evaluation only a finite number of Γ's are needed, so we can truncate the completeness relation (14.18) to terms $0 \leqslant b \leqslant b_{max}$. A more pragmatic attitude is to observe that the final results of the calculation are the 3-j and 6-j coefficients for the fully symmetric reps of $Sp(n)$, and that the spinster algebra (14.2) is a formal device for projecting only the fully symmetric reps from various Clebsch-Gordan series for $Sp(n)$.

The most striking result of this section is that the 3-j and 6-j coefficients are just the $SO(n)$ coefficients evaluated for $n \to -n$. The reason for this we already understand from chapter 13.

When we took the Grassmann extension of Clifford algebras in (14.2), it was not too surprising that the main effect was to interchange the role of symmetrization and antisymmetrization. All antisymmetric tensor reps of $SO(n)$ correspond to the symmetric rep of $Sp(n)$. What is more surprising is that if we take the expression we derived for the $SO(n)$ 3-j and 6-j coefficients and replace the dimension n by $-n$, we obtain exactly the corresponding result for $Sp(n)$. The negative dimension arises in these cases through the relation $\binom{-n}{a} = (-1)^a \binom{n+a-1}{a}$, which may be justified by analytic continuation of binomial coefficients by the Beta function.

14.2 RACAH COEFFICIENTS

So far, we have computed the 6-j coefficients for fully symmetric reps of $Sp(n)$. $Sp(2)$ plays a special role here; the symplectic invariant $f^{\mu\nu}$ has only one independent component, and it must be proportional to $\varepsilon^{\mu\nu}$. Hence, $Sp(2) \simeq SU(2)$. The observation that $SU(2)$ can be viewed as $SO(-2)$ was first made by Penrose [280], who used it to compute $SU(2)$ invariants using "binors." His method does not generalize to $SO(n)$, for which spinors are needed to project onto totally antisymmetric reps (for the case $n = 2$, this is not necessary as there are no other reps). For $SU(2)$, *all* reps are fully symmetric (Young tableaux consist of a single row), and our 6-j's are all the 6-j's needed for computing $SU(2) \simeq SO(3)$ group-theoretic factors. More pedantically: $SU(2) \simeq Spin(3) \simeq \widetilde{SO}(3)$. Hence, all the Racah [286] and Wigner coefficients, familiar from the atomic physics textbooks, are special cases of our spinor/spinster 6-j's. Wigner's 3-j symbol (5.14)

$$\begin{pmatrix} j_1 & j_2 & J \\ m_1 & m_2 & -M \end{pmatrix} \equiv \frac{(-1)^{j_1-j_2+M}}{\sqrt{2J+1}} \langle j_1 j_2 m_1 m_2 | JM \rangle \tag{14.25}$$

is really a clebsch with our 3-j as a normalization factor.

This may be expressed more simply in diagrammatic form:

$$\begin{pmatrix} j_1 & j_2 & J \\ m_1 & m_2 & -M \end{pmatrix} = \frac{i^{phase}}{\sqrt{\;2j_1,\ 2j_2,\ 2J\;}} \;\; 2j_1,\ 2j_2,\ 2J \tag{14.26}$$

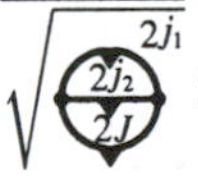

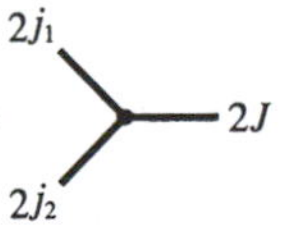

where we have not specified the phase convention on the right-hand side, as in the calculation of physical quantities such phases cancel. Factors of 2 appear because our integers $a, b, \ldots = 1, 2, \ldots$ count the numbers of $SU(2)$ 2-dimensional reps ($SO(3)$ spinors), while the usual $j_1, j_2, \ldots = \frac{1}{2}, 1, \frac{3}{2}, \ldots$ labels correspond to $SO(3)$ angular momenta.

It is easy to verify (up to a sign) the completeness and orthogonality properties of Wigner's 3-j symbols

$$\sum_{J,M} (2J+1) \begin{pmatrix} j_1 & j_2 & J \\ m_1 & m_2 & M \end{pmatrix} \begin{pmatrix} j_1 & j_2 & J \\ m_1 & m_2 & M \end{pmatrix} \sim \sum_J \frac{d_{2J}}{2j_1,\ 2j_2,\ 2J} \;\; 2j_1,\ 2j_2,\ 2J,\ 2j_1,\ 2j_2$$

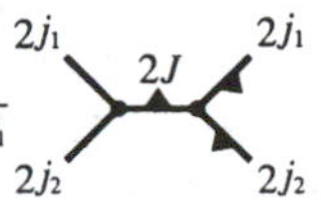

$$= \;\; 2j_1,\ 2j_2 \;\; \sim \delta_{m_1 m_1'} \delta_{m_2 m_2'} \tag{14.27}$$

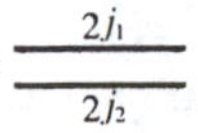

$$\sum_{m_1 m_2} \begin{pmatrix} j_1 & j_2 & J \\ m_1 & m_2 & M \end{pmatrix} \begin{pmatrix} j_1 & j_2 & J' \\ m_1 & m_2 & M' \end{pmatrix} \sim \frac{1}{\text{(birdtrack: } 2j_1, 2j_2, 2J)} \text{(birdtrack: } 2J, 2j_1, 2j_2, 2J) \, \delta_{JJ'}$$

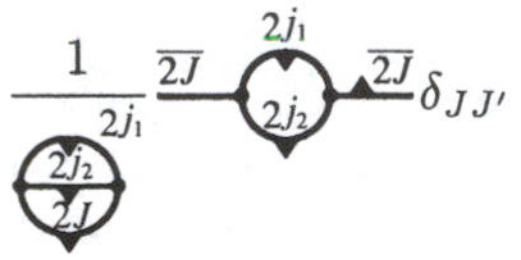

$$\sim \frac{\delta_{MM'} \delta_{JJ'}}{2J+1} \,. \qquad (14.28)$$

The expression (14.22) for our 3-j coefficient with $n = 2$ gives the expression usually written as Δ in Racah's formula for $\binom{j\ k\ l}{\alpha\ \gamma\ \gamma}$,

$$\frac{1}{\Delta(j,k,l)} = (-1)^{j+k+l} \text{(birdtrack: } 2j, 2k, 2l) = \frac{(j+k+l+1)!}{(j+k-l)!(k+l-j)!(l+j-k)!} \,. \qquad (14.29)$$

Wigner's 6-j coefficients (5.15) are the same as ours, except that the 3-vertices are normalized as in (14.26)

$$\begin{Bmatrix} j_1 & j_2 & j_3 \\ k_1 & k_2 & k_3 \end{Bmatrix} = \frac{1}{\sqrt{\text{(birdtracks: } 2j_1, 2k_2, 2k_3)\,(2j_2, 2k_1, 2k_3)\,(2j_3, 2k_1, 2k_2)\,(2j_1, 2j_2, 2j_3)}} \text{(birdtrack: } 2k_2, 2k_3, 2j_1, 2j_2, 2j_3, 2k_1) \,, \qquad (14.30)$$

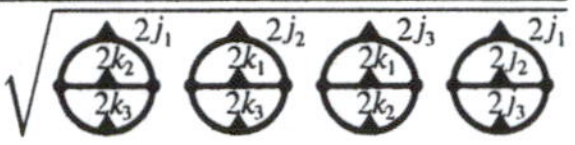

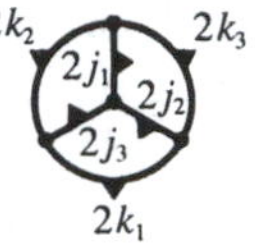

which gives Racah's formula using (14.23), with $n = 2$:

$$\begin{Bmatrix} j_1 & j_2 & j_3 \\ k_1 & k_2 & k_3 \end{Bmatrix} = [\Delta(j_1 k_2 k_3)\Delta(k_1 j_2 k_3)\Delta(k_1 k_2 j_3)\Delta(j_1 j_2 j_3)]^{1/2}$$

$$\times \sum_t \frac{(-1)^t (t+1)!}{t_1! t_2! t_3! t_4! t_5! t_6! t_7!} \,, \qquad \text{where}$$

$$\begin{aligned} t_1 &= t - j_1 - j_2 - j_3 \,, & t_5 &= j_1 + j_2 - k_1 + k_2 - t \,, \\ t_2 &= t - j_1 - k_2 - k_3 \,, & t_6 &= j_2 + j_3 + k_2 + k_3 - t \,, \\ t_3 &= t - k_1 - j_2 - k_3 \,, & t_7 &= j_3 + j_1 + k_3 + k_1 - t \,, \\ t_4 &= t - k_1 - k_2 - j_3 \,. & & \end{aligned} \qquad (14.31)$$

14.3 HEISENBERG ALGEBRAS

What are these "spinsters"? A trick for relating $SO(n)$ antisymmetric reps to $Sp(n)$ symmetric reps? That can be achieved without spinsters: indeed, Penrose [280] had observed many years ago that $SO(-2)$ yields Racah coefficients in a much more elegant manner than the usual angular momentum manipulations. In chapter 13, we have also proved that for any scalar constructed from tensor invariants, $SO(-n) \simeq Sp(n)$. This theorem is based on elementary properties of permutations and establishes the equivalence between 6-j coefficients for $SO(-n)$ and $Sp(n)$, without reference to spinsters or any other Grassmann extensions.

Nevertheless, spinsters are the natural supersymmetric extension of spinors, and the birdtrack derivation offers a different perspective from the literature discussions of metaplectic reps of the symplectic group [309, 322, 102, 193, 222]. They do not appear in the usual classifications, because they are infinite-dimensional reps of $Sp(n)$. However, they are not as unfamiliar as they might seem; if we write the

Grassmannian γ matrices for $Sp(2D)$ as $\gamma_\mu = (p_1, p_2, \dots p_D, x_1, x_2 \dots x_D)$ and choose $f_{\mu\nu}$ of form

$$f = \begin{pmatrix} 0 & \mathbf{1} \\ -\mathbf{1} & 0 \end{pmatrix}, \tag{14.32}$$

the defining commutator relation (14.2) is the defining relation for a Heisenberg algebra, except for a missing factor of i:

$$[p_i, x_j] = \delta_{ij}\mathbf{1}, \qquad i, j = 1, 2, \dots D. \tag{14.33}$$

If we include an extra factor of i into the definition of the "momenta" above, we find that spinsters resemble an antiunitary Grassmann-valued rep of the usual Heisenberg algebra. The Clifford algebra has its spinor reps, and the Heisenberg algebra has its infinite-dimensional Fock representation. The Fock space rep of the metaplectic group $Mp(n)$ is the double cover of the symplectic group $Sp(n)$, just as the spinors rep of the *Spin* group is the double cover of the rotation group $SO(n)$.

Chapter Fifteen

$SU(n)$ family of invariance groups

$SU(n)$ preserves the Levi-Civita tensor, in addition to the Kronecker δ of section 9.10. This additional invariant induces nontrivial decompositions of $U(n)$ reps. In this chapter, we show how the theory of $SU(2)$ reps (the quantum mechanics textbooks' theory of angular momentum) is developed by birdtracking; that $SU(3)$ is the unique group with the Kronecker delta and a rank-3 antisymmetric primitive invariant; that $SU(4)$ is isomorphic to $SO(6)$; and that for $n \geq 4$, only $SU(n)$ has the Kronecker δ and rank-n antisymmetric tensor primitive invariants.

15.1 REPS OF $SU(2)$

For $SU(2)$, we can construct an additional invariant matrix that would appear to induce a decomposition of $V \otimes V$ reps:

$$E_{b}^{a}{}_{,d}^{c} = \frac{1}{2}\varepsilon^{ac}\varepsilon_{bd} =$$ 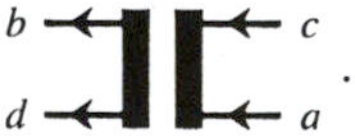 (15.1)

However, by (6.28) this can be written as a sum over Kronecker deltas and is not an independent invariant. So what does ε^{ac} do? It does two things; it removes the distinction between a particle and an antiparticle (if q_a transforms as a particle, then $\varepsilon^{ab}q_b$ transforms as an antiparticle), and it reduces the reps of $SU(2)$ to the fully symmetric ones. Consider $V \otimes V$ decomposition (7.4)

$$\boxed{1} \otimes \boxed{2} = \boxed{1|2} + \bullet$$

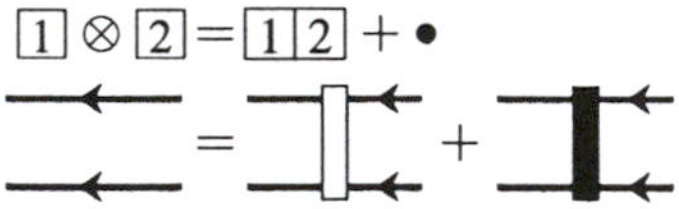 (15.2)

$$2^2 = \frac{2 \cdot 3}{2} + \frac{2 \cdot 1}{2}\,.$$

The antisymmetric rep is a singlet,

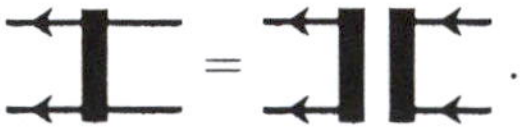 (15.3)

Now consider the $\otimes V^3$ and $\otimes V^4$ space decompositions, obtained by adding successive indices one at a time:

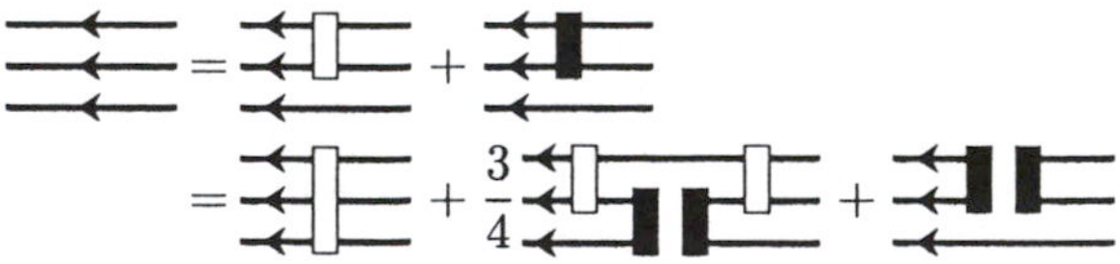

$\boxed{1} \times \boxed{2} \times \boxed{3} = \boxed{1|2|3} + \boxed{1} + \boxed{3}$

$$= \ldots + \frac{4}{3} \ldots + \ldots + \frac{3}{2} \ldots + \frac{4}{3} \ldots + \ldots$$

$$\boxed{1} \times \boxed{2} \times \boxed{3} \times \boxed{4} = \boxed{1|2|3|4} + \boxed{1|4} + \boxed{3|4} + \boxed{1|2} + \bullet + \bullet . \qquad (15.4)$$

This is clearly leading us into the theory of $SO(3)$ angular momentum addition (or $SU(2)$ spin, *i.e.*, both integer and half-integer irreps of the rotation group), described in any quantum mechanics textbook. We shall, anyway, persist a little while longer, just to illustrate how birdtracks can be used to recover some familiar results.

The projection operator for m-index rep is

$$\mathbf{P}_m = \ldots . \qquad (15.5)$$

The dimension is $\mathrm{tr}\,\mathbf{P}_m = 2(2+1)(2+2)\ldots(2+m-1)/m! = m+1$. In quantum mechanics textbooks m is set to $m = 2j$, where j is the spin of the rep. The projection operator (7.10) for the adjoint rep (spin 1) is

$$\ldots = \ldots - \frac{1}{2} \ldots . \qquad (15.6)$$

This can be rewritten as ... using (15.3). The quadratic casimir for the defining rep is

$$\ldots = \frac{3}{2} \ldots . \qquad (15.7)$$

Using

$$\ldots = \ldots - \frac{1}{2} \ldots = \frac{1}{2} \ldots , \qquad (15.8)$$

we can compute the quadratic casimir for any rep

$$\ldots = n \ldots$$

$$C_2(n) = \ldots = n^2 \ldots$$

$$= n \left\{ \ldots + (n-1) \ldots \right\}$$

$$= n\left(\frac{3}{2} + \frac{n-1}{2}\right) \ldots = \frac{n(n+2)}{2} . \qquad (15.9)$$

The Dynkin index for n-index rep is given by

$$\ell(n) = \frac{C_2(n)d_n}{C_2(2)d_2} = \frac{n(n+1)(n+2)}{24} . \tag{15.10}$$

We can also construct clebsches for various Kronecker products. For example, $\lambda_p \otimes \lambda_1$ is given by

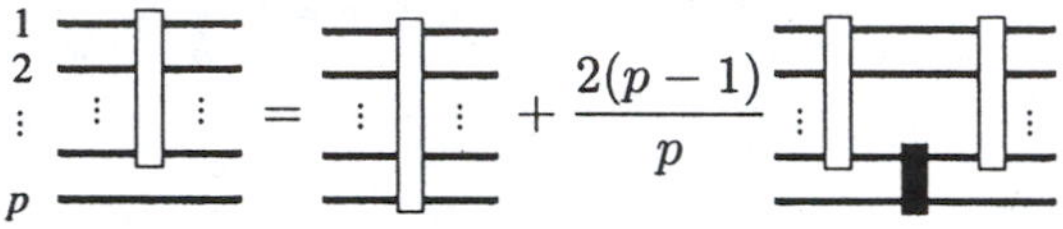

$$1,\ 2,\ \vdots,\ p \qquad = \qquad + \frac{2(p-1)}{p} \qquad \tag{15.11}$$

for any $U(n)$. For $SU(2)$ we have (15.3), so

$$\boxed{1}\boxed{2}\boxed{\cdots}\boxed{p\text{-}1} \times \boxed{p} = \boxed{1}\boxed{2}\boxed{\cdots}\boxed{p} + \boxed{1}\boxed{2}\boxed{\cdots}\boxed{p\text{-}2}$$

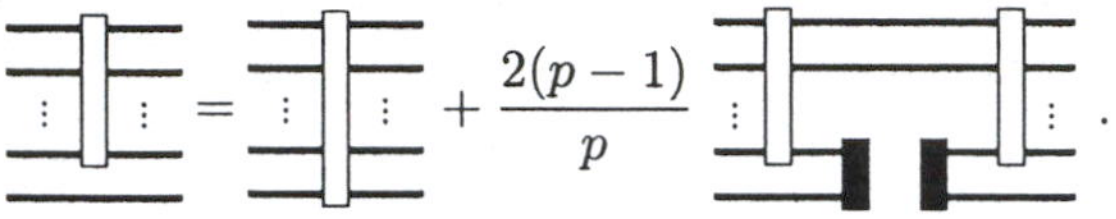

$$ = + \frac{2(p-1)}{p} \qquad . \tag{15.12}$$

Hence, the Clebsch-Gordan for $\lambda_p \otimes \lambda_1 \to \lambda_{p-1}$ is

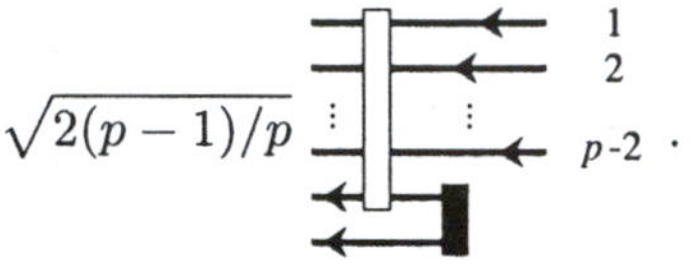

$$\sqrt{2(p-1)/p} \qquad 1,\ 2,\ p\text{-}2 \ . \tag{15.13}$$

As we have already given the complete theory of $SO(3)$ angular momentum in chapter 14, by giving explicit expressions for all Wigner 6-j coefficients (Racah coefficients), we will not pursue this further here.

Group weights have an amusing graph-theoretic interpretation for $SO(3)$. For a planar vacuum (no external legs) diagram weight W_G with normalization $\alpha = 2$, W_G is the number of ways of coloring the lines of the graph with three colors [280]. This, in turn, is related to the chromatic polynomials, Heawood's conjecture, and the 4-color problem [293, 267].

15.2 $SU(3)$ AS INVARIANCE GROUP OF A CUBIC INVARIANT

QCD hadrons are built from quarks and antiquarks, and with hadron spectrum consisting of the following

1. Mesons, each built from a quark and an antiquark.
2. Baryons, each built from three quarks or antiquarks in a fully antisymmetric color combination.
3. No exotic states, *i.e.*, no hadrons built from other combinations of quarks and antiquarks.

We shall show here that for such hadronic spectrum the color group can be only $SU(3)$.

In the group-theoretic language, the above three conditions are a list of the primitive invariants (color singlets) that define the color group:

1. One primitive invariant is δ^a_b, so the color group is a subgroup of $SU(n)$.
2. There is a cubic antisymmetric invariant f^{abc} and its dual f_{abc}.
3. There are no further primitive invariants. This means that any invariant tensor can be written in terms of the tree contractions of δ^b_a, f^{abc} and f_{bca}.

In the birdtrack notation,

$f^{abc} =$  $, f_{abc} =$ 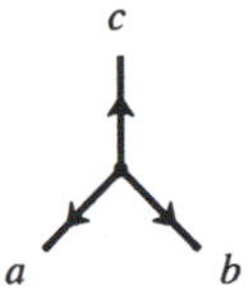 . (15.14)

f_{abc} and f^{abc} are fully antisymmetric:

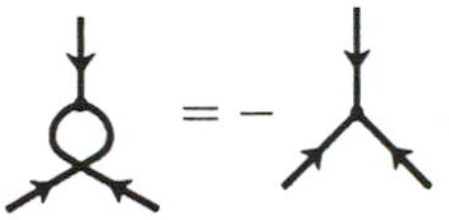 . (15.15)

We can already see that the defining rep dimension is at least three, $n \geq 3$, as otherwise f_{abc} would be identically zero. Furthermore, f's must satisfy a normalization condition,

$$f^{abc} f_{bdc} = \alpha \delta^a_d$$

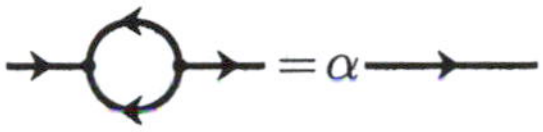 . (15.16)

(For convenience we set $\alpha = 1$ in what follows.) If this were not true, eigenvalues of the invariant matrix $F^a_d = f^{abc} f_{bdc}$ could be used to split the n-dimensional rep in a direct sum of lower-dimensional reps; but then n-dimensional rep would not be the defining rep.

$V \otimes V$ **states:** According to (7.4), they split into symmetric and antisymmetric subspaces. The antisymmetric space is reduced to $n + n(n-3)/2$ by the f^{abc} invariant:

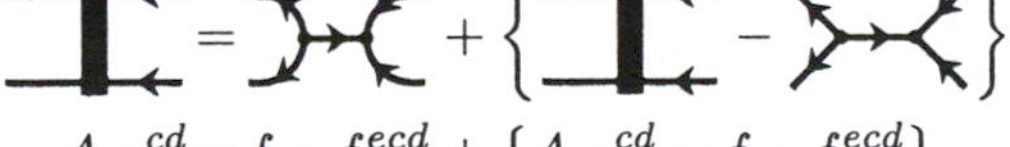

$$A_{ab}{}^{cd} = f_{abe} f^{ecd} + \left\{ A_{ab}{}^{cd} - f_{abe} f^{ecd} \right\} . \qquad (15.17)$$

On the symmetric subspace the $f_{abe} f^{ecd}$ invariant vanishes due to its antisymmetry, so this space is not split. The simplest invariant matrix on the symmetric subspace involves four f's:

$K_{ab,}{}^{cd} =$ 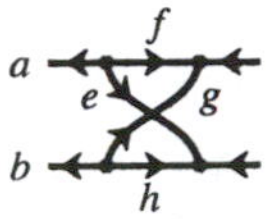 $= f_{aef} f_{bhg} f^{ceh} f^{dfg}$. (15.18)

As the symmetric subspace is not split, this invariant must have a single eigenvalue

$K_{ab,}{}^{cd} = \beta S_{ab,}{}^{cd} = \beta$ 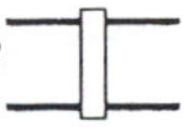 . (15.19)

Tracing $K_{ab,}{}^{ad}$ fixes $\beta = \frac{2}{n+1}$. The assumption, that k is not an independent invariant, means that we do not allow the existence of exotic $qq\overline{qq}$ hadrons. The requirement, that all invariants be expressible as trees of contractions of the primitives

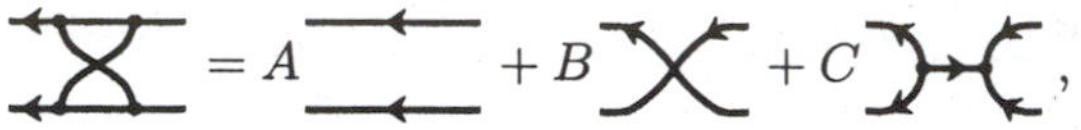

(15.20)

leads to the relation (15.19). The left-hand side is symmetric under index interchange $a \leftrightarrow b$, so $C = 0$ and $A = B$.

$V \otimes \overline{V}$ **states:** The simplest invariant matrix that we can construct from f's is

$$Q_{b}{}^{a}{}_{,c}{}^{d} = \frac{1}{\alpha}$$

$$= f^{aed} f_{bce} \, . \qquad (15.21)$$

By crossing (15.19), $\mathbf{Q}$ satisfies a characteristic equation,

$$\mathbf{Q}^2 = \frac{1}{n+1}\{\mathbf{1} + T\}$$

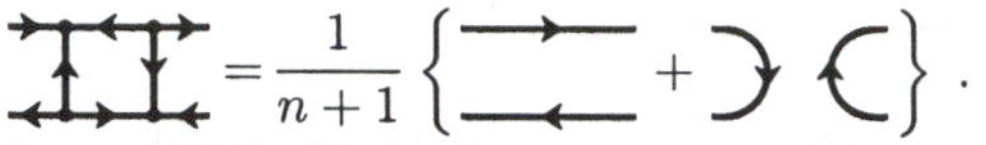

(15.22)

On the traceless subspace (7.8), this leads to

$$\left(\mathbf{Q}^2 - \frac{1}{n+1}\mathbf{1}\right)\mathbf{P}_2 = 0 \, , \qquad (15.23)$$

with eigenvalues $\pm 1/\sqrt{n+1}$. $V \otimes \overline{V}$ contains the adjoint rep, so at least one of the eigenvalues must correspond to the adjoint projection operator. We can compute the adjoint rep eigenvalue from the invariance condition (4.36) for $f^{\,bcd}$:

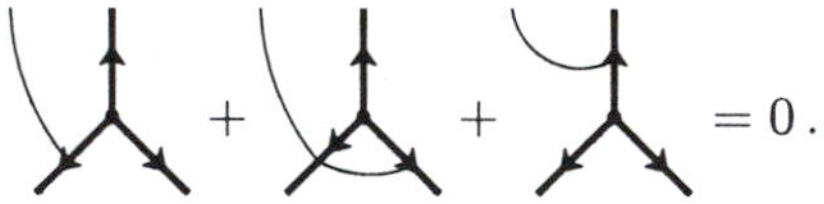

(15.24)

Contracting with f^{bcd}, we find

$$\mathbf{P}_A\mathbf{Q} = -\frac{1}{2}\mathbf{P}_A \, . \qquad (15.25)$$

Matching the eigenvalues, we obtain $1/\sqrt{n+1} = 1/2$, so $n = 3$: quarks can come in three colors only, and f_{abc} is proportional to the Levi-Civita tensor ε_{abc} of $SU(3)$. The invariant matrix $\mathbf{Q}$ is not an independent invariant; the $n(n-3)/2$-dimensional antisymmetric space (15.17) has dimension zero, so $\mathbf{Q}$ can be expressed in terms of Kronecker deltas:

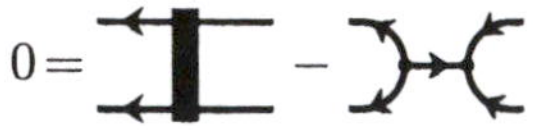

$$0 = A_{ab}{}^{cd} - Q_{a}{}^{c}{}_{,b}{}^{d} \, . \qquad (15.26)$$

We have proven that the only group that satisfies the conditions 1–3, at the beginning of this section, is $SU(3)$. Of course, it is well known that the color group of physical hadrons is $SU(3)$, and this result might appear rather trivial. That it is not so will become clear from the further examples of invariance groups, such as the G_2 family of the next chapter.

15.3 LEVI-CIVITA TENSORS AND $SU(n)$

In chapter 12, we have shown that the invariance group for a symplectic invariant f^{ab} is $Sp(n)$. In particular, for $f^{ab} = \varepsilon^{ab}$, the Levi-Civita tensor, the invariance group is $SU(2) = Sp(2)$. In the preceding section, we have proven that the invariance group of a skew-symmetric invariant f^{abc} is $SU(3)$, and that f^{abc} must be proportional to the Levi-Civita tensor. Now we shall show that for $f^{abc\ldots d}$ with r indices, the invariance group is $SU(r)$, and f is always proportional to the Levi-Civita tensor. (We consider here unitary transformations only; in general, the whole group $SL(3)$ preserves the Levi-Civita tensor.) $r = 2$ and $r = 3$ cases had to be treated separately, because it was possible to construct from f^{ab} and f^{abc} tree invariants on the $\overline{V} \otimes V \rightarrow \overline{V} \otimes V$ space, which could reduce the group $SU(n)$ to a subgroup. For f^{ab}, $n \geq 4$ this is, indeed, what happens: $SU(n) \rightarrow Sp(n)$, for n even.

For $r \geq 4$, we assume here that the primitive invariants are δ_a^b and the fully skew-symmetric invariant tensors

$f^{a_1 a_2 \ldots a_r} =$  , $f_{a_1 a_2 \ldots a_r} =$  , $r > 3$. (15.27)

A fully antisymmetric object can be realized only in $n \geq r$ dimensions. By the primitiveness assumption

 $= \alpha$ 

 $= \dfrac{2\alpha}{n-1}$ 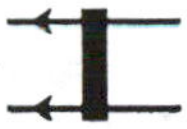 , *etc.*, (15.28)

i.e., various contractions of f's must be expressible in terms of δ's, otherwise there would exist additional primitives. (f invariants themselves have too many indices and cannot appear on the right-hand side of the above equations.)

The projection operator for the adjoint rep can be built only from $\delta_b^a \delta_d^c$ and $\delta_d^a \delta_b^c$. From section 9.10, we know that this can give us only the $SU(n)$ projection operator (7.8), but just for fun we feign ignorance and write

$\frac{1}{a}$  $= A \{$  $+ b$  $\}$. (15.29)

The invariance condition (6.56) on $f_{ab\ldots c}$ yields

$0 =$ 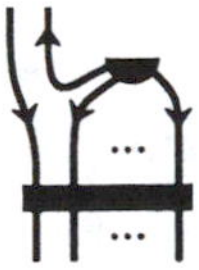 $+ b$  .

Contracting from the top, we get $0 = 1 + bn$. Antisymmetrizing all outgoing legs, we get

$0 =$ 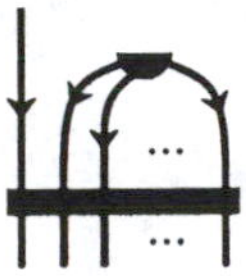 . (15.30)

Contracting with δ_b^a from the side, we get $0 = n - r$. As in (6.30), this defines the Levi-Civita tensor in n dimensions and can be rewritten as

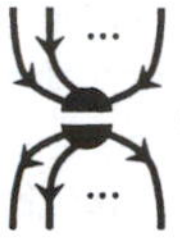

$$= n\alpha \quad . \tag{15.31}$$

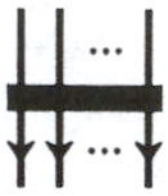

(The conventional Levi-Civita normalization is $n\alpha = n!$.) The solution $b = -1/n$ means that T_i is traceless, *i.e.*, the same as for the $SU(n)$ case considered in section 9.10. To summarize: The invariance condition forces $f_{abc...c}$ to be proportional to the Levi-Civita tensor (in n dimensions, a Levi-Civita tensor is the only fully antisymmetric tensor of rank n), and the primitives δ_b^a, $f_{ab...d}$ (rank n) have $SU(n)$ as their unique invariance algebra.

15.4 $SU(4)$–$SO(6)$ ISOMORPHISM

We have shown that if the primitive invariants are $\delta_b^{\prime a}$, $f_{ab...cd'}$, the corresponding Lie group is the defining rep of $SU(n)$, and $f_{ab...cd}$ is proportional to the Levi-Civita tensor. However, there are still interesting things to be said about particular $SU(n)$'s. As an example, we will establish the $SU(4) \simeq SO(6)$ isomorphism.

The antisymmetric $SU(4)$ rep is of dimension $d_A = 4 \cdot 3/2 = 6$. Let us introduce clebsches

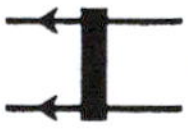

$$A_{ab,}{}^{cd} = \frac{1}{4}(\gamma^\mu)_{ab}(\gamma_\mu)^{cd}, \qquad \mu = 1, 2, \ldots, 6 \, . \tag{15.32}$$

1/4 normalization ensures that γ's will have the Dirac matrix normalization.

The Levi-Civita tensor induces a quadratic symmetric invariant on the 6-dimensional space

$$g_{\mu\nu} = \quad = $$  
$$= \frac{1}{4}(\gamma_\mu)^{ab}\epsilon_{bacd}(\gamma_\nu)^{dc} \, . \tag{15.33}$$

This invariant has an inverse:

$$g^{\mu\nu} = \quad = 6 \quad , \tag{15.34}$$

where the factor 6 is the normalization factor, fixed by the condition $g_{\mu\nu}g^{\nu\sigma} = \delta_\mu^\sigma$:

$$g_{\mu\nu}g^{\nu\sigma} =$$ 
$$= 6$$ 
$$= 6$$

$$= 6\,\frac{(n-3)}{4}\,\frac{(n-2)}{3}$$

$$= \;=\delta^{\sigma}_{\mu}\,. \qquad (15.35)$$

Here we have used (6.28), (15.32), and the orthonormality for clebsches:

$$(\gamma_\mu)^{ab}(\gamma^\mu)_{ba} = 4\delta^\nu_\mu\,. \qquad (15.36)$$

As we have shown in chapter 10, the invariance group for a symmetric invariant $g_{\mu\nu}$ is $SO(d_A)$. One can check that the generators for the 6-dimensional rep of $SU(4)$, indeed, coincide with the defining rep generators of $SO(6)$, and that the dimension of the Lie algebra is in both cases 15.

The invariance condition (6.56) for the Levi-Civita tensor is

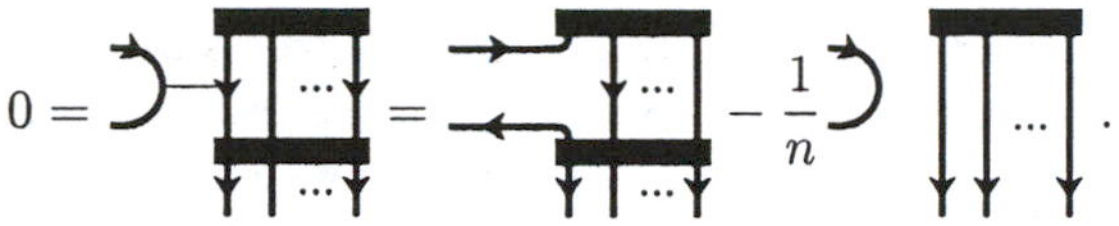

$$0 = \quad = \quad - \frac{1}{n} \quad . \qquad (15.37)$$

For $SU(4)$ we have

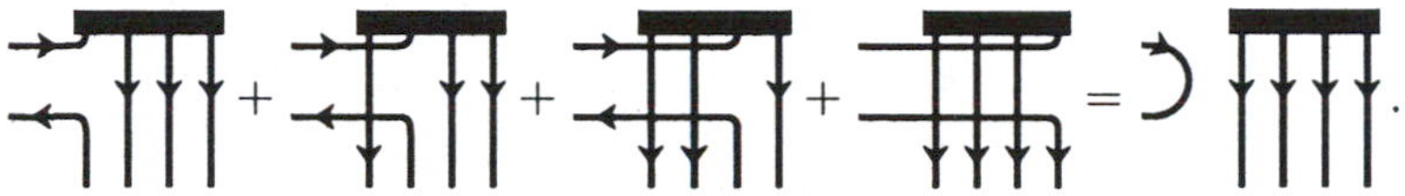

$$+ \quad + \quad + \quad = \quad . \qquad (15.38)$$

Contracting with $(\gamma_\mu)^{ab}(\gamma_\nu)^{cd}$, we obtain

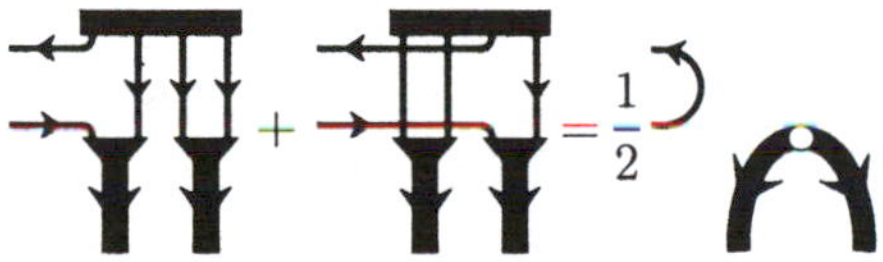

$$+ \quad = \frac{1}{2}$$

$$(\gamma_\mu)^{be}(\gamma_\nu)_{ab} + (\gamma_\mu)_{ad}(\gamma_\nu)^{de} = 2\delta^e_a g_{\mu\nu}\,. \qquad (15.39)$$

Here $(\gamma_\nu)_{ab} \equiv (\gamma_\nu)^{cd}\varepsilon_{dcab}$, and we recognize the Dirac equation (11.4). So the clebsches (15.32) are, indeed, the γ-matrices for $SO(6)$ (semi)spinor reps (11.65).

Chapter Sixteen

G_2 family of invariance groups

In this chapter, we begin the construction of all invariance groups that possess a symmetric quadratic and an antisymmetric cubic invariant in the defining rep. The resulting classification is summarized in figure 16.1. We find that the cubic invariant must satisfy either the Jacobi relation (16.7) or the alternativity relation (16.11). In the former case, the invariance group can be any semisimple Lie group in its adjoint rep; we pursue this possibility in the next chapter. The latter case is developed in this chapter; we find that the invariance group is either $SO(3)$ or the exceptional Lie group G_2. The problem of evaluation of $3n$-j coefficients for G_2 is solved completely by the reduction identity (16.14). As a by-product of the construction, we give a proof of Hurwitz's theorem (section 16.5) and demonstrate that the independent casimirs for G_2 are of order 2 and 6, by explicitly reducing the order 4 casimir in section 16.4. Here we are concerned only with the derivation of G_2. For a systematic discussion of G_2 invariants (in tensorial notation) we refer the reader to Macfarlane [221].

Consider the following list of primitive invariants:

1. δ^a_b, so the invariance group is a subgroup of $SU(n)$.

2. Symmetric $g^{ab} = g^{ba}$, $g_{ab} = g_{ba}$, so the invariance group is a subgroup of $SO(n)$. As in chapter 10, we take this invariant in its diagonal, Kronecker delta form δ_{ab}.

3. A cubic antisymmetric invariant f_{abc}.

Primitiveness assumption requires that all other invariants can be expressed in terms of the tree contractions of δ_{ab} , f_{abc}.

In the diagrammatic notation, one keeps track of the antisymmetry of the cubic invariant by reading the indices off the vertex in a fixed order:

$f_{abc} =$  $= -$  $= -f_{acb}\,.$ (16.1)

The primitiveness assumption implies that the double contraction of a pair of f's is proportional to the Kronecker delta. We can use this relation to fix the overall normalization of f's:

$$f_{abc}f_{cbd} = \alpha\, \delta_{ad}$$

 $= \alpha$ ——— . (16.2)

For convenience, we shall often set $\alpha = 1$ in what follows.

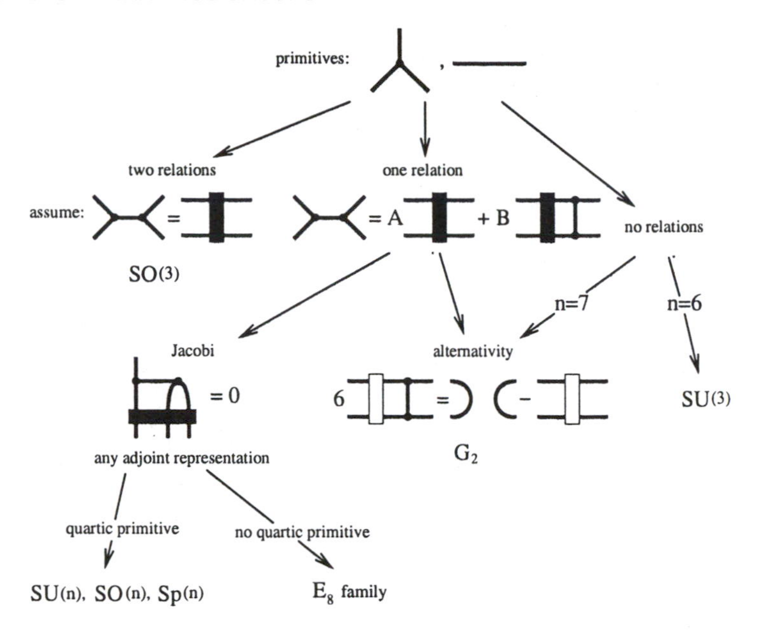

Figure 16.1 Logical organization of chapters 16–17. The invariance groups $SO(3)$ and G_2 are derived in this chapter, while the E_8 family is derived in chapter 17.

The next step in our construction is to identify all invariant matrices on $V \otimes V$ and construct the Clebsch-Gordan series for decomposition of 2-index tensors. There are six such invariants: the three distinct permutations of indices of $\delta_{ab}\delta_{cd}$, and the three distinct permutations of free indices of $f_{abe}f_{ecd}$. For reasons of clarity, we shall break up the discussion in two steps. In the first step, section 16.1, we assume that a linear relation between these six invariants exists. Pure symmetry considerations, together with the invariance condition, completely fix the algebra of invariants and restrict the dimension of the defining space to either 3 or 7. In the second step, section 16.3, we show that a relation *assumed* in the first step must exist because of the invariance condition.

Example. Consider "quarks" and "hadrons" of a Quantum Chromodynamics with the hadronic spectrum consisting of the following singlets:

1. Quark-antiquark mesons.

2. Mesons built of two quarks (or antiquarks) in a symmetric color combination.

3. Baryons built of three quarks (or antiquarks) in a fully antisymmetric color combination.

4. No exotics, *i.e.*, no hadrons built from other combinations of quarks and antiquarks.

As we shall now demonstrate, for this hadronic spectrum the color group is either $SO(3)$, with quarks of three colors, or the exceptional Lie group G_2, with quarks of seven colors.

16.1 JACOBI RELATION

If the above six invariant tensors are not independent, they satisfy a relation of form

$$0 = A \; + B \; + C \; + D \; + E \; + F \; . \quad (16.3)$$

Antisymmetrizing a pair of indices yields

$$0 = A' \; + E \; + F' \; , \quad (16.4)$$

and antisymmetrizing any three indices yields

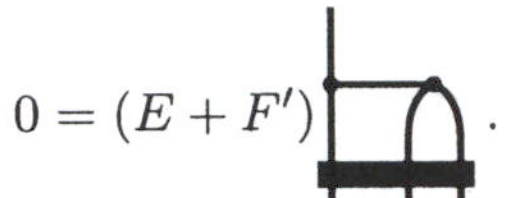

$$0 = (E + F') \; . \quad (16.5)$$

If the tensor itself vanishes, f's satisfy the *Jacobi relation* (4.49):

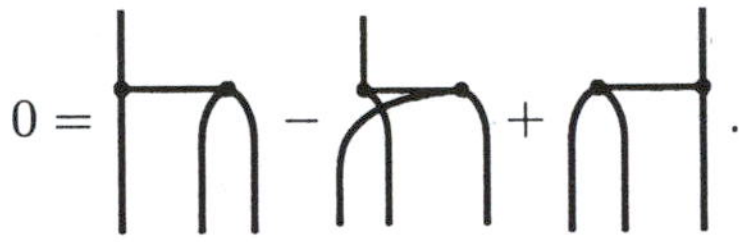

$$0 = \; - \; + \; . \quad (16.6)$$

If $A' \neq 0$ in (16.4), the Jacobi relation relates the second and the third term:

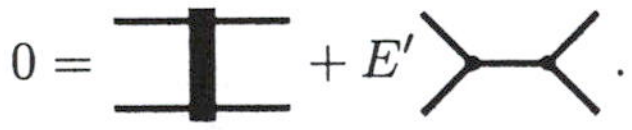

$$0 = \; + E' \; . \quad (16.7)$$

The normalization condition (16.2) fixes $E' = -1$:

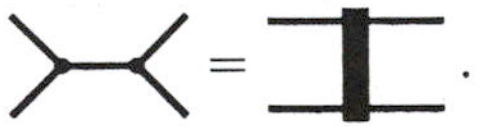

$$= \; . \quad (16.8)$$

Contracting the free ends of the top line with δ_{ab}, we obtain $1 = (n-1)/2$, so $n = 3$. We conclude that if pair contraction of f's is expressible in terms of δ's, the invariance group is $SO(3)$, and f_{abc} is proportional to the 3-index Levi-Civita tensor. To spell it out; in three dimensions, an antisymmetric rank-3 tensor can take only one value, $f_{abc} = \pm f_{123}$, that can be set equal to ± 1 by the normalization convention (16.2).

If $A' = 0$ in (16.4), the Jacobi relation is the only relation we have, and the adjoint rep of any simple Lie group is a possible solution. We return to this case in chapter 17.

16.2 ALTERNATIVITY AND REDUCTION OF f-CONTRACTIONS

If the Jacobi relation does not hold, we must have $E = -F'$ in (16.5), and (16.4) takes the form

$$+ \; = A'' \; . \quad (16.9)$$

Contracting with δ_{ab} fixes $A'' = 3/(n-1)$. Symmetrizing the top two lines and rotating the diagrams by 90^0, we obtain the *alternativity relation*:

$$= \frac{1}{n-1}\left\{ \quad - \quad \right\} . \qquad (16.10)$$

The name comes from the octonion interpretation of this formula (see section 16.4). Adding the two equations, we obtain

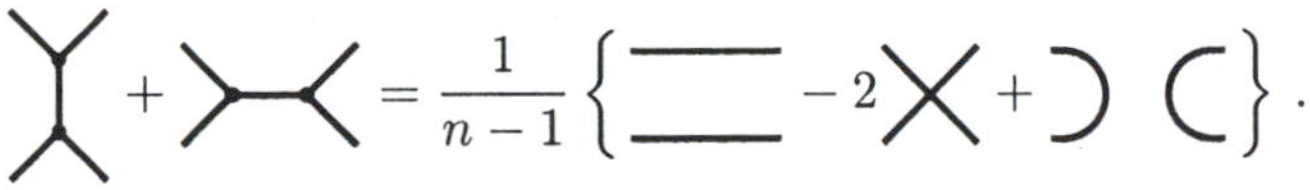

$$+ \quad = \frac{1}{n-1}\left\{ \quad - 2 \quad + \quad \right\} . \qquad (16.11)$$

By (16.9), the invariant  is reducible on the antisymmetric subspace. By (16.10), it is also reducible on the symmetric subspace. The only independent $f \cdot f$ invariant is  , which, by the normalization (16.2), is already the projection operator that projects the antisymmetric 2-index tensors onto the n-dimensional defining space. The Clebsch-Gordan decomposition of $V \otimes V$ follows:

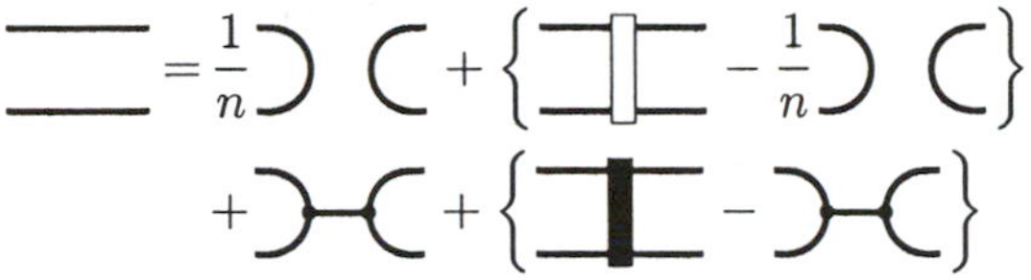

$$= \frac{1}{n} \quad + \left\{ \quad - \frac{1}{n} \quad \right\}$$

$$+ \quad + \left\{ \quad - \quad \right\}$$

$$n^2 = 1 + \frac{(n-1)(n+2)}{2} + n + \frac{n(n-3)}{2} . \qquad (16.12)$$

The dimensions of the reps are obtained by tracing the corresponding projection operators.

The adjoint rep  of $SO(n)$ is now split into two reps. Which one is the new adjoint rep? We determine this by considering (6.56), the invariance condition for f_{abc}. If we take  to be the projection operator for the adjoint rep, we again get the Jacobi condition (16.6), with $SO(3)$ as the only solution. However, if we demand that the last term in (16.12) is the adjoint projection operator

$$\frac{1}{a} \quad = \quad - \quad , \qquad (16.13)$$

the invariance condition takes the form

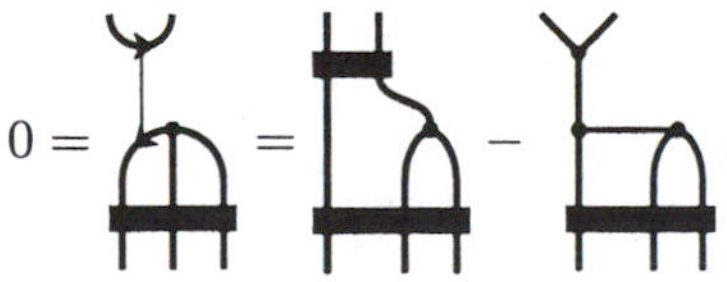

$$0 = \quad = \quad - \quad . \qquad (16.14)$$

The last term can be simplified by (6.19) and (16.9):

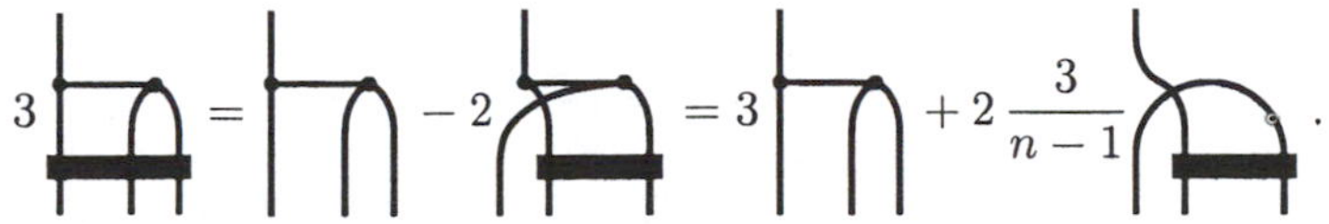

$$3 \quad = \quad - 2 \quad = 3 \quad + 2 \frac{3}{n-1} \quad .$$

Substituting back into (16.14) yields

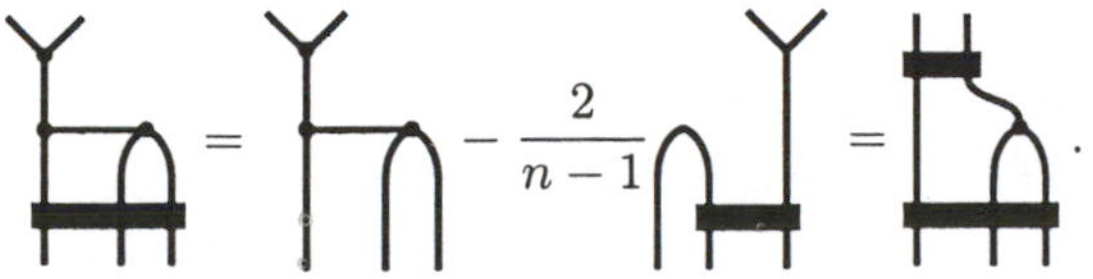

$$= \quad - \frac{2}{n-1} \quad = \quad .$$

Expanding the last term and redrawing the equation slightly, we have

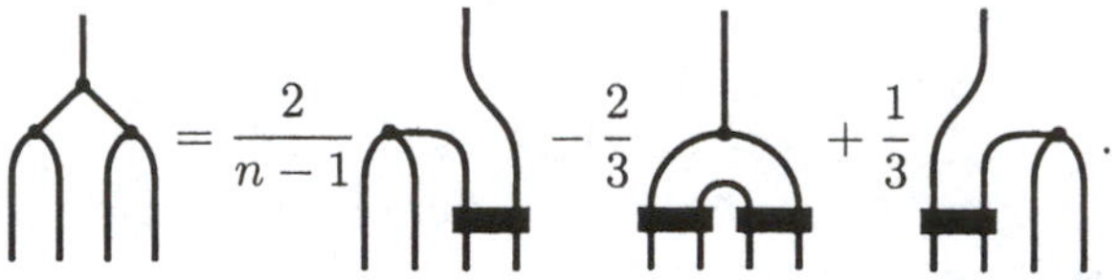

This equation is antisymmetric under interchange of the left and the right index pairs. Hence, $2/(n-1) = 1/3$, and the invariance condition is satisfied *only* for $n = 7$. Furthermore, the above relation gives us the G_2 *reduction identity*

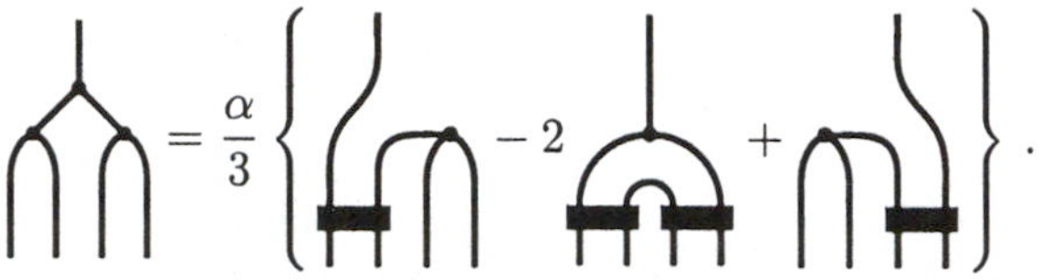

(16.15)

This identity is the key result of this chapter: it enables us to recursively reduce *all* contractions of products of δ-functions and pairwise contractions $f_{abc}f_{cde}$, and thus completely solves the problem of evaluating any casimir or $3n$-j coefficient of G_2.

The invariance condition (16.14) for f_{abc} implies that

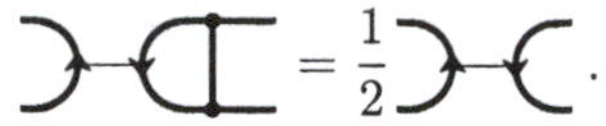

(16.16)

The "triangle graph" for the defining rep can be computed in two ways, either by contracting (16.10) with f_{abc}, or by contracting the invariance condition (16.14) with δ_{ab}:

(16.17)

So, the alternativity and the invariance conditions are consistent if $(n-3)(n-7) = 0$, *i.e.*, *only* for three or seven dimensions. In the latter case, the invariance group is the exceptional Lie group G_2, and the above derivation is also a proof of Hurwitz's theorem (see section 16.4).

In this way, symmetry considerations together with the invariance conditions suffice to determine the algebra satisfied by the cubic invariant. The invariance condition fixes the defining dimension to $n = 3$ or 7. Having assumed only that *a* cubic antisymmetric invariant exists, we find that if the cubic invariant is not a structure constant, it can be realized only in seven dimensions, and its algebra is *completely* determined. The identity (16.15) plays the role analogous to one the Dirac relation $\{\gamma_\mu, \gamma_\nu\} = 2g_{\mu\nu}\mathbf{1}$ plays for evaluation of traces of products of Dirac gamma-matrices, described above in chapter 11. Just as the Dirac relation obviates the need for explicit reps of γ's, (16.14) reduces any $f \cdot f \cdot f$ contraction to a sum of terms linear in f and obviates any need for explicit construction of f's.

The above results enable us to compute any group-theoretic weight for G_2 in two steps. First, we replace all adjoint rep lines by the projection operators $\mathbf{P}_A$ (16.13). The resulting expression contains Kronecker deltas and chains of contractions of f_{abc}, which can then be reduced by systematic application of the reduction identity (16.15).

The above 1975 diagrammatic derivation of the Hurwitz theorem was one of the first nontrivial applications of the birdtrack technology [73, 74, 82]. More recently, the same diagrammatic proof of Hurwitz's theorem has been given by Dominic Boos [27], based on the algebraic proof by Markus Rost [299].

16.3 PRIMITIVITY IMPLIES ALTERNATIVITY

The step that still remains to be proven is the assertion that the alternativity relation (16.10) follows from the primitiveness assumption. We complete the proof in this section. The proof is rather inelegant and should be streamlined (an exercise for the reader).

If no relation (16.3) between the three $f \cdot f$ contractions is assumed, then by the primitiveness assumption the adjoint rep projection operator $\mathbf{P}_A$ is of the form

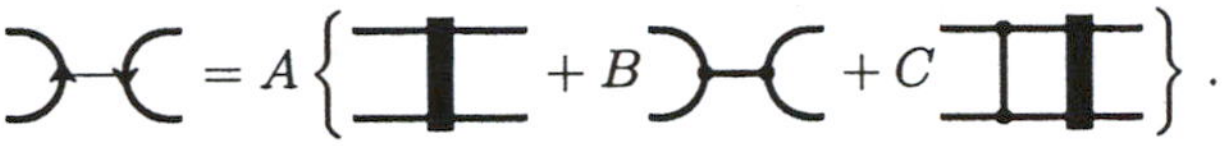

$$= A\left\{ \quad + B \quad + C \quad \right\} . \qquad (16.18)$$

Assume that the Jacobi relation does not hold; otherwise, this immediately reduces to $SO(3)$. The generators must be antisymmetric, as the group is a subgroup of $SO(n)$. Substitute the adjoint projection operator into the invariance condition (6.56) (or (16.14)) for f_{abc}:

$$0 = \quad + B \quad + C \quad . \qquad (16.19)$$

Resymmetrize this equation by contracting with  . This is evaluated substituting (6.19) and using the relation (6.61):

$$= 0 . \qquad (16.20)$$

The result is

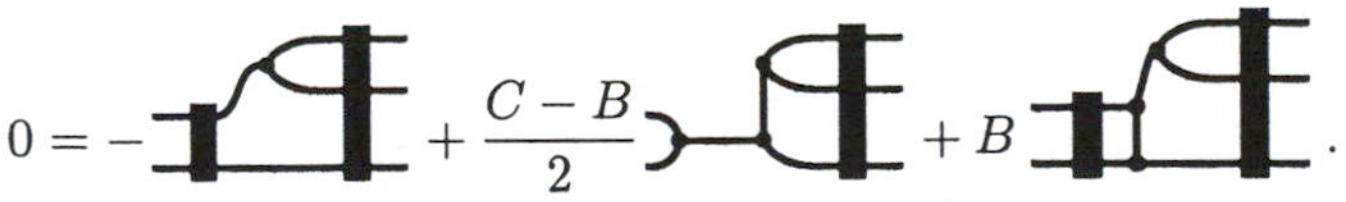

$$0 = - \quad + \frac{C-B}{2} \quad + B \quad . \qquad (16.21)$$

Multiplying (16.19) by B, (16.21) by C, and subtracting, we obtain

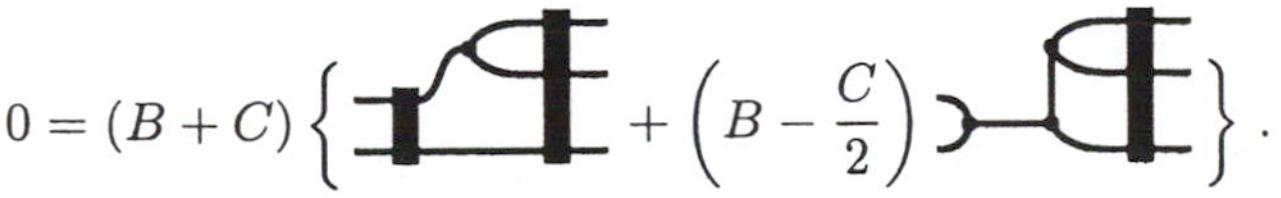

$$0 = (B + C)\left\{ \quad + \left(B - \frac{C}{2}\right) \quad \right\} . \qquad (16.22)$$

We treat the case $B + C = 0$ below, in (16.26).

If $B + C \neq 0$, by contracting with f_{abc} we get $B - C/2 = -1$, and

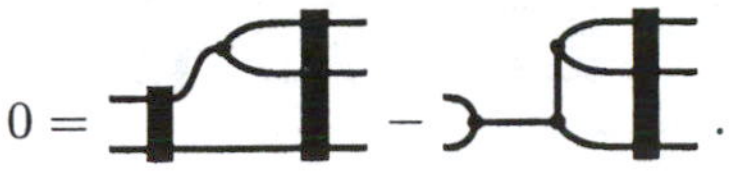

$$0 = \quad - \quad . \qquad (16.23)$$

To prove that this is equivalent to the alternativity relation, we contract with  , expand the 3-leg antisymmetrization, and obtain

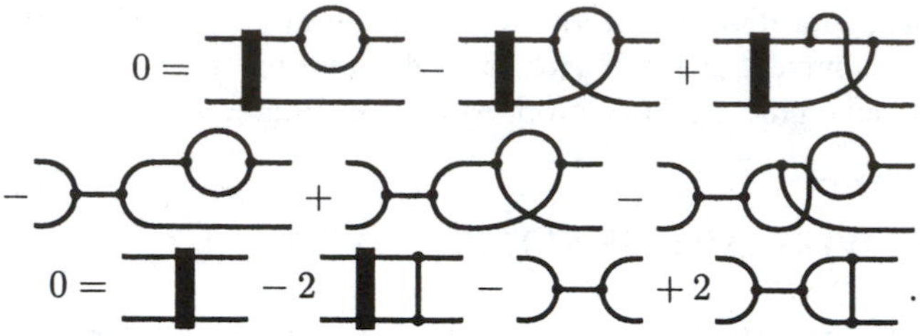

$$0 = \quad - \quad + $$
$$- \quad + \quad - $$
$$0 = \quad - 2 \quad - \quad + 2 \quad . \qquad (16.24)$$

The triangle subdiagram can be computed by adding (16.19) and (16.21)

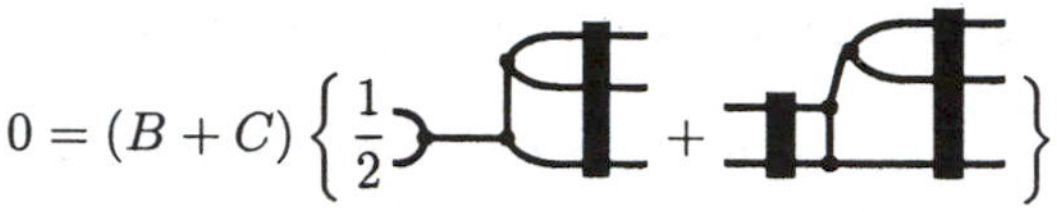

$$0 = (B + C)\left\{\frac{1}{2} \quad + \quad \right\}$$

and contracting with  . The result is

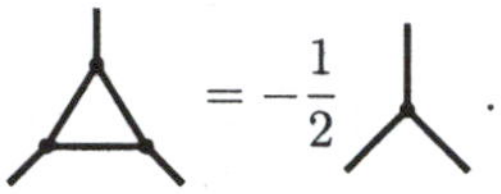

$$= -\frac{1}{2} \quad . \qquad (16.25)$$

Substituting into (16.24), we recover the alternativity relation (16.10). Hence, we have proven that the primitivity assumption implies the alternativity relation for the case $B + C \neq 0$ in (16.22).

If $B + C = 0$, (16.19) takes the form

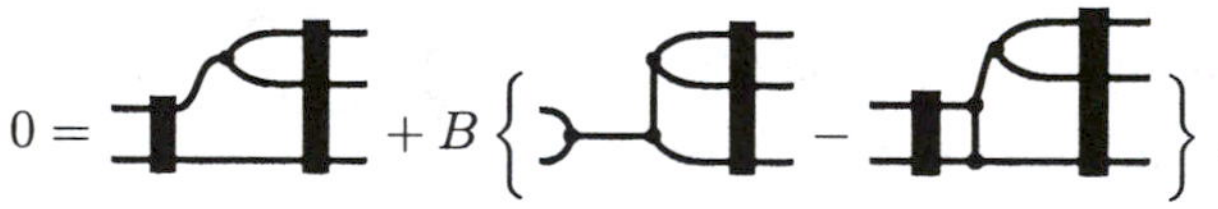

$$0 = \quad + B\left\{ \quad - \quad \right\} . \qquad (16.26)$$

Using the normalization (7.38) and orthonormality  conditions, we obtain

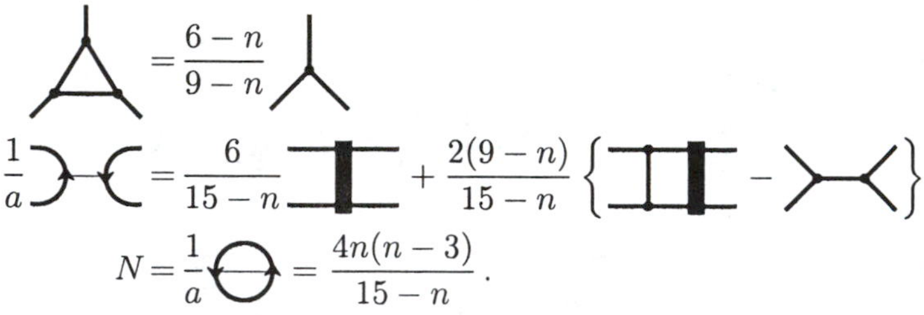

$$= \frac{6-n}{9-n} \qquad (16.27)$$

$$\frac{1}{a} = \frac{6}{15-n} \quad + \frac{2(9-n)}{15-n}\left\{ \quad - \quad \right\} \qquad (16.28)$$

$$N = \frac{1}{a} = \frac{4n(n-3)}{15-n} . \qquad (16.29)$$

The remaining antisymmetric rep

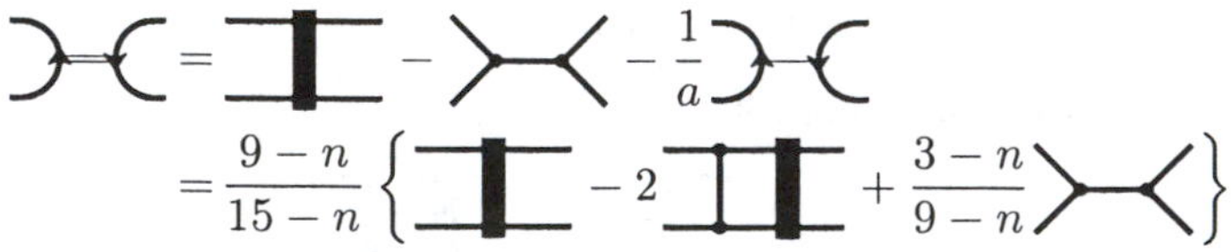

$$= \quad - \quad - \frac{1}{a}$$
$$= \frac{9-n}{15-n}\left\{ \quad - 2 \quad + \frac{3-n}{9-n} \right\} \qquad (16.30)$$

has dimension

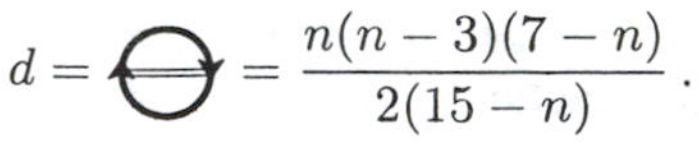

$$d = \quad = \frac{n(n-3)(7-n)}{2(15-n)} . \qquad (16.31)$$

The dimension cannot be negative, so $d \leq 7$. For $n = 7$, the projection operator (16.30) vanishes identically, and we recover the alternativity relation (16.10).

The Diophantine condition (16.31) has two further solutions: $n = 5$ and $n = 6$.

The $n = 5$ is eliminated by examining the decomposition of the traceless symmetric subspace in (16.12), induced by the invariant $\mathbf{Q} =$ . By the primitiveness assumption, $\mathbf{Q}^2$ is reducible on the symmetric subspace

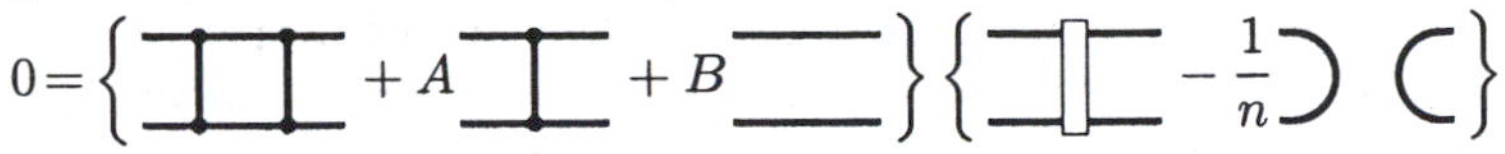

$$0 = (\mathbf{Q}^2 + A\mathbf{Q} + B\mathbf{1})\mathbf{P}_2 \,.$$

Contracting the top two indices with δ_{ab} and $(T_i)_{ab}$, we obtain

$$\left(\mathbf{Q}^2 - \frac{1}{2}\frac{3-n}{9-n}\mathbf{Q} - \frac{5}{2}\frac{6-n}{(2+n)(9-n)}\mathbf{1}\right)\mathbf{P}_2 = 0 \,. \qquad (16.32)$$

For $n = 5$, the roots of this equation are irrational and the dimensions of the two reps, induced by decomposition with respect to $\mathbf{Q}$, are not integers. Hence, $n = 5$ is not a solution. The $n = 6$ case appears to be related to Westbury's *sextonians* [340, 208, 209, 341] a 6-dimensional alternative algebra, intermediate between the complex quaternions and octonions. I leave the proof of that as an exercise to the reader.

16.4 CASIMIRS FOR G_2

In this section, we prove that the independent casimirs for G_2 are of order 2 and 6, as indicated in table 7.1. As G_2 is a subgroup of $SO(7)$, its generators are antisymmetric, and only even-order casimirs are nonvanishing.

The quartic casimir, in the notation of (7.9),

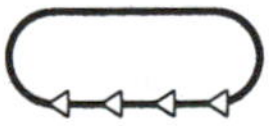

$$= \operatorname{tr} X^4 = \sum_{ijkl} x_i x_j x_k x_l \operatorname{tr}(T_i T_j T_k T_l) \,,$$

can be reduced by manipulating it with the invariance condition (6.56)

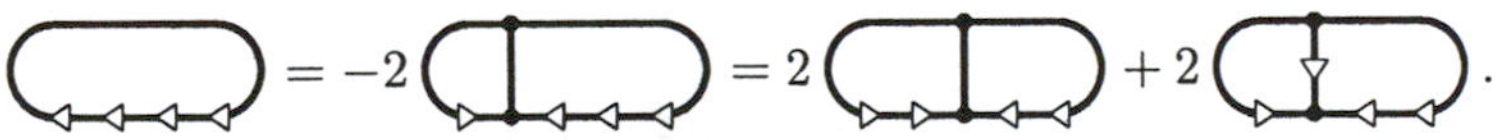

The last term vanishes by further manipulation with the invariance condition

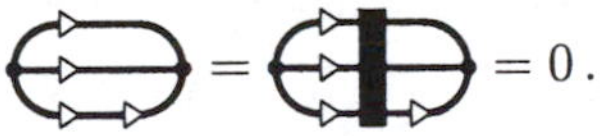

(16.33)

The remaining term is reduced by the alternativity relation (16.10)

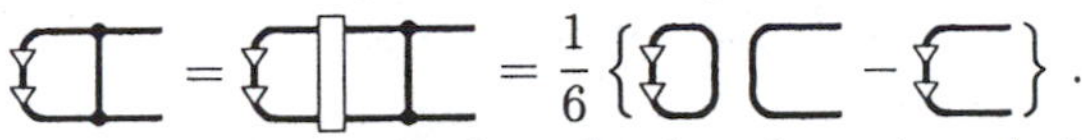

This yields the explicit expression for the reduction of quartic casimirs in the defining rep of G_2:

$$\operatorname{tr} X^4 = \frac{1}{4}\left(\operatorname{tr} X^2\right)^2 \,. \qquad (16.34)$$

As the defining rep is 7-dimensional, the characteristic equation (7.10) reduces the casimirs of order 8 or higher. Hence, the independent casimirs for G_2 are of order 2 and 6.

16.5 HURWITZ'S THEOREM

Throughout this text the field over which the defining vector space V is defined is either $\mathbb{R}$, the field of real numbers, or $\mathbb{C}$, the field of complex numbers. Neither quaternions (a skew field or division ring), nor octonions (a nonassociative algebra) form a field.

Frobenius's theorem states that the only associative real division algebras are the real numbers, the complex numbers, and the quaternions. In order to interpret the results obtained above, we need to define *normed algebras*.

Definition (Curtis [70]). A normed algebra A is an $(n+1)$-dimensional vector space over a field F with a product xy such that

$$\begin{aligned} (i)\ x(cy) &= (cx)y = c(xy)\,, \qquad c \in F \\ (ii)\ x(y+z) &= xy + xz\,, \qquad x, y, z \in A \\ (x+y)z &= xz + yz, \end{aligned}$$

and a nondegenerate quadratic norm that permits composition

$$(iii) \qquad N(xy) = N(x)N(y)\,, \qquad N(x) \in F. \tag{16.35}$$

Here F will be the field of real numbers. Let $\{\mathbf{e}_0, \mathbf{e}_1, \ldots, \mathbf{e}_n\}$ be a basis of A over F:

$$x = x_0\mathbf{e}_0 + x_1\mathbf{e}_1 + \ldots + x_n\mathbf{e}_n\,, \qquad x_a \in F\,, \quad \mathbf{e}_a \in A\,. \tag{16.36}$$

It is always possible to choose $\mathbf{e}_o = \mathbf{I}$ (see Curtis [70]). The product of remaining bases must close the algebra:

$$\mathbf{e}_a\mathbf{e}_b = -d_{ab}\mathbf{I} + f_{abc}\mathbf{e}_c\,, \qquad d_{ab}, f_{abc} \in F \qquad a, \ldots, c = 1, 2, \ldots, n\,. \tag{16.37}$$

The norm in this basis is

$$N(x) = x_0^2 + d_{ab}x_a x_b. \tag{16.38}$$

From the symmetry of the associated inner product (Tits [325]),

$$(x, y) = (y, x) = -\frac{N(x+y) - N(x) - N(y)}{2}, \tag{16.39}$$

it follows that $-d_{ab} = (\mathbf{e}_a, \mathbf{e}_b) = (\mathbf{e}_b, \mathbf{e}_a)$ is symmetric, and it is always possible to choose bases $\mathbf{e}_a$ such that

$$\mathbf{e}_a\mathbf{e}_b = -\delta_{ab} + f_{abc}\mathbf{e}_c. \tag{16.40}$$

Furthermore, from

$$\begin{aligned} -(xy, x) &= \frac{N(xy+x) - N(x)N(y)}{2} = N(x)\frac{N(y+1) - N(y) - 1}{2} \\ &= N(x)(y, 1), \end{aligned} \tag{16.41}$$

it follows that $f_{abc} = (\mathbf{e}_a, \mathbf{e}_b, \mathbf{e}_c)$ is fully antisymmetric. [In Tits's notation [325], the multiplication tensor f_{abc} is replaced by a cubic antisymmetric form (a, a', a''), his equation (14)]. The composition requirement (16.35) expressed in terms of bases (16.36) is

$$\begin{aligned} 0 &= N(xy) - N(x)N(y) \\ &= x_a x_b y_c y_d \left(\delta_{ac}\delta_{bd} - \delta_{ab}\delta_{cd} + f_{ace}f_{cbd}\right). \end{aligned} \qquad (16.42)$$

To make a contact with section 16.2, we introduce diagrammatic notation (factor $i\sqrt{6/\alpha}$ adjusts the normalization to (16.2))

$$f_{abc} = i\sqrt{\frac{6}{\alpha}}$$  . (16.43)

Diagrammatically, (16.42) is given by

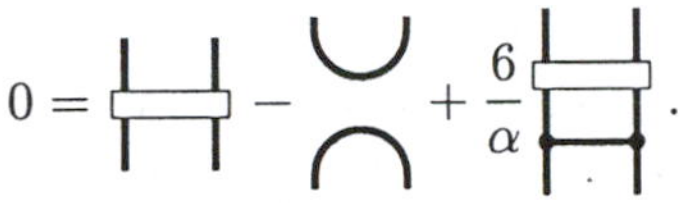 (16.44)

This is precisely the alternativity relation (16.10) we have proven to be nontrivially realizable only in three and seven dimensions. The trivial realizations are $n = 0$ and $n = 1$, $f_{abc} = 0$. So we have inadvertently proven

Hurwitz's theorem [165, 166, 70, 169]: *$(n{+}1)$-dimensional normed algebras over reals exist only for $n = 0, 1, 3, 7$ (real, complex, quaternion, octonion).*

We call (16.10) the *alternativity* relation, because it can also be obtained by substituting (16.40) into the alternativity condition for octonions [304]

$$\begin{aligned} [xyz] &\equiv (xy)z - x(yz), \\ [xyz] &= [zxy] = [yzx] = -[yxz]. \end{aligned} \qquad (16.45)$$

Cartan [43] was first to note that $G_2(7)$ is the isomorphism group of octonions, *i.e.*, the group of transformations of octonion bases (written here in the infinitesimal form)

$$\mathbf{e}'_a = (\delta_{ab} + iD_{ab})\mathbf{e}_b,$$

which preserve the octonionic multiplication rule (16.40). The reduction identity (16.15) was first derived by Behrends *et al.* [18], in index notation: see their equation (V.21) and what follows. Tits also constructed the adjoint rep projection operator for $G_2(7)$ by defining the derivation on an octonion algebra as

$$Dz = \langle x, y\rangle z = -\frac{1}{2}((x \cdot y) \cdot z) + \frac{3}{2}[(y, z)x - (x, z)y]$$

[Tits 1966, equation (23)], where

$$\mathbf{e}_a \cdot \mathbf{e}_b \equiv f_{abc}\mathbf{e}_c, \qquad (16.46)$$

$$(\mathbf{e}_a, \mathbf{e}_b) \equiv -\delta_{ab}. \qquad (16.47)$$

Substituting $x = x_a\mathbf{e}_a$, we find

$$(Dz)_d = -3x_a y_b \left(\frac{1}{2}\delta_{ab}\delta_{bd} + \frac{1}{6}f_{abe}f_{ecd}\right) z_c. \qquad (16.48)$$

The term in the brackets is just the $G_2(7)$ adjoint rep projection operator $\mathbf{P}_A$ in (16.13), with normalization $\alpha = -3$.

Chapter Seventeen

E_8 family of invariance groups

In this chapter we continue the construction of invariance groups characterized by a symmetric quadratic and an antisymmetric cubic primitive invariant. In the preceding chapter we proved that the cubic invariant must either satisfy the alternativity relation (16.11), or the Jacobi relation (4.48), and showed that the first case has $SO(3)$ and G_2 as the only interesting solutions.

Here we pursue the second possibility and determine all invariance groups that preserve a symmetric quadratic (4.28) and an antisymmetric cubic primitive invariant (4.46),

(17.1)

with the cubic invariant satisfying the Jacobi relation (4.48)

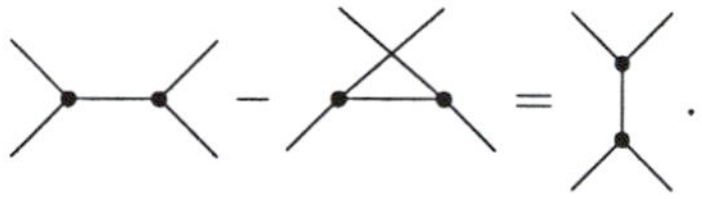

(17.2)

Our task is twofold:

1. Enumerate all Lie algebras defined by the primitives (17.1). The key idea here is the primitiveness assumption (3.39). By requiring that the list of (17.1) is the full list of primitive invariants, *i.e.*, that any invariant tensor can be expressed as a linear sum over the tree invariants constructed from the quadratic and the cubic invariants, we are classifying those invariance groups for which *no quartic primitive* invariant exists in the adjoint rep (see figure 16.1).

2. Demonstrate that we can compute all $3n$-j coefficients (or casimirs, or vacuum bubbles); the ones up to 12-j are listed in table 5.1. Due to the antisymmetry (17.1) of structure constants and the Jacobi relation (17.2), we need to concentrate on evaluation of only the even-order symmetric casimirs, a subset of (7.13):

   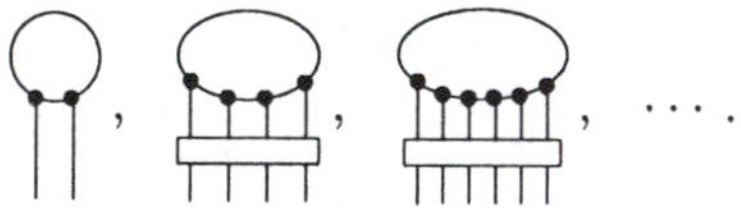

   (17.3)

   Here cheating a bit and peeking into the list of the Betti numbers (table 7.1) offers some moral guidance: the orders of Dynkin indices for the E_8 group are 2, 8, 12, 14, 18, 20, 24, 30. In other words, there is no way manual birdtracking is going to take us to the end of this road.

We accomplish here most of 1: the Diophantine conditions (17.13)–(17.19) and (17.38)–(17.40) yield all of the E_8 family Lie algebras, and no stragglers, but we fail to prove that there exist no further Diophantine conditions, and that all of these groups actually exist. We are much further from demonstrating 2: the projection operators (17.15), (17.16), (17.31)–(17.33) for the E_8 family enable us to evaluate diagrams with internal loops of length 5 or smaller, but we have no proof that *any* vacuum bubble can be so evaluated. Should we be intimidated by existence of Dynkin indices of order 30? Not necessarily: we saw that any classical Lie group vacuum bubble can be iteratively reduced to a polynomial in n, regardless of the number of its Dynkin indices. But for F_4, E_6, E_7, and E_8 such algorithms remain unknown.

As, by assumption, the defining rep satisfies the Jacobi relation (17.2), the defining rep is in this case also A, the adjoint rep of some Lie group. Hence, in this chapter we denote the dimension of the defining rep by N, the cubic invariant by the Lie algebra structure constants $-iC_{ijk}$, and draw the invariants with the thin (adjoint) lines, as in (17.1) and (17.2).

The assumption that the defining rep is irreducible means in this case that the Lie group is simple, and the quadratic casimir (Cartan-Killing tensor) is proportional to the identity

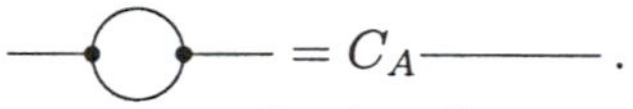
(17.4)

In this chapter we shall choose normalization $C_A = 1$. The Jacobi relation (17.2) reduces a loop with three structure constants

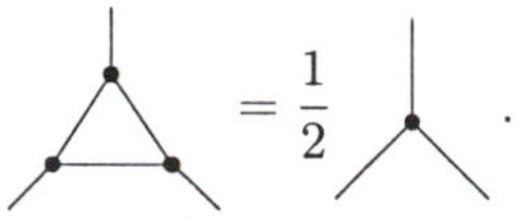
(17.5)

Remember diagram (1.1)? The one diagram that launched this whole odyssey? In order to learn how to reduce such 4-vertex loops we turn to the decomposition of the $A \otimes A$ space.

In what follows, we will generate quite a few irreducible reps. In order to keep track of them, we shall label each family of such reps (for example, the eigenvalues $\lambda_{\blacksquare}$, $\lambda_{\square\square}$ in (17.12)) by the generalized Young tableau (or Dynkin label) notation for the E_8 irreducible reps (section 17.4). A review of related literature is given in section 21.2.

17.1 TWO-INDEX TENSORS

The invariance group of the quadratic invariant (17.1) alone is $SO(n)$, so as in table 10.1, $A \otimes A$ decomposes into singlet, symmetric, and antisymmetric subspaces.

Of the three possible tree invariants in $A \otimes A \to A \otimes A$ constructed from the cubic invariant (17.1), only two are linearly independent because of the Jacobi relation (17.2). The first one induces a decomposition of antisymmetric $A \otimes A$ tensors into two subspaces:

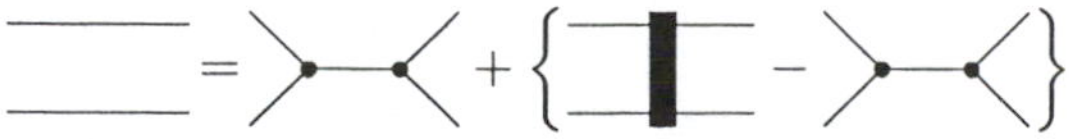

$$+\frac{1}{N} \quad + \left\{ \quad - \frac{1}{N} \quad \right\} \tag{17.6}$$

$$\mathbf{1} = \mathbf{P}_{\square} + \mathbf{P}_{\boxminus} + \mathbf{P}_{\bullet} + \mathbf{P}_{s} \,.$$

As the other invariant matrix in $A \otimes A \to A \otimes A$ we take

$$\mathbf{Q}_{ij,kl} = \;. \tag{17.7}$$

By the Jacobi relation (17.2), $\mathbf{Q}$ has zero eigenvalue on the antisymmetric subspace

$$\mathbf{Q}\mathbf{P}_{\boxminus} = \quad \mathbf{P}_{\boxminus} = \frac{1}{2} \quad \mathbf{P}_{\boxminus} = \frac{1}{2}\mathbf{P}_{\square}\mathbf{P}_{\boxminus} = 0 \,, \tag{17.8}$$

so $\mathbf{Q}$ can decompose only the symmetric subspace $\mathrm{Sym}^2 A$.

The assumption that there exists no primitive quartic invariant is the *defining relation* for the E_8 family. By the primitiveness assumption, the 4-index loop invariant $\mathbf{Q}^2$ is not an independent invariant, but is expressible in terms of any full linearly independent set of the 4-index tree invariants $\mathbf{Q}_{ij,k\ell}$, $C_{ijm}C_{mk\ell}$, and δ_{ij}'s constructed from the primitive invariants (17.1),

$$+ A \quad + B \quad + C \quad + D \quad + E \quad = 0 \,.$$

Rotate by 90^0 and compare. That eliminates two coefficients. Flip any pair of adjacent legs and use the Jacobi relation (17.2) (*i.e.*, the invariance condition). Only one free coefficient remains:

$$- \frac{1}{6}\left\{ \quad + \quad \right\} - \frac{q}{2}\left\{ \quad + \quad + \quad \right\} = 0 \,. \tag{17.9}$$

Now, trace over a pair of adjacent legs, and evaluate 2- and 3-loops using (17.4) and (17.5). This expresses the parameter q in terms of the adjoint dimension, and (17.9) yields the characteristic equation for $\mathbf{Q}$ restricted to the traceless symmetric subspace,

$$\left(\mathbf{Q}^2 - \frac{1}{6}\mathbf{Q} - \frac{5}{3(N+2)}\mathbf{1}\right)\mathbf{P}_s = 0 \,. \tag{17.10}$$

An eigenvalue of $\mathbf{Q}$ satisfies the characteristic equation

$$\lambda^2 - \frac{1}{6}\lambda - \frac{5}{3(N+2)} = 0 \,,$$

so the adjoint dimension can be expressed as

$$N + 2 = \frac{5}{3\lambda(\lambda - 1/6)} = 60\left\{\frac{6 - \lambda^{-1}}{6} - 2 + \frac{6}{6 - \lambda^{-1}}\right\} . \tag{17.11}$$

As we shall seek for values of λ such that the adjoint rep dimension N is an integer, it is natural to reparametrize the two eigenvalues as

$$\lambda_{\blacksquare} = -\frac{1}{m-6} \,, \qquad \lambda_{\square\square} = \frac{1}{6}\frac{m}{m-6} \,, \tag{17.12}$$

a form that will lend itself to Diophantine analysis. In terms of the parameter m, the dimension of the adjoint representation is given by

$$N = -2 + 60\,(m/6 - 2 + 6/m)\,, \tag{17.13}$$

and the two eigenvalues map into each other under $m/6 \to 6/m$. Substituting

$$\lambda_{\Box\Box} - \lambda_{\blacksquare} = \frac{1}{6}\frac{m+6}{m-6} \tag{17.14}$$

into (3.48), we obtain the corresponding projection operators:

$$\mathbf{P}_{\blacksquare} = \quad = -\frac{6(m-6)}{m+6}\left\{ \quad - \frac{1}{6}\frac{m}{m-6} \quad \right\}\mathbf{P}_s \tag{17.15}$$

$$\mathbf{P}_{\Box\Box} = \quad = \frac{6(m-6)}{m+6}\left\{ \quad + \frac{1}{m-6} \quad \right\}\mathbf{P}_s\,. \tag{17.16}$$

In order to compute the dimensions of the two subspaces, we evaluate

$$\mathrm{tr}\,\mathbf{P}_s\mathbf{Q} = \quad - \frac{1}{N} \quad = -\frac{N+2}{2} \tag{17.17}$$

and obtain

$$d_{\Box\Box} = \mathrm{tr}\,\mathbf{P}_{\Box\Box} = \frac{(N+2)(1/\lambda_{\blacksquare} + N - 1)}{2(1 - \lambda_{\Box\Box}/\lambda_{\blacksquare})}\,. \tag{17.18}$$

Dimension $d_{\blacksquare}$ is obtained by interchanging $\lambda_{\blacksquare}$ and $\lambda_{\Box\Box}$. Substituting (17.13), (17.12) leads to

$$d_{\Box\Box} = \frac{5(m-6)^2(5m-36)(2m-9)}{m(m+6)}$$

$$d_{\blacksquare} = \frac{270(m-6)^2(m-5)(m-8)}{m^2(m+6)}\,. \tag{17.19}$$

To summarize, in absence of a primitive 4-index invariant, $A \otimes A$ decomposes into five irreducible reps

$$\mathbf{1} = \mathbf{P}_{\Box} + \mathbf{P}_{\boxminus} + \mathbf{P}_{\bullet} + \mathbf{P}_{\Box\Box} + \mathbf{P}_{\blacksquare}\,. \tag{17.20}$$

The decomposition is parametrized by rational values of m, and is possible only for integer N and $d_{\blacksquare}$ that satisfy the Diophantine conditions (17.13), (17.19).

This happened so quickly that the reader might have missed it: our homework problem is done. What we have accomplished by (17.9) is the reduction of the adjoint rep 4-vertex loop in (1.1) for, as will turn out, *all* exceptional Lie algebras.

17.2 DECOMPOSITION OF $\mathrm{Sym}^3 A$

Now that you have aced the homework assignment (1.1), why not go for extra credit: can you disentangle vacuum bubbles whose shortest loop is of length 6,

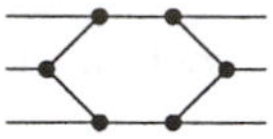

$$= ? \qquad (17.21)$$

If you have an elegant solution, let me know. But what follows next is cute enough.

The general strategy for decomposition of higher-rank tensor products is as follows; the equation (17.10) reduces $\mathbf{Q}^2$ to $\mathbf{Q}$, $\mathbf{P}_r$ weighted by the eigenvalues $\lambda_{\square\square}$, $\lambda_{\blacksquare}$. For higher-rank tensor products, we shall use the same result to decompose symmetric subspaces. We shall refer to a decomposition as "uninteresting" if it brings no new Diophantine condition. As $\mathbf{Q}$ acts only on the symmetric subspaces, decompositions of antisymmetric subspaces will always be uninteresting, as was already the case in (17.8). We illustrate this by working out the decomposition of $\mathrm{Sym}^3 A$.

The invariance group of the quadratic invariant (17.1) alone is $SO(N)$, with the seven reps Clebsch-Gordan decomposition of the $SO(N)$ 3-index tensors (table 10.4): one fully symmetric, one fully antisymmetric, two copies of the mixed symmetry rep, and three copies of the defining rep. As the Jacobi relation (17.2) trivializes the action of $\mathbf{Q}$ on any antisymmetric pair of indices, the only serious challenge that we face is reducing $\otimes A^3$ within the fully symmetric $\mathrm{Sym}^3 A$ subspace.

As the first step, project out the A and $A \otimes A$ content of $\mathrm{Sym}^3 A$:

$$\mathbf{P}_{\square} = \frac{3}{N+2} \qquad (17.22)$$

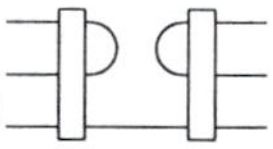

$$\mathbf{P}_{\boxminus} = \frac{6(N+1)(N^2-4)}{5(N^2+2N-5)} \, . \qquad (17.23)$$

$\mathbf{P}_{\square}$ projects out $\mathrm{Sym}^3 A \to A$, and $\mathbf{P}_{\boxminus}$ projects out the antisymmetric subspace (17.6) $\mathrm{Sym}^3 A \to V_{\boxminus}$. The ugly prefactor is a normalization, and will play no role in what follows. We shall decompose the remainder of the $\mathrm{Sym}^3 A$ space

$$\mathbf{P}_r = S - \mathbf{P}_{\square} - \mathbf{P}_{\boxminus} = \qquad (17.24)$$

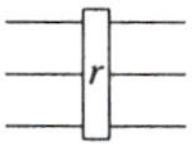

by the invariant tensor $\mathbf{Q}$ restricted to the $\mathbf{P}_r$ remainder subspace

$$\mathbf{Q} = \quad , \qquad \hat{\mathbf{Q}} = \qquad \hat{\mathbf{Q}} = \mathbf{P}_r \mathbf{Q} \mathbf{P}_r \, . \qquad (17.25)$$

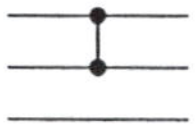

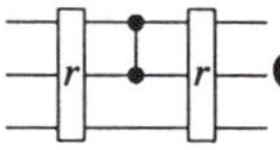

We can partially reduce $\hat{\mathbf{Q}}^2$ using (17.10), but symmetrization leads also to a new invariant tensor,

$$\hat{\mathbf{Q}}^2 = \frac{1}{3} \qquad + \frac{2}{3} \qquad . \qquad (17.26)$$

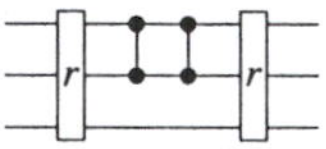

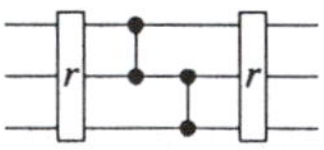

A calculation that requires applications of the Jacobi relation (17.2), symmetry identities (6.63) such as

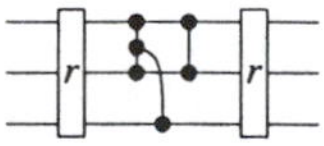

$$= 0 \, , \qquad (17.27)$$

and relies on the fact that $\mathbf{P}_r$ contains no A, $A \otimes A$ subspaces yields

$$\hat{\mathbf{Q}}^3 = \frac{1}{3} [\text{birdtrack}] + \frac{2}{3} [\text{birdtrack}] . \tag{17.28}$$

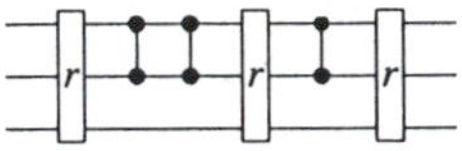

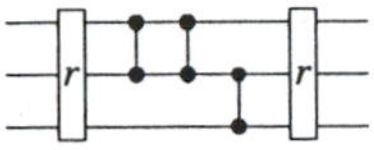

Reducing by (17.10) and using $\lambda_{\blacksquare} + \lambda_{\square\square} = 1/6$ leads to

$$\hat{\mathbf{Q}}^3 = \frac{1}{6} \left\{ \frac{1}{3} \hat{\mathbf{Q}}^2 + \frac{2}{3} [\text{birdtrack}] \right\} - \lambda_{\blacksquare} \lambda_{\square\square} \hat{\mathbf{Q}} . \tag{17.29}$$

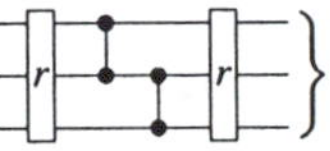

The extra tensor can be eliminated by (17.26), and the result is a cubic equation for $\hat{\mathbf{Q}}$:

$$0 = \left(\hat{\mathbf{Q}} - \frac{1}{18} \mathbf{1} \right) \left(\hat{\mathbf{Q}} - \lambda_{\blacksquare} \mathbf{1} \right) \left(\hat{\mathbf{Q}} - \lambda_{\square\square} \mathbf{1} \right) \mathbf{P}_r . \tag{17.30}$$

The projection operators for the corresponding three subspaces are given by (3.48)

$$\begin{aligned} \mathbf{P}_{\square} &= \frac{1}{(1/18 - \lambda_{\blacksquare})(1/18 - \lambda_{\square\square})} \left(\hat{\mathbf{Q}} - \lambda_{\blacksquare} \mathbf{1} \right) \left(\hat{\mathbf{Q}} - \lambda_{\square\square} \mathbf{1} \right) \mathbf{P}_r \\ &= - \frac{162\,(m-6)^2}{(m+3)(m+12)} \left(\hat{\mathbf{Q}}^2 - \frac{1}{6} \hat{\mathbf{Q}} - \frac{6m}{(2-6m)^2} \mathbf{1} \right) \mathbf{P}_r , \end{aligned} \tag{17.31}$$

$$\begin{aligned} \mathbf{P}_{\square\blacksquare} &= \frac{1}{(\lambda_{\blacksquare} - 1/18)(\lambda_{\blacksquare} - \lambda_{\square\square})} \left(\hat{\mathbf{Q}} - \frac{1}{18} \mathbf{1} \right) \left(\hat{\mathbf{Q}} - \lambda_{\square\square} \mathbf{1} \right) \mathbf{P}_r \\ &= \frac{54\,(m-6)^2}{(m+3)(m+6)} \left(\hat{\mathbf{Q}}^2 - \frac{m-24}{18(m-6)} \hat{\mathbf{Q}} + \frac{1}{18(m-6)} \mathbf{1} \right) \mathbf{P}_r , \end{aligned} \tag{17.32}$$

$$\begin{aligned} \mathbf{P}_3 &= \frac{1}{(\lambda_{\square\square} - 1/18)(\lambda_{\square\square} - \lambda_{\blacksquare})} \left(\hat{\mathbf{Q}} - \frac{1}{18} \mathbf{1} \right) \left(\hat{\mathbf{Q}} - \lambda_{\blacksquare} \mathbf{1} \right) \mathbf{P}_r \\ &= \frac{108\,(m-6)^2}{(m+6)(m+12)} \left(\hat{\mathbf{Q}}^2 - \frac{2(m-3)}{9(m-6)} \hat{\mathbf{Q}} + \frac{m}{108(m-6)} \mathbf{1} \right) \mathbf{P}_r . \end{aligned} \tag{17.33}$$

The presumption is (still to be proved for a general tensor product) that the interesting reductions only occur in the symmetric subspaces, always via the $\mathbf{Q}$ characteristic equation (17.10). As the overall scale of $\mathbf{Q}$ is arbitrary, there is only one rational parameter in the problem, either $\lambda_{\square\square}/\lambda_{\blacksquare}$ or m, or whatever is convenient. Hence all dimensions and $3n$-j coefficients (casimirs, Dynkin indices, vacuum bubbles) will be ratios of polynomials in m.

To proceed, we follow the method outlined in appendix A. On $\mathbf{P}_{\square}$, $\mathbf{P}_{\boxminus}$ subspaces $S\mathbf{Q}$ has eigenvalues

$$S\mathbf{Q}\mathbf{P}_{\square} = [\text{birdtrack}] = \frac{1}{3} [\text{birdtrack}] \qquad \rightarrow \lambda_{\square} = 1/3 \tag{17.34}$$

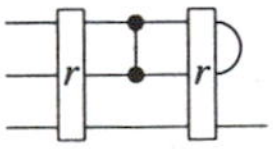

$$S\mathbf{Q}\mathbf{P}_{\boxminus} = [\text{birdtrack}] = \frac{1}{6} [\text{birdtrack}] \qquad \rightarrow \lambda_{\boxminus} = 1/6 , \tag{17.35}$$

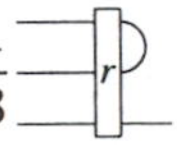

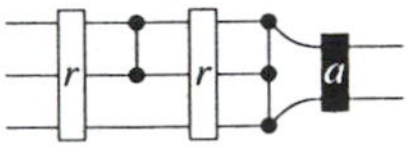

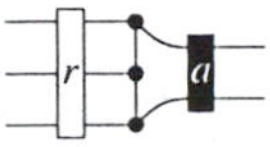

so the eigenvalues are $\lambda_{\square} = 1/3$, $\lambda_{\boxminus} = 1/6$, $\lambda_3 = 1/18$, $\lambda_{\square\blacksquare} = \lambda_{\square\square}$, $\lambda_{\square} = \lambda_{\square\square}$. The dimension formulas (A.8) require evaluation of

$$\mathrm{tr}\, S\mathbf{Q} = [\text{birdtrack}] = - \frac{N(N+2)}{6} \tag{17.36}$$

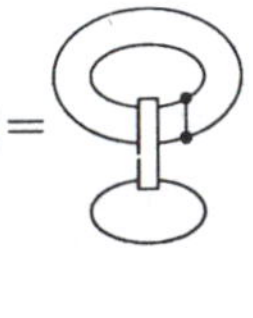

$$\mathrm{tr}(S\mathbf{Q})^2 = \quad = \frac{N(3N+16)}{36}. \tag{17.37}$$

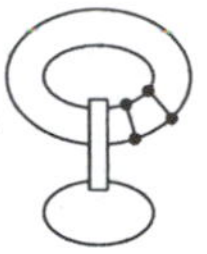

Substituting into (A.8) we obtain the dimensions of the three new reps:

$$d_{\square} = \frac{27(m-5)(m-8)(2m-15)(2m-9)(5m-36)(5m-24)}{m^2(3+m)(12+m)} \tag{17.38}$$

$$d_{\square\blacksquare} = \frac{10(m-6)^2(m-5)(m-1)(2m-9)(5m-36)(5m-24)}{3m^2(6+m)(12+m)} \tag{17.39}$$

$$d_3 = \frac{5(m-5)(m-8)(m-6)^2(2m-15)(5m-36)}{m^3(3+m)(6+m)}(36-m). \tag{17.40}$$

17.3 DIOPHANTINE CONDITIONS

As N in (17.13) is an integer, allowed m are rationals $m = P/Q$ built from Q any combination of subfactors of the denominator $360 = 1 \cdot 2^3 \cdot 3^2 \cdot 5$, and the numerator $P = 1, 2,$ or 5, where P and Q are relative primes. The solutions are symmetric under interchange $m/6 \leftrightarrow 6/m$, so we need to check only the 23 rationals $m \geq 6$. The Diophantine conditions (17.13), (17.19), and (17.38) are satisfied only for $m = 5$, 8, 9, 10, 12, 18, 20, 24, 30, and 36. The solutions that survive the Diophantine conditions form the E_8 *family*, listed in table 17.1. The formulas (17.15), (17.16) yield, upon substitution of N, $\lambda_{\square\square}$ and $\lambda_{\blacksquare}$, the $A\otimes A$ Clebsch-Gordan series for the E_8 family (table 17.2).

Particularly interesting is the $(36-m)$ factor in the d_3 formula (17.40): positivity of a dimension excludes $m > 36$ solutions, and vanishing of the corresponding projection operator (17.33) for $m = 36$ implies a birdtrack identity valid only for E_8, the presumed key to the homework assignment (17.21). For inspiration, go through the derivation of (18.37), the analogous 6-loop reduction formula for E_6. According to ref. [294], the smallest vacuum bubble that has no internal loop with fewer than six edges has fourteen vertices and is called the "Coxeter graph."

Birdtracks yield the E_8 family, but they do not tie it into the Cartan-Killing theory. For that we refer the reader to the very clear [29] and thorough exposition by Deligne [89]. All the members of the family are immediately identifiable, with exception of the $m = 30$ case. The $m = 30$ solution was found independently by Landsberg and Manivel [209], who identify the corresponding column in table 17.1 as a class of nonreductive algebras. Here this set of solutions will be eliminated by (19.42), which says that it does not exist as a semisimple Lie algebra for the F_4 subgroup of E_8.

The main result of all this heavy birdtracking is that $N > 248$ is excluded by the positivity of d_3, and $N = 248$ is special, as $\mathbf{P}_3 = 0$ implies existence of a tensorial identity on the $\mathrm{Sym}^3 A$ subspace specific to E_8. That dimensions should all factor into terms linear in m is altogether not obvious.

m	5	8	9	10	12	15	18	24	30	36
		A_1	A_2	G_2	D_4	F_4	E_6	E_7	.	E_8
N	0	3	8	14	28	52	78	133	190	248
d_3	0	0	1	7	56	273	650	1,463	1,520	0
$d_{\blacksquare}$	0	−3	0	64	700	4,096	11,648	40,755	87,040	147,250
$d_{\square\blacksquare}$	0	0	27	189	1,701	10,829	34,749	152,152	392,445	779,247

Table 17.1 All solutions of Diophantine conditions (17.13), (17.19), and (17.38).

17.4 DYNKIN LABELS AND YOUNG TABLEAUX FOR E_8

A rep of E_8 is characterized by 8 Dynkin labels $(a_1a_2a_3a_4a_5a_6a_7a_8)$. The correspondence between the E_8 Dynkin diagram from table 7.6, Dynkin labels, irreducible tensor Young tableaux, and the dimensions [294] of the lowest reps is

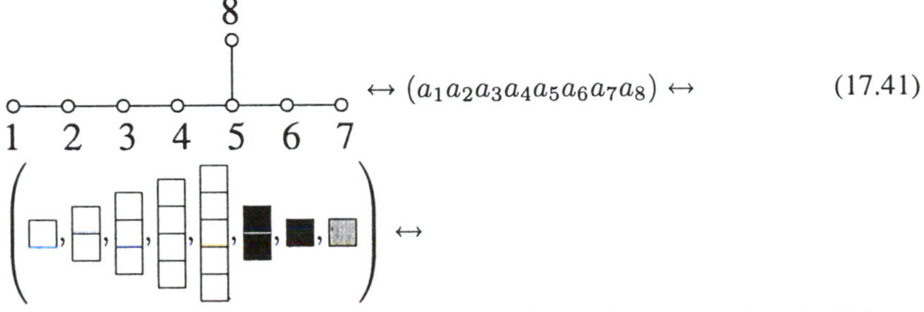

$$\leftrightarrow (a_1a_2a_3a_4a_5a_6a_7a_8) \leftrightarrow \qquad (17.41)$$

(248, 30380, 2450240, 146325270, 6899079264, 6696000, 3875, 147250)

Label a_1 counts the number of not antisymmetrized defining (= adjoint) representation indices. Labels a_2 through a_5 count the number of antisymmetric doublets, triplets, quadruplets, and quintuplets, respectively. Label a_7 counts the number of not antisymmetrized ■ indices, and a_6 the number of its antisymmetrized doublets. The label a_8 counts the number of ▒.

	$\square \otimes \square$	=	$\square$	+	$\begin{smallmatrix}\square\\\square\end{smallmatrix}$	+	$\bullet$	+	$\square\square$	+	$\blacksquare$
Dynkin labels											
E_8	$(10000000) \otimes (10000000)$	=	(10000000)	+	(01000000)	+	(00000000)	+	(20000000)	+	(00000010)
E_7	$(1000000) \otimes (1000000)$	=	(1000000)	+	(0100000)	+	(0000000)	+	(2000000)	+	(0000100)
E_6	$(000001) \otimes (000001)$	=	(000001)	+	(001000)	+	(000000)	+	(000002)	+	(100010)
F_4	$(1000) \otimes (1000)$	=	(1000)	+	(0100)	+	(0000)	+	(2000)	+	(0010)
D_4	$(0100) \otimes (0100)$	=	(0100)	+	(1010)	+	(0000)	+	(0200)	+	$(2000) + (0001)$
G_2	$(10) \otimes (10)$	=	(10)	+	(03)	+	(00)	+	(20)	+	(02)
A_2	$(11) \otimes (11)$	=	(11)	+	$(12) + (21)$	+	(00)	+	(22)	+	(11)
A_1	$(2) \otimes (2)$	=	(2)	+	(0)	+	(4)	+	(4)		
Dimensions	N^2	=	N	+	$\frac{N(N-3)}{2}$	+	1	+	$d_{\square\square}$	+	$d_{\blacksquare}$
E_8	248^2	=	248	+	30,380	+	1	+	27,000	+	3,875
E_7	133^2	=	133	+	8,645	+	1	+	7,371	+	1,539
E_6	78^2	=	78	+	2,925	+	1	+	2,430	+	650
F_4	52^2	=	52	+	1,274	+	1	+	1,053	+	324
D_4	28^2	=	28	+	350	+	1	+	300	+	$35 + 35 + 35$
G_2	14^2	=	14	+	77	+	1	+	77	+	27
A_2	8^2	=	8	+	$10 + \overline{10}$	+	1	+	27	+	8
A_1	3^2	=	3	+	0	+	1	+	5	+	0

Table 17.2 E_8 family Clebsch-Gordan series for $A \otimes A$. Corresponding projection operators are given in (17.6), (17.15), and (17.16).

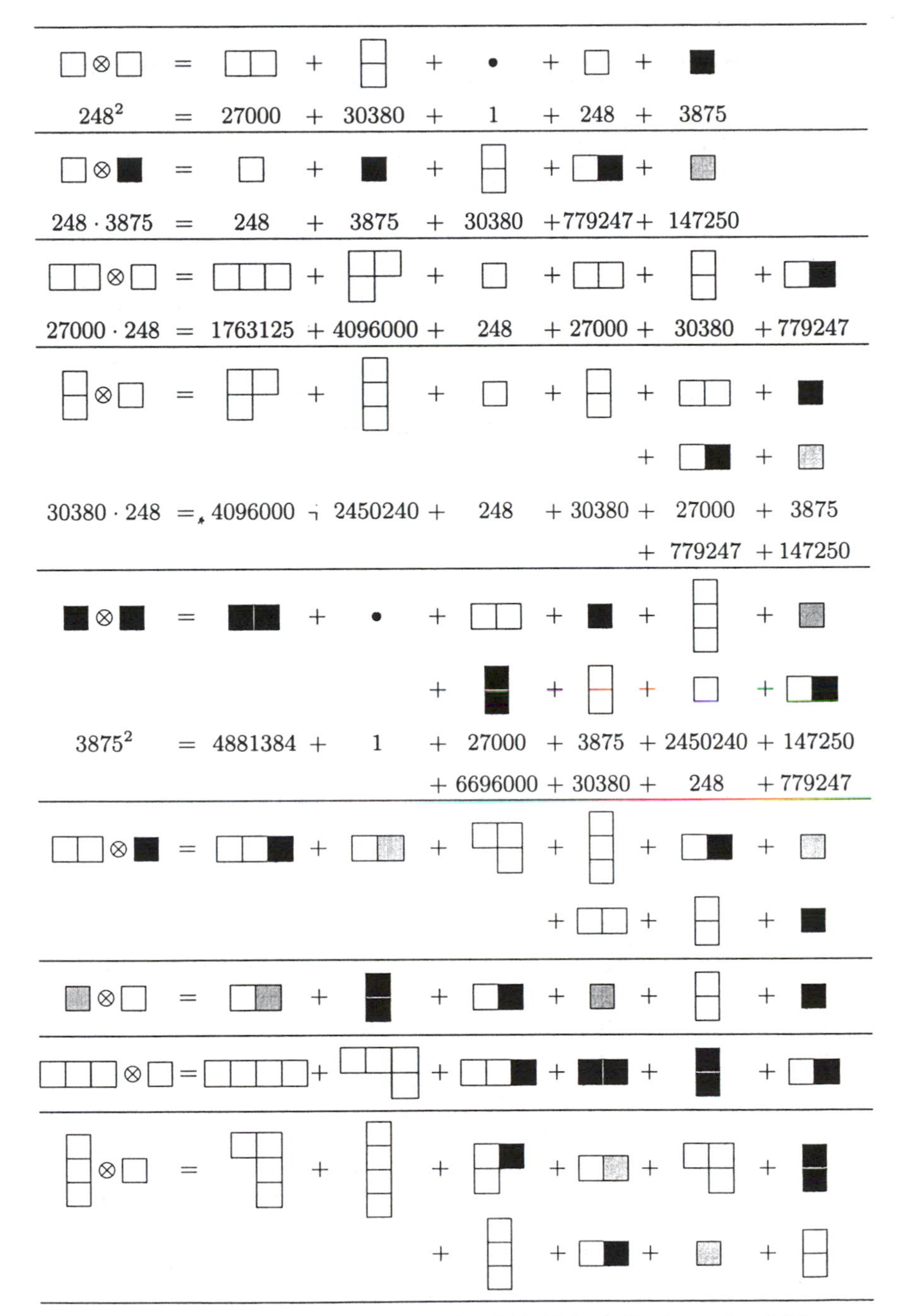

Table 17.3 Some of the low-dimensional E_8 Clebsch-Gordan series [352].

Chapter Eighteen

E_6 family of invariance groups

In this chapter, we determine all invariance groups whose primitive invariant tensors are δ^a_b and fully symmetric d_{abc}, d^{abc}. The reduction of $V \otimes V$ space yields a rule for evaluation of the loop contraction of four d-invariants (18.9). The reduction of $V \otimes \bar{V}$ yields the first Diophantine condition (18.13) on the allowed dimensions of the defining rep. The reduction of $V \otimes V \otimes V$ tensors is straightforward, but the reduction of $A \otimes V$ space yields the second Diophantine condition (d_4 in table 18.4) and limits the defining rep dimension to $n \leq 27$. The solutions of the two Diophantine conditions form the E_6 *family* consisting of E_6, A_5, $A_2 + A_2$, and A_2. For the most interesting $E_6, n = 27$ case, the cubic casimir (18.44) vanishes. This property of E_6 enables us to evaluate loop contractions of 6 d-invariants (18.37), reduce $V \otimes A$ tensors (table 18.5), and investigate relations among the higher-order casimirs of E_6 in section 18.8. In section 18.7 we introduce a Young tableau notation for any rep of E_6 and exemplify its use in construction of the Clebsch-Gordan series (table 18.6).

18.1 REDUCTION OF TWO-INDEX TENSORS

By assumption, the primitive invariants set that we shall study here is

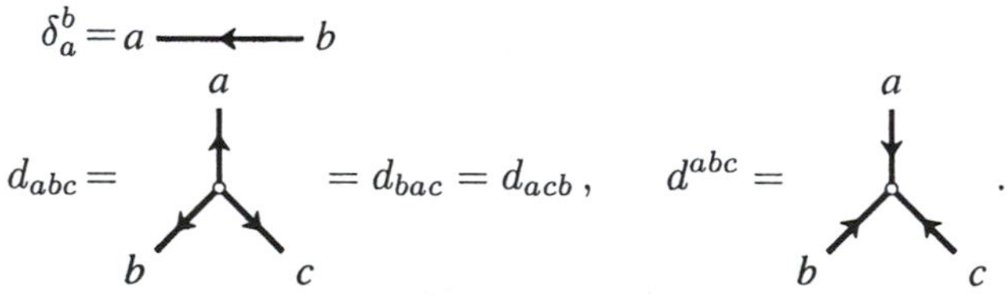

(18.1)

Irreducibility of the defining n-dimensional rep implies

$$d_{abc}d^{bcd} = \alpha\delta^d_a$$

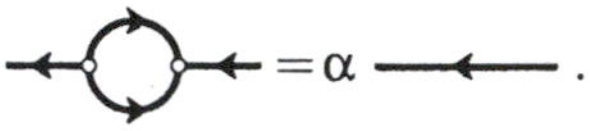

(18.2)

The value of α depends on the normalization convention. For example, Freudenthal [130] takes $\alpha = 5/2$. Kephart [187] takes $\alpha = 10$. We find it convenient to set it to $\alpha = 1$.

We can immediately write the Clebsch-Gordan series for the 2-index tensors. The symmetric subspace in (9.4) is reduced by the $d_{abc}d^{cde}$ invariant:

(18.3)

The rep dimensions and Dynkin indices are given in table 18.1.

E_6	$(000010) \otimes (000010)$	$=$ (000100)	$\oplus$	(100000)	$\oplus$	(000020)
A_5	$(00010) \otimes (00010)$	$=$ (00101)	$\oplus$	(01000)	$\oplus$	(00020)
A_2	$(02) \otimes (02)$	$=$ (12)	$\oplus$	(20)	$\oplus$	(04)
dimension	n^2	$= n(n-1)/2$	$+$	n	$+$	$n(n-1)/2$
E_6	27^2	$=$ 351	$+$	27	$+$	351
A_5	15^2	$=$ 105	$+$	15	$+$	105
$A_2 + A_2$	9^2	$=$ 36	$+$	9	$+$	36
A_2	6^2	$=$ 15	$+$	6	$+$	15
index	$2n\ell$	$= (n-2)\ell$	$+$	ℓ	$+$	$(n+1)\ell$
E_6	$2 \cdot 27 \cdot \frac{1}{4}$	$=$ $\frac{25}{4}$	$+$	$\frac{1}{4}$	$+$	7
A_5	$2 \cdot 15 \cdot \frac{1}{3}$	$=$ $\frac{13}{3}$	$+$	$\frac{1}{3}$	$+$	$\frac{16}{3}$
$A_2 + A_2$	$2 \cdot 9 \cdot \frac{1}{2}$	$=$ $\frac{7}{2}$	$+$	$\frac{1}{2}$	$+$	5
A_2	$2 \cdot 6 \cdot \frac{5}{6}$	$=$ $\frac{10}{3}$	$+$	$\frac{5}{6}$	$+$	$\frac{35}{6}$

Table 18.1 E_6 family Clebsch-Gordan series Dynkin labels, dimensions, and Dynkin indices for $V \otimes V$. The defining rep Dynkin index ℓ is computed in (18.14).

By the primitiveness assumption, any $V^2 \otimes \bar{V}^2$ invariant is a linear combination of all tree invariants that can be constructed from the primitives:

$$= a \quad + b \quad + c \quad . \tag{18.4}$$

In particular,

$$\frac{1}{\alpha^2} \quad = \frac{1}{\alpha^2} \quad = \frac{A}{\alpha} \quad + B \quad . \tag{18.5}$$

One relation on constants A, B follows from a contraction with δ_a^b:

$$\frac{1}{\alpha^2} \quad = \frac{A}{\alpha} \quad + B$$

$$1 = A + B\frac{n+1}{2} .$$

The other relation follows from the invariance condition (6.53) on d_{abc}:

$$\frac{1}{\alpha} \quad = -\frac{1}{2} \quad . \tag{18.6}$$

label	$\square \otimes \blacksquare$	=	$\bullet$	$\oplus$	$\boxtimes$	$\oplus$	$\blacksquare\square$
E_6	$(000010) \otimes (100000)$	=	(000000)	$\oplus$	(000001)	$\oplus$	(100010)
A_5	$(00010) \otimes (01000)$	=	(00000)	$\oplus$	(10001)	$\oplus$	(01010)
A_2	$(02) \otimes (20)$	=	(00)	$\oplus$	(11)	$\oplus$	(22)
dimension	n^2	=	1	+	$\frac{4n(n-1)}{n+9}$	+	$\frac{(n+3)^2(n-1)}{n+9}$
E_6	27^2	=	1	+	78	+	650
A_5	15^2	=	1	+	35	+	189
$A_2 + A_2$	9^2	=	1	+	16	+	64
A_2	6^2	=	1	+	8	+	27
index	$2n\ell$	=	0	+	1	+	$\frac{2(n+3)^2}{n+9}\ell$
E_6	$2 \cdot 27 \cdot \frac{1}{4}$	=	0	+	1	+	$50 \cdot \frac{1}{4}$
A_5	$2 \cdot 15 \cdot \frac{1}{3}$	=	0	+	1	+	$27 \cdot \frac{1}{3}$
$A_2 + A_2$	$2 \cdot 9 \cdot \frac{1}{2}$	=	0	+	1	+	$16 \cdot \frac{1}{2}$
A_2	$2 \cdot 6 \cdot \frac{5}{6}$	=	0	+	1	+	$\frac{54}{6}$

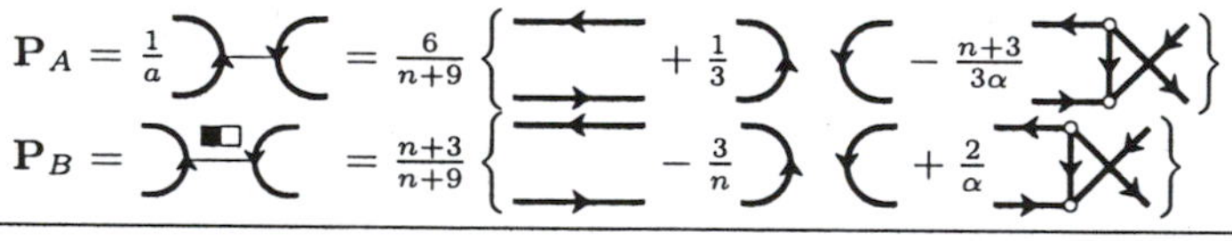

Table 18.2 E_6 family Clebsch-Gordan series for $V \otimes \overline{V}$. The defining rep Dynkin index ℓ is computed in (18.14).

Contracting (18.5) with $(T_i)_a^b$, we obtain

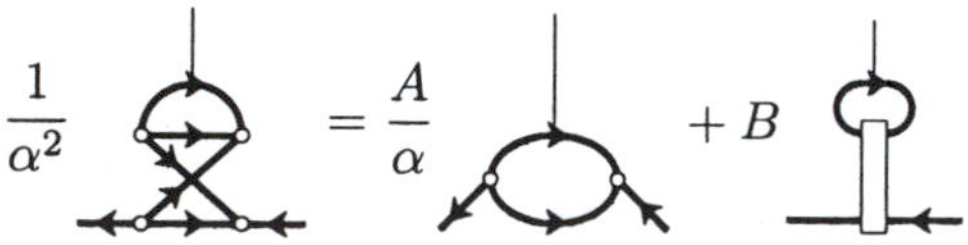

$$\frac{1}{4} = -\frac{A}{2} + \frac{B}{2}, \qquad A = -\frac{n-3}{2(n+3)}, \qquad B = \frac{3}{n+3}. \tag{18.7}$$

18.2 MIXED TWO-INDEX TENSORS

Let us apply the above result to the reduction of $V \otimes \bar{V}$ tensors. As always, they split into a singlet and a traceless part (9.54). However, now there exists an additional invariant matrix

$$\mathbf{Q}_{b\,c}^{a\,d} = \quad , \tag{18.8}$$

which, according to (18.5) and (18.7), satisfies the characteristic equation

$$= A \quad + \frac{B}{2}\left\{ \quad + \quad \right\}$$

$$\mathbf{Q}^2 = -\frac{1}{2}\frac{n-3}{n+3}\mathbf{Q} + \frac{1}{2}\frac{3}{n+3}(\mathbf{T}+1). \tag{18.9}$$

On the traceless $V \otimes \bar{V}$ subspace, the characteristic equation for $\mathbf{Q}$ takes the form

$$\mathbf{P}_2\left(\mathbf{Q}+\frac{1}{2}\right)\left(\mathbf{Q}-\frac{3}{n+3}\right) = 0\,, \tag{18.10}$$

where $\mathbf{P}_2$ is the traceless projection operator (9.54). The associated projection operators (3.48) are

$$\mathbf{P}_A = \frac{\mathbf{Q}-\frac{3}{n+3}}{-\frac{1}{2}-\frac{3}{n+3}}\mathbf{P}_2\,, \qquad \mathbf{P}_B = \frac{\mathbf{Q}+\frac{1}{2}}{\frac{3}{n+3}+\frac{1}{2}}\mathbf{P}_2\,. \tag{18.11}$$

Their birdtracks form and their dimensions are given in table 18.2.

$\mathbf{P}_A$, the projection operator associated with the eigenvalue $-\frac{1}{2}$, is the adjoint rep projection operator, as it satisfies the invariance condition (18.6). To compute the dimension of the adjoint rep, we use the relation

$$- \quad = \frac{4}{n+9}\left\{ \quad - \quad \right\}, \tag{18.12}$$

that follows trivially from the form of the projection operator $\mathbf{P}_A$ in table 18.2. The dimension is computed by taking trace (3.52),

$$N = \quad = \frac{4n(n-1)}{n+9}\,. \tag{18.13}$$

The 6-j coefficient, needed for the evaluation of the Dynkin index (7.27), can also be evaluated by substituting (18.12) into

$$= \quad + \frac{4}{n+9}\left\{0 - \quad \right\}$$

$$= N\left(1-\frac{4}{n+9}\right).$$

The Dynkin index for the E_6 family defining rep is

$$\ell = \frac{1}{6}\frac{n+9}{n-3}\,. \tag{18.14}$$

18.3 DIOPHANTINE CONDITIONS AND THE E_6 FAMILY

The expressions for the dimensions of various reps (see tables in this chapter) are ratios of polynomials in n, the dimension of the defining rep. As the dimension of a rep should be a nonnegative integer, these relations are the Diophantine conditions on the allowed values of n. The dimension of the adjoint rep (18.13) is one such condition; the dimension of V_4 from table 18.4 another. Furthermore, the positivity of the dimension d_4 restricts the solutions to $n \leq 27$. This leaves us with six solutions: $n = 3, 6, 9, 15, 21, 27$. As we shall show in chapter 21, of these solutions only $n = 21$ is spurious; the remaining five solutions are realized as the E_6 row of the Magic Triangle (figure 1.1).

In the Cartan notation, the corresponding Lie algebras are A_2, $A_2 + A_2$, A_5, and E_6. We do not need to prove this, as for E_6 Springer has already proved the existence of a cubic invariant, satisfying the relations required by our construction, and for the remaining Lie algebras the cubic invariant is easily constructed (see section 18.9). We call these invariance groups the E_6 *family* and list the corresponding dimensions, Dynkin labels, and Dynkin indices in the tables of this chapter.

18.4 THREE-INDEX TENSORS

The $V \otimes V \otimes V$ tensor subspaces of $U(n)$, listed in table 9.1, are decomposed by invariant matrices constructed from the cubic primitive d_{abc} in the following manner.

18.4.1 Fully symmetric $V \otimes V \otimes V$ tensors

We substitute expansion from table 18.1 into the symmetric projection operator

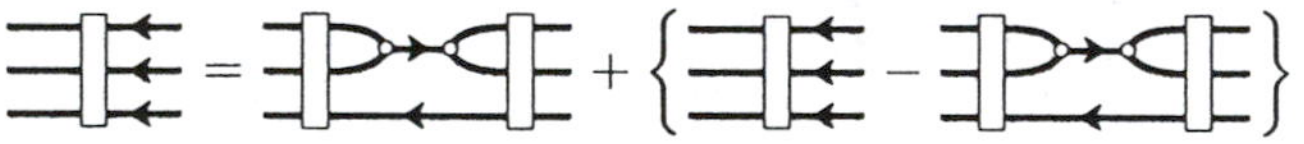

$$= + \left\{ - \right\} .$$

The $V \otimes \bar{V}$ subspace is decomposed by the expansion of table 18.2:

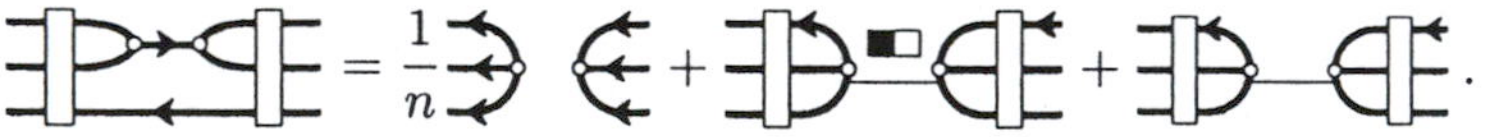

$$= \frac{1}{n} + + . \qquad (18.15)$$

The last term vanishes by the invariance condition (6.53). To get the correct projector operator normalization for the second term, we compute

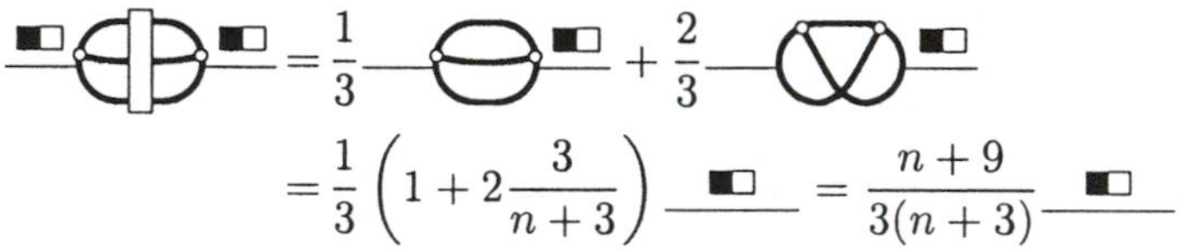

$$= \frac{1}{3} + \frac{2}{3}$$
$$= \frac{1}{3}\left(1 + 2\frac{3}{n+3}\right) = \frac{n+9}{3(n+3)} . \qquad (18.16)$$

Here, the second term is given by the ■□-subspace eigenvalue (18.10) of the invariant matrix **Q** from (18.8). The resulting decomposition is given in table 18.3.

18.4.2 Mixed symmetry $V \otimes V \otimes V$ tensors

The invariant $d_{abe}(T_i)_c^{\ e}$ satisfies

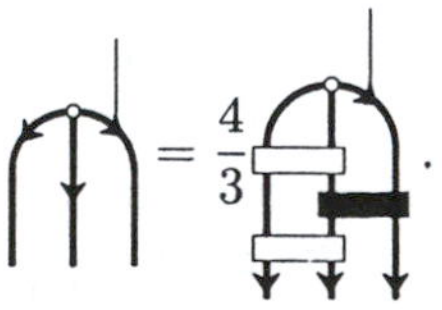

$$= \frac{4}{3} . \qquad (18.17)$$

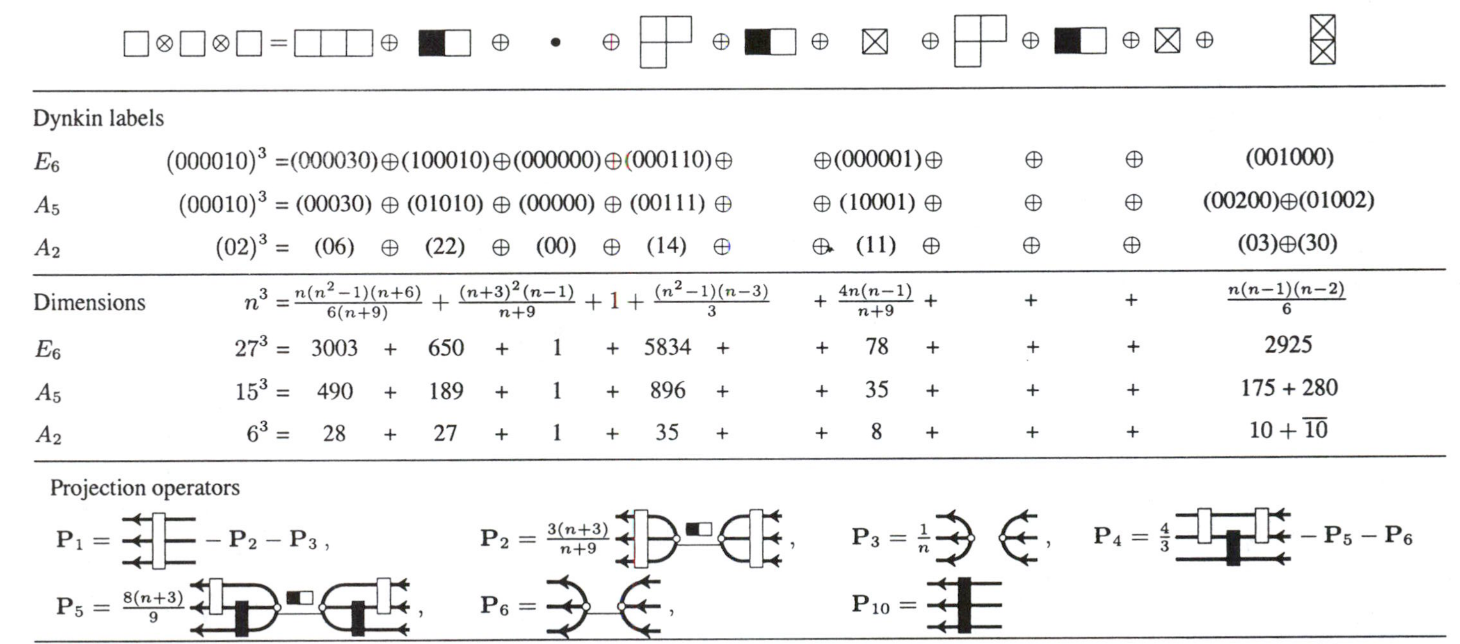

Dynkin labels											
E_6	$(000010)^3 =$	$(000030)\oplus$	$(100010)\oplus$	$(000000)\oplus$	$(000110)\oplus$	$\oplus$	$(000001)\oplus$	$\oplus$	$\oplus$		(001000)
A_5	$(00010)^3 =$	$(00030)\oplus$	$(01010)\oplus$	$(00000)\oplus$	$(00111)\oplus$	$\oplus$	$(10001)\oplus$	$\oplus$	$\oplus$		$(00200)\oplus(01002)$
A_2	$(02)^3 =$	$(06)\oplus$	$(22)\oplus$	$(00)\oplus$	$(14)\oplus$	$\oplus$	$(11)\oplus$	$\oplus$	$\oplus$		$(03)\oplus(30)$
Dimensions	$n^3 =$	$\frac{n(n^2-1)(n+6)}{6(n+9)} +$	$\frac{(n+3)^2(n-1)}{n+9} +$	$1 +$	$\frac{(n^2-1)(n-3)}{3}$	$+$	$\frac{4n(n-1)}{n+9} +$	$+$	$+$		$\frac{n(n-1)(n-2)}{6}$
E_6	$27^3 =$	$3003 +$	$650 +$	$1 +$	$5834 +$	$+$	$78 +$	$+$	$+$		2925
A_5	$15^3 =$	$490 +$	$189 +$	$1 +$	$896 +$	$+$	$35 +$	$+$	$+$		$175 + 280$
A_2	$6^3 =$	$28 +$	$27 +$	$1 +$	$35 +$	$+$	$8 +$	$+$	$+$		$10 + \overline{10}$

Projection operators

$\mathbf{P}_1 = \ldots - \mathbf{P}_2 - \mathbf{P}_3$, $\mathbf{P}_2 = \frac{3(n+3)}{n+9} \ldots$, $\mathbf{P}_3 = \frac{1}{n} \ldots$, $\mathbf{P}_4 = \frac{4}{3} \ldots - \mathbf{P}_5 - \mathbf{P}_6$

$\mathbf{P}_5 = \frac{8(n+3)}{9} \ldots$, $\mathbf{P}_6 = \ldots$, $\mathbf{P}_{10} = \ldots$

Table 18.3 E_6 family Clebsch-Gordan series for $V \otimes V \otimes V$, with ■□ defined in table 18.2. The dimensions and Dynkin labels of repeated reps are listed only once.

This follows from the invariance condition (6.53):

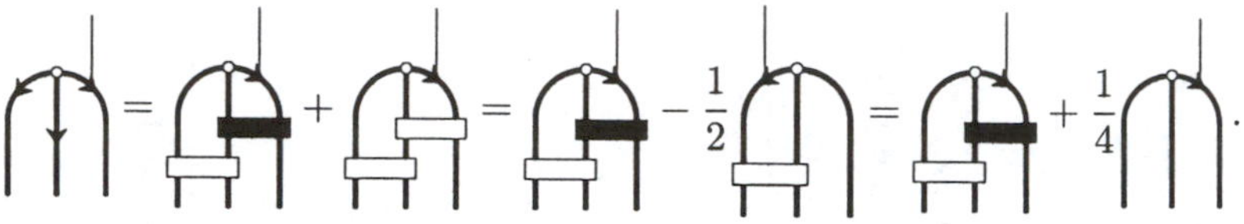

$$= \quad + \quad = \quad - \frac{1}{2} \quad = \quad + \frac{1}{4} \quad .$$

Hence, the adjoint subspace lies in the mixed symmetry subspace, projected by (9.10). Substituting expansions of tables 18.2 and 18.3, we obtain

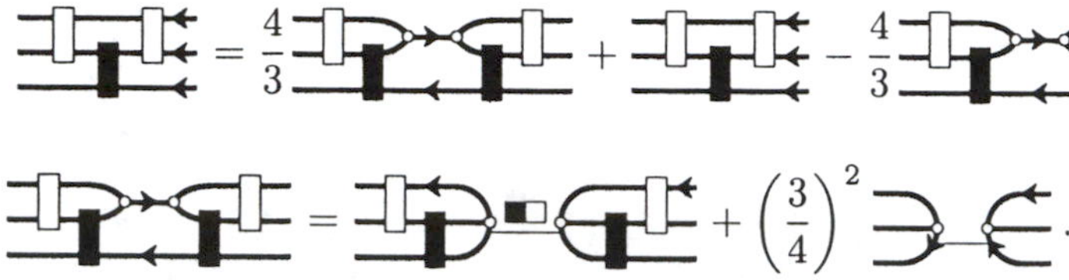

$$= \frac{4}{3} \quad + \quad - \frac{4}{3}$$

$$= \quad + \left(\frac{3}{4}\right)^2 \quad . \qquad (18.18)$$

The corresponding decomposition is listed in table 18.3. The other mixed symmetry subspace from table 9.1 decomposes in the same way.

18.4.3 Fully antisymmetric $V \otimes V \otimes V$ tensors

All invariant matrices on $\otimes V^3 \to \otimes V^3$, constructed from d_{abc} primitives, are symmetric in at least a pair of indices. They vanish on the fully antisymmetric subspace, hence, the fully antisymmetric subspace in table 9.1 is irreducible for E_6.

18.5 DEFINING $\otimes$ ADJOINT TENSORS

We turn next to the determination of the Clebsch-Gordan series for $V \otimes A$ reps. As always, this series contains the n-dimensional rep

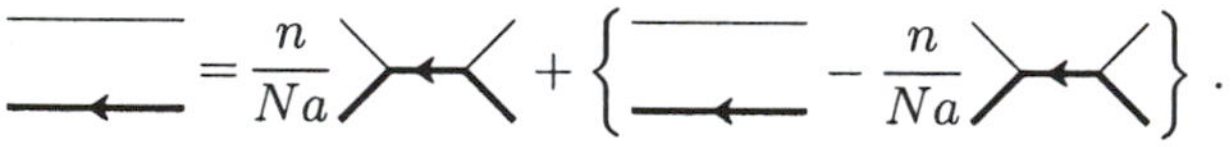

$$= \frac{n}{Na} \quad + \left\{ \quad - \frac{n}{Na} \quad \right\} .$$

$$1 = \quad \mathbf{P}_1 \quad + \quad \mathbf{P}_5 \qquad (18.19)$$

Existence of the invariant tensor

$$\qquad (18.20)$$

implies that $V \otimes A$ also contains a projection onto the $V \otimes V$ space. The symmetric rep in (18.3) does not contribute, as the d_{abc} invariance reduces (18.20) to a projection onto the V space:

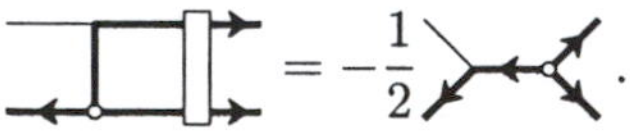

$$= -\frac{1}{2} \quad . \qquad (18.21)$$

The antisymmetrized part of (18.20),

$$\mathbf{R} = \quad , \qquad \mathbf{R}^\dagger = \quad , \qquad (18.22)$$

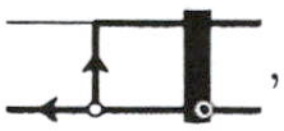

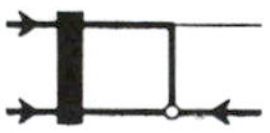

projects out the $V \otimes V$ antisymmetric intermediate state, as in (18.3):

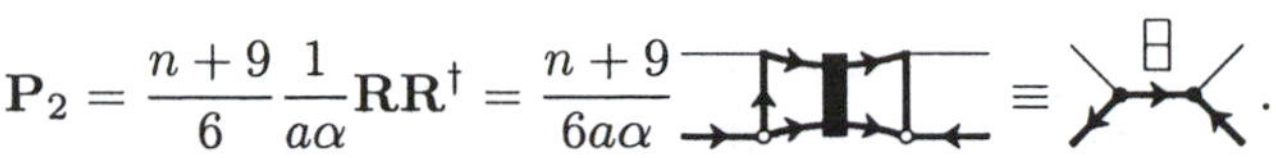

$$\mathbf{P}_2 = \frac{n+9}{6}\frac{1}{a\alpha}\mathbf{R}\mathbf{R}^\dagger = \frac{n+9}{6a\alpha} \quad \equiv \quad . \tag{18.23}$$

Here the normalization factor is evaluated by substituting the adjoint projection operator $\mathbf{P}_A$ (table 18.2) into

$$\mathbf{R}^\dagger\mathbf{R} = \quad = \frac{6}{n+9}a\alpha \quad . \tag{18.24}$$

In this way, $\mathbf{P}_5$ in (18.19) reduces to $\mathbf{P}_5 = \mathbf{P}_2 + \mathbf{P}_c$,

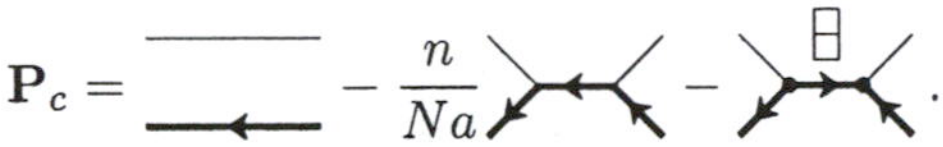

$$\mathbf{P}_c = \quad - \frac{n}{Na} \quad - \quad . \tag{18.25}$$

However, $\mathbf{P}_c$ subspace is also reducible, as there exists still another invariant matrix on $V \otimes A$ space:

$$\mathbf{Q} = \frac{1}{a} \quad . \tag{18.26}$$

We compute $\mathbf{Q}^2\mathbf{P}_c$ by substituting the adjoint projection operator and dropping the terms that belong to projections onto V and $V \otimes V$ spaces:

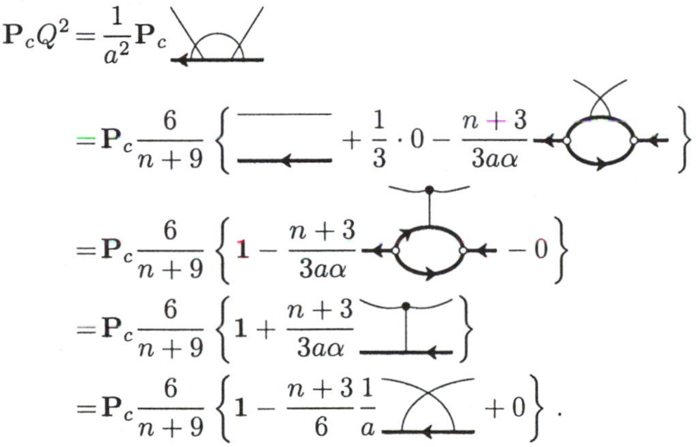

$$\begin{aligned}
\mathbf{P}_c Q^2 &= \frac{1}{a^2}\mathbf{P}_c \\
&= \mathbf{P}_c \frac{6}{n+9}\left\{ \quad + \frac{1}{3}\cdot 0 - \frac{n+3}{3a\alpha} \quad \right\} \\
&= \mathbf{P}_c \frac{6}{n+9}\left\{ \mathbf{1} - \frac{n+3}{3a\alpha} \quad - 0 \right\} \\
&= \mathbf{P}_c \frac{6}{n+9}\left\{ \mathbf{1} + \frac{n+3}{3a\alpha} \quad \right\} \\
&= \mathbf{P}_c \frac{6}{n+9}\left\{ \mathbf{1} - \frac{n+3}{6}\frac{1}{a} \quad + 0 \right\} .
\end{aligned} \tag{18.27}$$

The resulting characteristic equation is surprisingly simple:

$$\mathbf{P}_c(\mathbf{Q}+1)\left(\mathbf{Q} - \frac{6}{n+9}\right) = 0 . \tag{18.28}$$

The associated projection operators and rep dimensions are listed in table 18.4.

The rep V_4 has dimension zero for $n = 27$, singling out the exceptional group $E_6(27)$. Vanishing dimension implies that the corresponding projection operator (4.22) vanishes identically. This could imply a relation between the contractions of primitives, such as the G_2 alternativity relation implied by the vanishing of (16.30). To investigate this possibility, we expand $\mathbf{P}_4$ from table 18.4.

We start by using the invariance conditions and the adjoint projection operator $\mathbf{P}_A$ from table 18.2 to evaluate

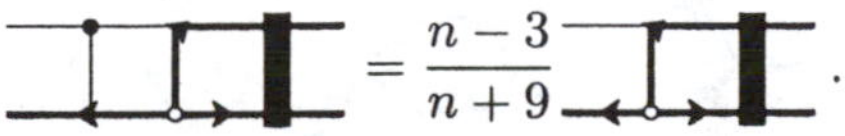

$$\quad = \frac{n-3}{n+9} \quad . \tag{18.29}$$

This yields

$$= \frac{n-3}{n+9} \qquad (18.30)$$

$$\mathbf{P}_4 = \frac{n+9}{n+15}\left\{\frac{1}{4} - + \frac{6}{n+9} - \frac{n+3}{n+9}\right\}.$$

Next, motivated by the hindsight of the next section, we rewrite $\mathbf{P}_2$ in terms of the cubic casimir (7.44). First we use invariance and Lie algebra (4.47) to derive the relation

$$= - \frac{1}{4}. \qquad (18.31)$$

We use the adjoint projection operator (18.11) to replace the $d_{abc}d^{cde}$ pair in the first term,

$$= \frac{1}{n+3}\left\{-\frac{n+9}{2} + 3 + \right\}. \qquad (18.32)$$

In terms of the cubic casimir (7.44), the $\mathbf{P}_2$ projection operator is given by

$$= \frac{n+9}{6(n+3)}\left\{-\frac{n+9}{4} - \frac{n-3}{4} + \frac{3}{2} + \right\}. \qquad (18.33)$$

Substituting back into (18.30), we obtain

$$\mathbf{P}_4 = \frac{n+9}{n+15}\left\{\frac{27-n}{6}\left(\frac{1}{n+9} - \frac{1}{4}\right) + \frac{n+9}{24}\right\}. \qquad (18.34)$$

We shall show in the next section that the cubic casimir, in the last term, vanishes for $n = 27$. Hence, each term in this expansion vanishes separately for $n = 27$, and no new relation follows from the vanishing of $\mathbf{P}_4$. Too bad.

However, the vanishing of the cubic casimir for $n = 27$ does lead to several important relations, special to the E_6 algebra. One of these is the reduction of the loop contraction of 6 d_{abc}'s. For E_6 (18.33) becomes

$$E_6: \quad = \frac{1}{5}\left\{ + \frac{3}{2} - 6 \right\}. \qquad (18.35)$$

The left-hand side of this equation is related to a loop of 6 d_{abc}'s (after substituting the adjoint projection operators):

$$E_6: \quad = 6 - \frac{3}{2}. \qquad (18.36)$$

The right-hand side of (18.35) contains no loop contractions. Substituting the adjoint operators in both sides of (18.35), we obtain a tree expansion for loops of length 6:

$$E_6: \quad \frac{1}{\alpha^3} \quad = \tag{18.37}$$

$$\frac{1}{500}\left\{ \frac{3}{2}\{\ +\ \} - \ + \ + \ + \ - \frac{5}{\alpha}\{\ +\ +\ +\ +\ \} + \frac{10}{\alpha}\{\ +\ +\ \} + \frac{15}{\alpha} \ - \frac{50}{\alpha^2}\{\ +\ +\ +\ +\ \} \right\}$$

At the time of writing this report, we lack a proof that we can compute any scalar invariant built from d_{abc} contractions. However, the scalar invariants that we might be unable to compute are of very high order, bigger than anything listed in table 5.1, as their shortest loop must be of length 8 or longer, with no less than 30 vertices in a vacuum bubble. (See table 2 in ref. [294] for the minimal number of vacuum bubble vertices for a given shortest loop, or "girth.")

The Dynkin indices (table 18.4) are computed using (7.29) with $\lambda =$ defining rep, $\mu =$ adjoint rep, $\rho = \lambda_3, \lambda_4$

$$\ell_\rho = \left(\frac{\ell}{n} + \frac{1}{N}\right) d_\rho - \frac{2\ell}{N} \; . \tag{18.38}$$

The value of the 6-j coefficient follows from (18.28), the eigenvalues of the exchange operator $\mathbf{Q}$.

18.6 TWO-INDEX ADJOINT TENSORS

$A \otimes A$ has the usual starting decomposition (17.7). As in section 9.1, we study the index interchange and the index contraction invariants $\mathbf{Q}$ and $\mathbf{R}$:

$$\mathbf{Q} = \quad , \quad \mathbf{R} = \quad . \tag{18.39}$$

	$A\otimes V$	=	V_1	$\oplus$	V_2	$\oplus$	V_3	$\oplus$	V_4
Dynkin labels	⊠ ⊗ □	=	□	$\oplus$	■■	$\oplus$	□⊠		
E_6	(000001) ⊗ (000010)	=	(000010)	$\oplus$	(010000)	$\oplus$	(000011)		
A_5	(10001) ⊗ (00010)	=	(00010)	$\oplus$	(10100)	$\oplus$	(10011)	$\oplus$	(00002)
A_2	(11) ⊗ (02)	=	(02)	$\oplus$	(21)	$\oplus$	(13)	$\oplus$	(10)
Dimensions	nN	=	n	+	$\frac{n(n-1)}{2}$	+	$\frac{4n(n+1)(n-3)}{n+15}$	+	$\frac{n(n-1)(n-3)(27-n)}{2(n+9)(n+15)}$
E_6	$27 \cdot 78$	=	27	+	351	+	1728	+	0
A_5	$15 \cdot 35$	=	15	+	105	+	384	+	21
$A_2 + A_2$	$9 \cdot 16$	=	9	+	36	+	90	+	9
A_2	$6 \cdot 8$	=	6	+	15	+	24	+	3
Dynkin indices	$n + N\ell$	=	ℓ	+	$(n-2)\ell$	+	$\frac{5(n+1)(n+9)}{3(n+15)}$	+	$\frac{(n-5)(27-n)}{6(n+15)}$
E_6	$27 + \frac{78}{4}$	=	$\frac{1}{4}$	+	$\frac{25}{4}$	+	40	+	0
A_5	$15 + \frac{35}{3}$	=	$\frac{1}{3}$	+	$\frac{13}{3}$	+	$\frac{64}{3}$	+	$\frac{2}{3}$
$A_2 + A_2$	$9 + \frac{16}{2}$	=	$\frac{1}{2}$	+	$\frac{7}{2}$	+	$\frac{25}{2}$	+	$\frac{1}{2}$
A_2	$6 + \frac{8 \cdot 5}{6}$	=	$\frac{5}{6}$	+	$\frac{10}{3}$	+	$\frac{25}{3}$	+	$\frac{1}{6}$

Projection operators

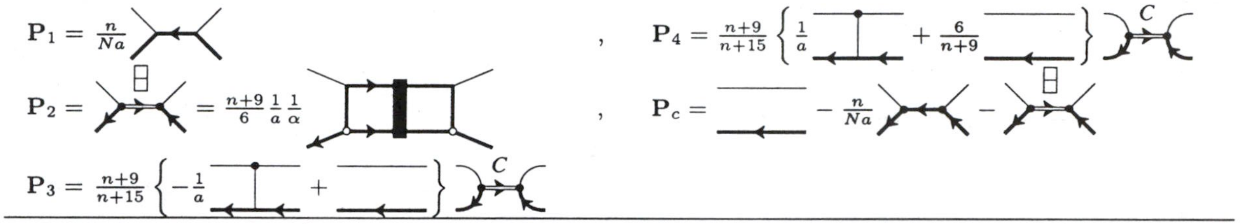

P₁ = n/Na
P₂ = = (n+9)/6 1/a 1/α
P₃ = (n+9)/(n+15) { −1/a + }
C
P₄ = (n+9)/(n+15) { 1/a + 6/(n+9) }
C
Pc = − n/Na −

Table 18.4 E_6 family Clebsch-Gordan series for $A\otimes V$.

The decomposition induced by **R** follows from table 18.2; it decomposes the symmetric subspace $\mathbf{P}_s$

$$\mathbf{P}_s\mathbf{R}\mathbf{P}_s = \frac{1}{a^3} \quad + \frac{1}{a^2} \quad . \tag{18.40}$$

By (9.80) **R** has no effect on the antisymmetric subspaces $\mathbf{P}_A, \mathbf{P}_a$. The corresponding projection operators are normalized by evaluating

$$\frac{1}{a^3} \quad = \frac{(27-n)(n+1)}{2(n+9)^2}$$

$$\frac{1}{a^2} \quad = \frac{12(n-3)}{(n+9)^2} \quad . \tag{18.41}$$

Such relations are evaluated by substituting the Clebsch-Gordan series of table 18.2 into , which yields

$$= \frac{16}{(n+9)^2}\left\{ \quad + (n-2) \quad + \frac{(n+1)(n+9)}{16} \quad \right\} .$$

Equation (18.41) then follows by substitution into

$$= \quad - \frac{C_A}{4} \quad = -\frac{a^2}{2}\frac{(n+1)(n-27)}{(n+9)^2} \quad . \tag{18.42}$$

This implies that the norm of the cubic casimir (7.44) is given by

$$\frac{1}{N}d_{ijk}d_{ijk} = \frac{1}{N} \quad = 4\frac{1}{N} \quad = 2a^3\frac{(n+1)(27-n)}{(n+9)^2} . \tag{18.43}$$

Positivity of the norm restricts $n \leq 27$. For E_6 ($n = 27$), the cubic casimir vanishes identically:

$$E_6 : \quad = 0 . \tag{18.44}$$

18.6.1 Reduction of antisymmetric three-index tensors

Consider the clebsch for projecting the antisymmetric subspace of $V \otimes V \otimes V$ onto $A \otimes A$. By symmetry, it projects only onto the antisymmetric subspace of $A \otimes A$:

$$= \quad . \tag{18.45}$$

Furthermore, it does not contribute to the adjoint subspace:

$$= - \quad + \quad = 0 . \tag{18.46}$$

That both terms vanish can easily be checked by substituting the adjoint projection operator (table 18.2). Furthermore, by substituting (18.37) we have

$$E_6 \quad n = 27 : \quad = \frac{1}{30} \quad . \tag{18.47}$$

This means that for E_6 reps ⊠ and fully antisymmetrized 3-index tensors are equivalent.

	$A\otimes A =$	V_1	$\oplus$	V_2	$\oplus$	V_3	$\oplus$	V_4	$\oplus$	V_5	$\oplus$	V_6	$\oplus$	V_7
label	⊠ ⊗ ⊠ =	•	$\oplus$	·	$\oplus$	■□	$\oplus$	⊠⊠	$\oplus$	⊠	$\oplus$		⊠ over ⊠	
E_6	$(000001)^2 =$	(000000)	+	·	+	(100010)	+	(000002)	+	(000001)	+		(00100)	
A_5	$(10001)^2 =$	(00000)	+	(10001)	+	(01010)	+	(20002)	+	(10001)	+	(01002)	+	(20010)
A_2	$(11)^2 =$	(00)	+	(11)		+		(22)	+	(11)	+	(03)	+	(30)
dimension	$N^2 =$	1	+	$N(1-\delta_{n,27})$	+	$\frac{(n+3)^2(n-1)}{n+9}$	+		+	N	+			
E_6	$78^2 =$	1	+	0	+	650	+	2430	+	78	+		2925	
A_5	$35^2 =$	1	+	35	+	189	+	405	+	35	+	280	+	280
$A_2 + A_2$	$16^2 =$	1	+	16	+		+		+	16	+	52	+	52
A_2	$8^2 =$	1	+	8	+	0	+	27	+	8	+	10	+	10

Projection operators for $E_6 (n = 27)$:

$\mathbf{P}_1 = \frac{1}{78}$ [diagram] , $\mathbf{P}_4 =$ [diagram] $- \mathbf{P}_1 - \mathbf{P}_3$, $\mathbf{P}_6 =$ [diagram] $- \mathbf{P}_5$

$\mathbf{P}_3 = \frac{650}{\text{[diagram]}}$ [diagram] , $\mathbf{P}_5 = \frac{78}{\text{[diagram]}}$ [diagram]

Table 18.5 E_6 family Clebsch-Gordan series of $A\otimes A$.

18.7 DYNKIN LABELS AND YOUNG TABLEAUX FOR E_6

A rep of E_6 is characterized by six Dynkin labels $(a_1a_2a_3a_4a_5a_6)$. The corresponding Dynkin diagram is given in table 7.6. The relation of the Dynkin labels to the Young tableaux (section 7.9) is less obvious than in the case of $SU(n)$, $SO(n)$, and $Sp(n)$ groups, because for E_6 they correspond to tensors made traceless also with respect to the cubic invariant d_{abc}.

The first three labels a_1, a_2, a_3 have the same significance as for the $U(n)$ Young tableaux. a_1 counts the number of (not antisymmetrized) contravariant indices (columns of one box ■). a_2 counts the number of antisymmetrized contravariant index pairs (columns of two boxes ▮). a_3 is the number of antisymmetrized covariant index triples. That is all as expected, as the symmetric invariant d_{abc} cannot project anything from the antisymmetric subspaces. That is why the antisymmetric reps in table 18.1 and table 18.3 have the same dimension as for $SU(27)$.

However, according to (18.47), an antisymmetric contravariant index triple is equivalent to an antisymmetric pair of adjoint indices. Hence, contrary to the $U(n)$ intuition, this rep is *real*. We can use the clebsches from (18.47) to turn any set of $3p$ antisymmetrized contravariant indices into p adjoint antisymmetric index pairs. For example, for $p = 2$ we have

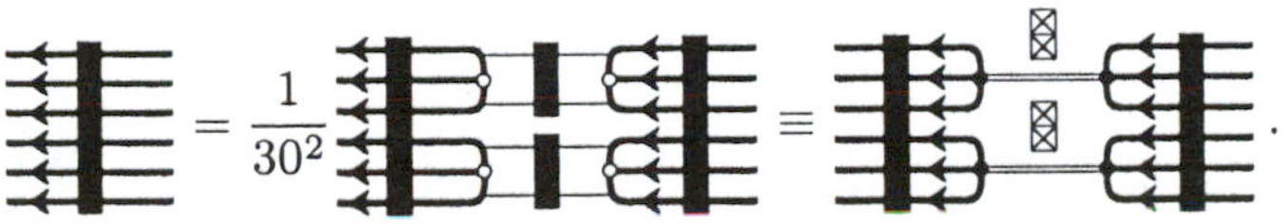

$$= \frac{1}{30^2} \qquad \equiv \qquad . \qquad (18.48)$$

Hence, a column of more than two boxes is always reduced modulo 3 to a_3 antisymmetric adjoint pairs (in the above example $a_3 = p$), that we shall denote by columns of two crossed boxes ⊠⊠.

In the same fashion, the antisymmetric covariant index n-tuples contribute to a_3, the number of antisymmetric adjoint pairs ⊠⊠, a_4 antisymmetrized covariant index pairs ⊟, and a_5 (not antisymmetrized) covariant indices □.

Finally, taking a trace of a covariant-contravariant index pair implies removing both a singlet *and* an adjoint rep. We shall denote the adjoint rep by ⊠. The number of (not antisymmetrized) adjoint indices is given by a_6. For example, an $SU(n)$ tensor $x^a_b \in V \otimes \bar{V}$ decomposes into three reps of table 18.2. The first one is the singlet (000000), that we denote by •. The second one is the adjoint subspace (0000001) = ⊠. The remainder is labeled by the number of covariant indices $a_1 = 1$, and contravariant indices $a_5 = 1$, yielding (100010) = ■□ rep.

Any set of $2p$ antisymmetrized adjoint indices is equivalent to p *symmetrized* pairs by the identity

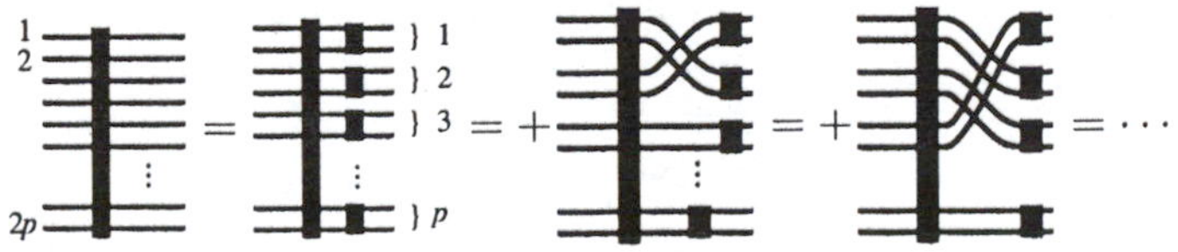

$$= \quad \begin{matrix} \}\,1 \\ \}\,2 \\ \}\,3 \\ \vdots \\ \}\,p \end{matrix} \quad = + \qquad = + \qquad = \cdots \qquad (18.49)$$

This reduces any column of three ⊠⊠⊠ or more antisymmetric indices. We conclude that any irreducible E_6 tensor can, therefore, be specified by six numbers $a_1, a_2, \ldots a_6$.

An E_6 tensor is made irreducible by projecting out all invariant subspaces. We do this by identifying all invariant tensors with right indices and symmetries and constructing the corresponding projection operators, as exemplified by tables 18.1 through 18.5. If we are interested only in identifying the terms in a Clebsch-Gordan series, this can be quickly done by listing all possible nonvanishing invariant projections (many candidates vanish by symmetry or the invariance conditions) and checking whether their dimensions (from the Patera-Sankoff tables [273]) add up. Examples are given in table 18.6.

To summarize, the correspondence between the E_6 Dynkin diagram from table 7.6, Dynkin labels, irreducible tensors, and the dimensions of the lowest corresponding reps is

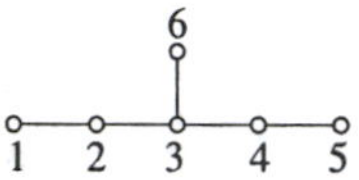

$$\leftrightarrow (a_1, a_2, a_3, a_4, a_5, a_6) \leftrightarrow$$

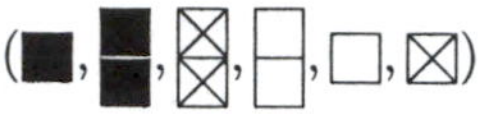

$$(\blacksquare, \ldots) \leftrightarrow (27, 351, 2925, 351, 27, 78) \tag{18.50}$$

$a_1 =$ number of not antisymmetrized contravariant indices

$a_2 =$ number of antisymmetrized contravariant pairs

$a_3 =$ number of antisymmetrized adjoint index pairs

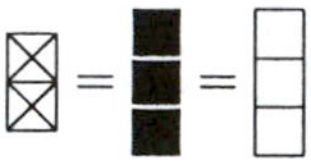

$a_4 =$ number of antisymmetrized covariant pairs

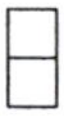

$a_5 =$ number of not antisymmetrized covariant indices

$a_6 =$ number of not antisymmetrized adjoint indices

For example, the Young tableau for the rep (2,1,3,2,1,2) can be drawn as

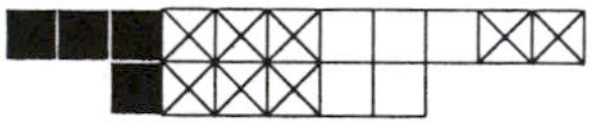

$$. \tag{18.51}$$

The difference in the number of the covariant and contravariant indices

$$a_1 + 2a_2 - 2a_4 - a_5 \qquad (\text{mod } 3) \tag{18.52}$$

is called *triality*. Modulo 3 arises because of the conversion of antisymmetric triplets into the real antisymmetric adjoint pairs by (18.47). The triality is a useful check of correctness of a Clebsch-Gordan series, as all subspaces in the series must have the same triality.

18.8 CASIMIRS FOR E_6

In table 7.1 we have listed the orders of independent casimirs for E_6 as 2, 5, 6, 8, 9, 12. Here we shall use our construction of $E_6(27)$ to partially prove this statement.

27 · 27	=	351	+ 351	+ 27				
27 · 27	=	650	+ 1	+ 78				
351 · 27	=	5824	+ 2925	+ 650	+ 78			
27 · 351	=	7371	+ 27	+ 1728	+ 351			
27 · 78	=	1728	+ 27	+ 351				
78 · 78	=	2925	+ 2430	+ 1	+ 78	+ 650		
351 · 27	=	5824	+ 3003	+ 650				
27 · 351	=	7722	+ 27	+ 1728				
650 · 27	=	7722	+ 7371	+ 351	+ 351	+ 1728	+ 27	
331 · 78	=	17550	+ 351	+ 351	+ 27	+ 7311	+ 1728	
2925 · 27	=	51975	+ 1728	+ 17550	+ 7371	+ 351		

Table 18.6 Examples of the E_6 Clebsch-Gordan series in terms of the Young tableaux. Various terms in the expansion correspond to projections on various subspaces, indicated by the Clebsch-Gordan coefficients listed on the right. See tables 18.1 through 18.5 for explicit projection operators.

By the hermiticity of T_i, the fully symmetric tensor d_{ijk} from (18.43) is real, and

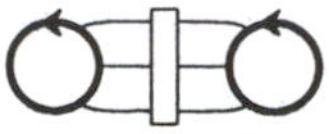

$$= (d_{ijk})^2 \geq 0. \tag{18.53}$$

By (18.43), this equals

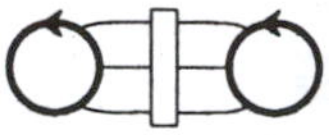

$$= \frac{a^3}{2} \frac{(n+1)(27-n)}{(n+9)^2} N. \tag{18.54}$$

The cubic casimir d_{ijk} *vanishes identically* for E_6.

Next we prove that the *quartic casimir* for E_6 is reducible. From table 18.1 expression for the adjoint rep projection operator we have

$$= \frac{3}{n+3} \left\{ -\frac{n+9}{6} \quad + \frac{1}{3} \quad + \quad \right\}, \tag{18.55}$$

which yields

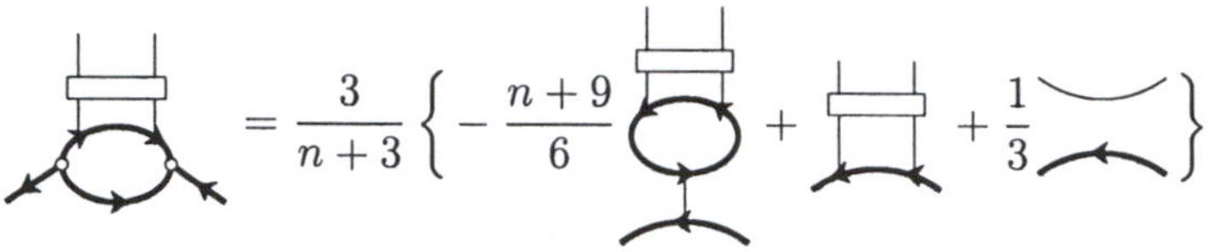

$$= \frac{3}{n+3} \left\{ -\frac{n+9}{6} \quad + \quad + \frac{1}{3} \quad \right\}. \tag{18.56}$$

Now the quartic casimir. By the invariance (6.53)

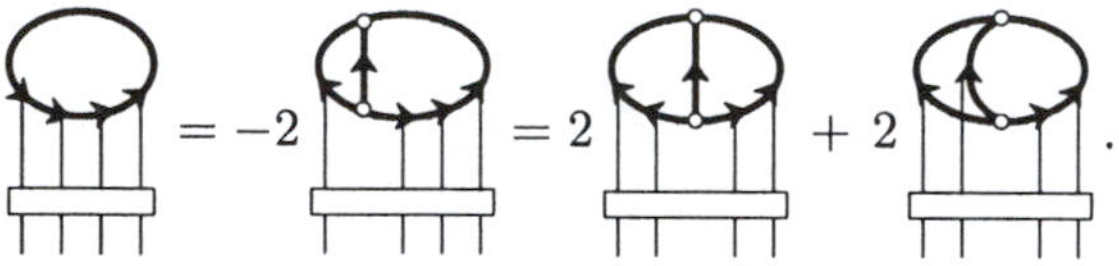

$$= -2 \quad = 2 \quad + 2 \quad . \tag{18.57}$$

The second term vanishes by the invariance (6.53):

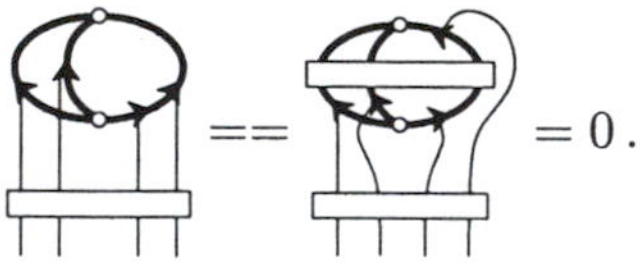

$$== \quad = 0\,. \tag{18.58}$$

Substituting (18.32), we obtain

$$= -\frac{n+9}{n-3} \quad + \frac{2}{n-3} \quad . \tag{18.59}$$

For E_6 the cubic casimir vanishes, and consequently the quartic casimir is a square of the quadratic casimir:

$$E_6 : \operatorname{tr} X^4 = \frac{1}{12} (\operatorname{tr} X^2)^2\,. \tag{18.60}$$

The *quintic* casimir $\operatorname{tr} X^5$ for E_6 must be irreducible, as it cannot be expressed as a power of $\operatorname{tr} X^2$. We leave it as an exercise to the reader to prove that $\operatorname{tr} X^6$ is irreducible.

The reducibility of $\operatorname{tr} X^7$ can be demonstrated by similar birdtrack manipulations, but as the higher irreducible casimirs are beyond manual calculation (according to table 7.1 the Betti numbers for E_6 are 2, 5, 6, 8, 9, 128, 9, 12), this task is better left to a computer [294].

18.9 SUBGROUPS OF E_6

Why is $A_2(6)$ in the E_6 family? The symmetric 2-index rep (9.2) of $SU(3)$ is 6-dimensional. The symmetric cubic invariant (18.2) can be constructed using a pair of Levi-Civita tensors,

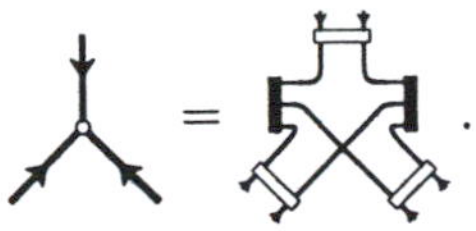

$$= \quad . \tag{18.61}$$

Contractions of several d_{abc}'s can be reduced using the projection operator properties (6.28) of Levi-Civita tensors, yielding expressions such as

$$A_2(6): \quad \frac{1}{\alpha} = \frac{1}{3}\left\{ + - 2 \right\}, \tag{18.62}$$

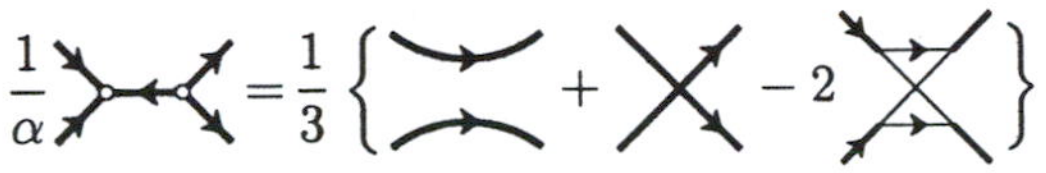

$$\frac{1}{\alpha} = \frac{4}{5}\left\{ - \frac{1}{3} \right\}, \quad etc.. \tag{18.63}$$

The reader can check that, for example, the Springer relation (18.65) is satisfied.

Why is $A_5(15)$ in the E_6 family? The antisymmetric 2-index rep (9.3) of $A_5 = SU(6)$ is 15-dimensional. The symmetric cubic invariant (18.2) is constructed using the Levi-Civita invariant (6.27) for $SU(6)$:

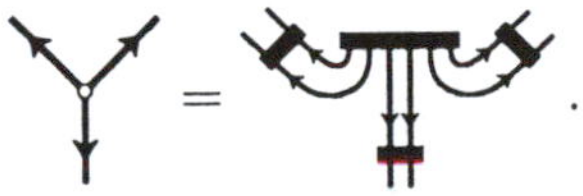

$$= \quad . \tag{18.64}$$

The reader is invited to check the correctness of the primitiveness assumption (18.5). All other results of this chapter then follow.

Is $A_2 + A_2(9)$ in the E_6 family? Exercise for the reader: unravel the $A_2 + A_2$ 9-dimensional rep, construct the d_{abc} invariant.

18.10 SPRINGER RELATION

Substituting $\mathbf{P}_A$ into the invariance condition (6.53), one obtains the *Springer relation* [314, 315]

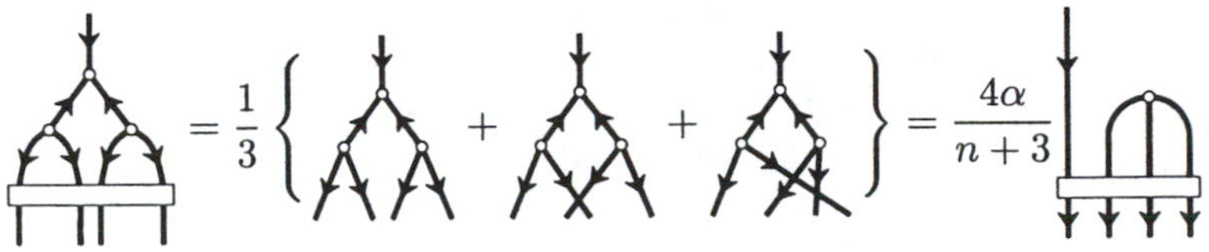

$$= \frac{1}{3}\left\{ + + \right\} = \frac{4\alpha}{n+3} \quad . \tag{18.65}$$

The Springer relation can be used to eliminate one of the three possible contractions of three d_{abc}'s. For the G_2 family it was possible to reduce *any* contraction of three

f_{abc}'s by (16.15); however, a single chain of three d_{abc}'s *cannot* be reducible. If it were, symmetry would dictate a reduction relation of the form

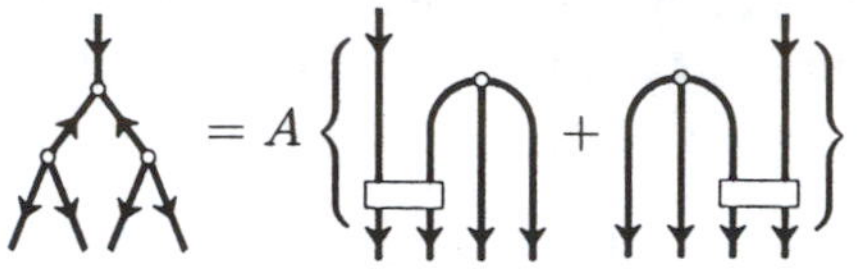

$$= A \left\{ \quad + \quad \right\} . \tag{18.66}$$

Contracting with d_{abc} one finds that contractions of pairs of d_{abc}'s should also be reducible:

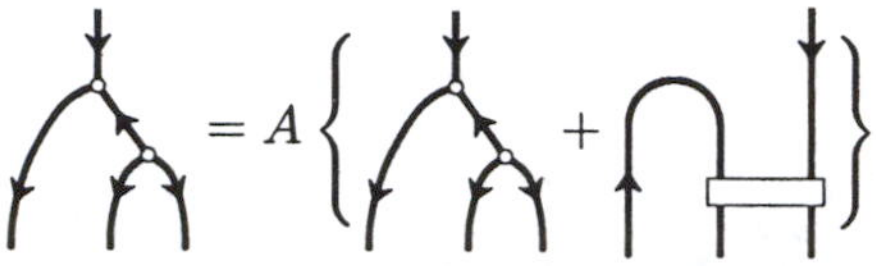

$$= A \left\{ \quad + \quad \right\} . \tag{18.67}$$

Contractions of this relation with d_{abc} and δ^a_b yields $n = 1$, *i.e.*, reduction relation (18.66) can be satisfied only by a trivial 1-dimensional defining rep.

18.11 SPRINGER'S CONSTRUCTION OF E_6

In the preceding sections we have given a self-contained derivation of the E_6 family, in notation unfamiliar to the handful of living experts on this subject. Here we translate our results into the more established algebraic notation, and identify the relations already given in the literature.

Definition (Springer [314, 315]). Let V, $\bar{V}$ be finite-dimensional vector spaces paired by an inner product $\langle \bar{x}, x \rangle$ (see section 3.1.2). Assume existence of symmetric trilinear forms $\langle x, y, z \rangle$, $\langle \bar{x}, \bar{y}, \bar{z} \rangle$. If $x, y \in V$, there exists by duality $x \times y \in \bar{V}$ such that

$$3\langle x, y, z \rangle = \langle x \times y, z \rangle \,, \tag{18.68}$$

with the $\bar{x} \times \bar{y} \in V$ product defined similarly. Assume that the $\times$ product satisfies *Springer relation* [130]

$$(x \times x) \times (x \times x) = \langle x, x, x \rangle \, x \tag{18.69}$$

(together with the corresponding formula for $x \to \bar{x}$). Springer proves that the exceptional simple Jordan algebra of $[3 \times 3]$ hermitian matrices x with octonionic matrix elements [129, 130, 304, 168] satisfies these assumptions, and that the characteristic equation for $[3 \times 3]$ matrix x yields the relation (18.69). Our purpose here is not to give an account of Freudenthal theory, but to aid the reader in relating the birdtrack notation to Freudenthal-Springer octonionic formulation. The reader is referred to the cited literature for the full exposition and proofs.

The nonassociative multiplication rule for elements x can be written in an orthonormal basis $x = x_a \mathbf{e}^a$, $\bar{x} = x^a \mathbf{e}_a$,

$$\langle \mathbf{e}_a, \mathbf{e}^b \rangle = \delta^b_a \,, \qquad a, b = 1, 2, \ldots\ldots, 27 \,. \tag{18.70}$$

Expand x, $\bar{x}$ and define [150]

$$\mathbf{e}^a \times \mathbf{e}^b = d^{abc}\mathbf{e}_c \,. \tag{18.71}$$

Expressed in this basis, (18.69) is the Springer relation (18.65), with $\alpha = 5/2$. Freudenthal and Springer prove that (18.69) is satisfied if d^{abc} is related to the Jordan product

$$\mathbf{e}^a \cdot \mathbf{e}^b = \hat{d}^{abc}\mathbf{e}_c$$

by

$$d^{abc} \equiv \hat{d}^{abc} - \frac{1}{2}[\delta^{ab}\,\mathrm{tr}(\mathbf{e}^c) + \delta^{ac}\,\mathrm{tr}(\mathbf{e}^b) + \delta^{bc}\,\mathrm{tr}(\mathbf{e}^a)] + \frac{1}{2}\,\mathrm{tr}(\mathbf{e}^a)\,\mathrm{tr}(\mathbf{e}^b)\,\mathrm{tr}(\mathbf{e}^c)\,.$$

The defining $n = 27$ representation of E_6 is the group of isomorphisms that leave $\langle \bar{x}, y\rangle = \delta_a^b x^a y_b$ and $\langle x, y, z\rangle = d^{abc}x_a y_b z_c$ invariant. The "derivation" (4.25) $V^2 \otimes \tilde{V} = V \otimes A \to V$ is given by Freudenthal, equation (1.21) in ref. [129]:

$$Dz \equiv [x, \bar{y}]\,z = 2\bar{y} \times (x \times z) - \frac{1}{2}\langle \bar{y}, z\rangle\, x - \frac{1}{6}\langle \bar{y}, x\rangle\, z\,.$$

Expressed in the basis (18.70), this is the adjoint projection operator $\mathbf{P}_A$ (table 18.2),

$$(Dz)_d = -3\,x_a y^b (P_A)_{b\,d}^{a\,c}\, z_c \,. \tag{18.72}$$

The invariance of the x-product is given by Freudenthal as

$$\langle Dx, x \times x\rangle = 0.$$

Expressed in the basis (18.70) this is the invariance condition (6.53) for d_{abc}.

Chapter Nineteen

F_4 family of invariance groups

In this chapter we classify and construct all invariance groups whose primitive invariant tensors are a symmetric bilinear d_{ab}, and a symmetric trilinear d_{abc}, satisfying the relation (19.16).

Take as primitives a symmetric quadratic invariant d_{ab} and a symmetric cubic invariant d_{abc}. As explained in chapter 12, we can use d_{ab} to lower all indices. In the birdtrack notation, we drop the open circles denoting symmetric 2-index invariant tensor d^{ab}, and we drop arrows on all lines:

$d^{ab} = d_{ab} =$,

$d_{abc} = d_{bac} = d_{acb} =$ 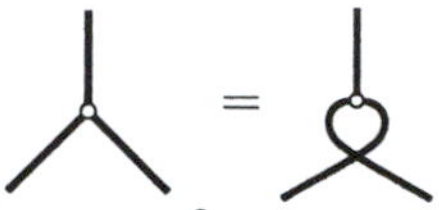 . (19.1)

The defining n-dimensional rep is by assumption irreducible, so

$d_{abc}d_{bcd} =$ 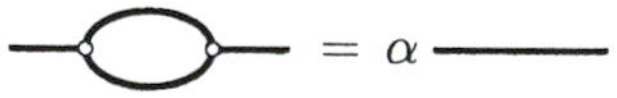 $= \alpha\, \delta_{ad}$ (19.2)

$d_{abb} =$  $= 0\,.$ (19.3)

Were (19.3) nonvanishing, we could use  to project out a 1-dimensional subspace, violating the assumption that the defining rep is irreducible. The value of α depends on the normalization convention (Schafer [304] takes $\alpha = 7/3$).

19.1 TWO-INDEX TENSORS

d_{abc} is a clebsch for $V \otimes V \to V$, so without any calculation the $V \otimes V$ space is decomposed into four subspaces:

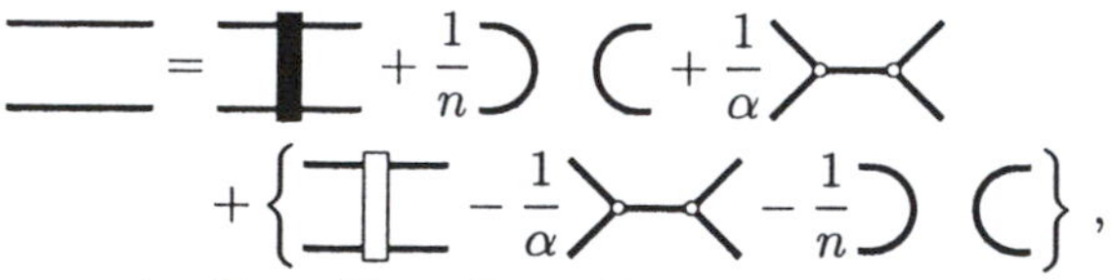

$$\mathbf{1} = \mathbf{P}_{\boxminus} + \mathbf{P}_{\bullet} + \mathbf{P}_{\square} + \mathbf{P}_{3}\,. \tag{19.4}$$

We turn next to the decompositions induced by the invariant matrix

$\mathbf{Q}_{ab,cd} = \frac{1}{\alpha}$  . (19.5)

I shall assume that **Q** does not decompose the symmetric subspace, *i.e.*, that its symmetrized projection can be expressed as

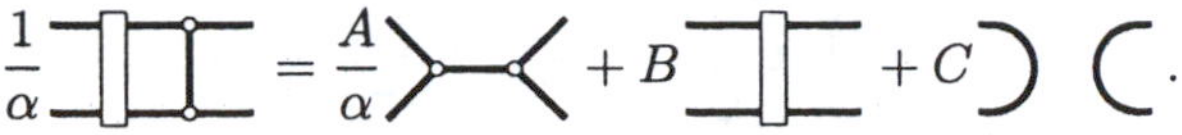

(19.6)

Together with the list of primitives (19.1), this assumption *defines* the F_4 family. This corresponds to the assumption (16.3) in the construction of G_2. I have not been able to construct the F_4 family without this assumption.

Invariance groups with primitives d_{ab}, d_{abc} that do not satisfy (19.6) do exist. The familiar example [73, 41] is the adjoint rep of $SU(n)$, $n \geq 4$, where d_{abc} is the Gell-Mann symmetric tensor (9.87).

Let us first dispose of the possibility that the invariant 4-tensors in (19.6) satisfy additional relationships. Symmetrizing (19.6) in all legs, we obtain

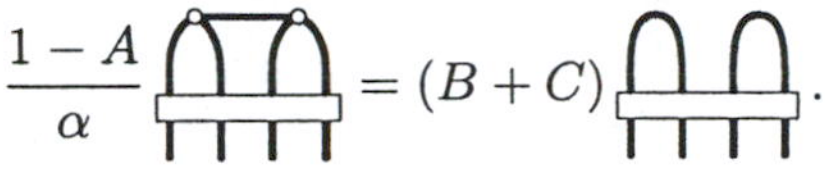

(19.7)

Neither of the tensors can vanish, as contractions with δ's would lead to

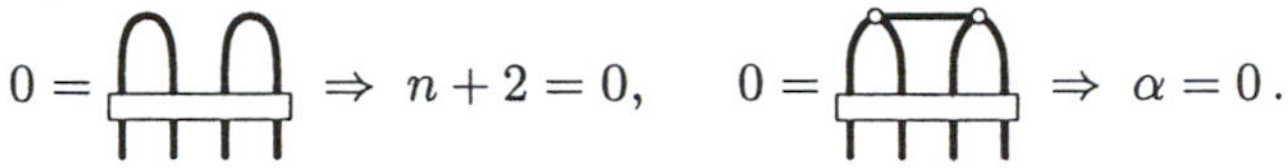

(19.8)

If the coefficients were to vanish, $1 - A = B + C = 0$, we would have

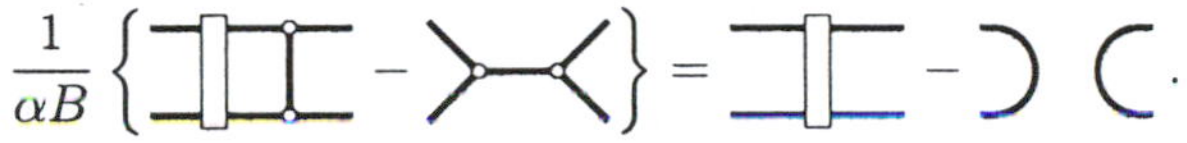

(19.9)

Antisymmetrizing the top two legs, we find that

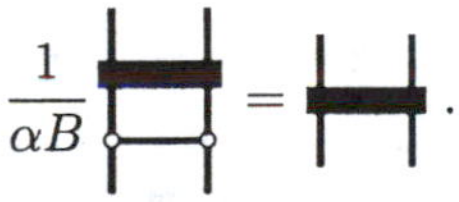

(19.10)

In this case the invariant matrix **Q** of (19.5) can be eliminated,

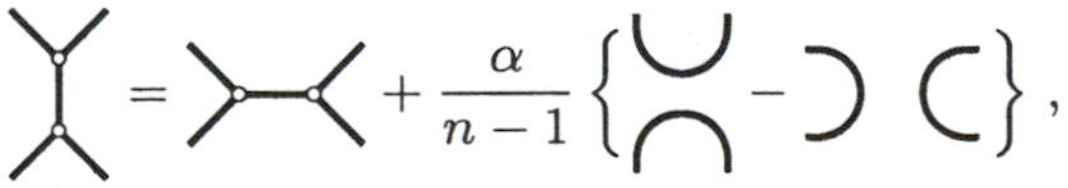

(19.11)

and does not split the antisymmetric part of (19.4). In that case the adjoint rep of $SO(n)$ would also be the adjoint rep for the invariance group of d_{abc}. However, the invariance condition

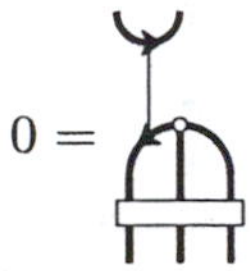

(19.12)

cannot in this case be satisfied for any positive dimension n:

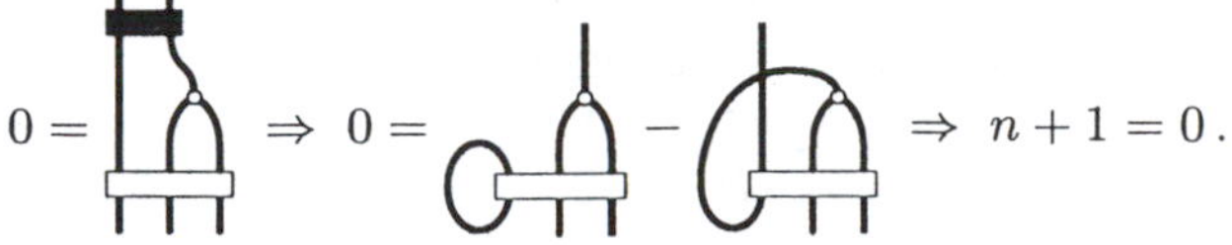

(19.13)

Hence the coefficients in (19.7) are nonvanishing, and there are no additional relations beyond (19.6). The coefficients are fixed by tracing with δ_{ab}:

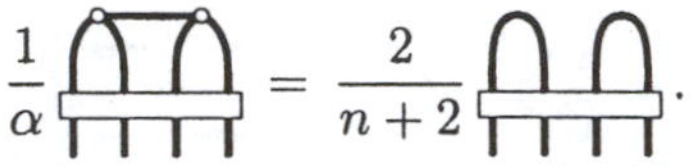
(19.14)

Expanding the symmetrization operator, we can write this relation as

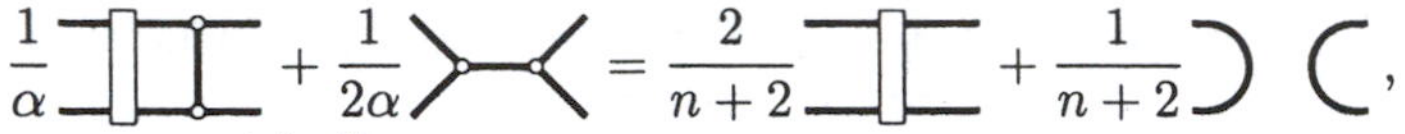
(19.15)

or, more symmetrically, as

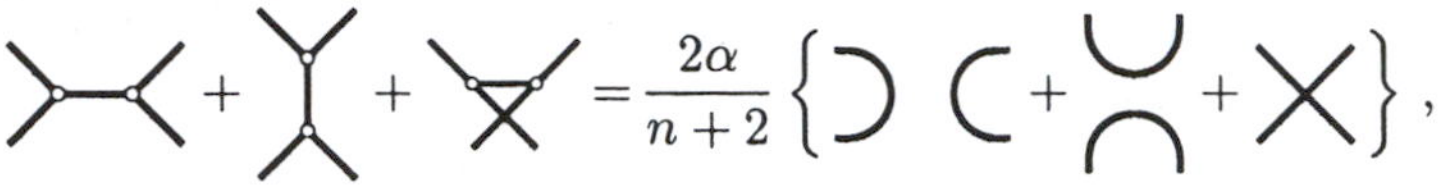

$$d_{abe}d_{ecd} + d_{ade}d_{ebc} + d_{ace}d_{ebd} = \frac{2\alpha}{n+2}(\delta_{ab}\delta_{cd} + \delta_{ad}\delta_{bc} + \delta_{ac}\delta_{bd}) \,. \quad (19.16)$$

In section 19.3, we shall show that this relation can be interpreted as the characteristic equation for [3×3] octonionic matrices. This is the *defining relation* for the F_4 family, equivalent to the assumption (19.6).

The eigenvalue of the invariant matrix **Q** on the n-dimensional subspace can now be computed from (19.15):

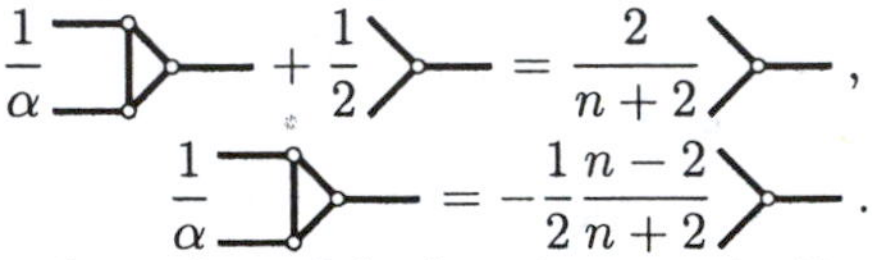
(19.17)

Let us now turn to the action of the invariant matrix **Q** on the antisymmetric subspace in (19.4). We evaluate $\mathbf{Q}^2$ with the help of (19.16) and the identity (6.60), replacing the top $d_{abe}d_{ecd}$ pair by

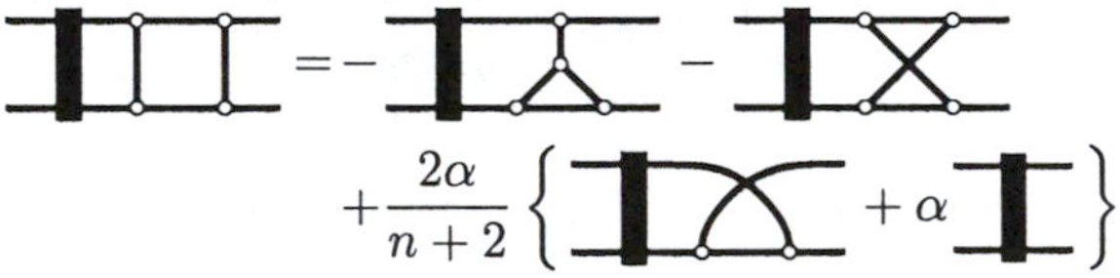

$$0 = A\left(\mathbf{Q}^2 - \frac{1}{2}\frac{n-6}{n+2}\mathbf{Q} - \frac{2}{n+2}\mathbf{1}\right) \,. \quad (19.18)$$

The roots are $\lambda_{\boxtimes} = -1/2$, $\lambda_{\boxminus} = 4/(n+2)$, and the associated projectors yield the adjoint rep and the antisymmetric rep

(19.19)

(19.20)

$\mathbf{P}_{\boxtimes}$ is the projection operator for the adjoint rep, as it satisfies the invariance condition (19.12). The dimensions of the two representations are

$$N = \operatorname{tr}\mathbf{P}_{\boxtimes} = \frac{3n(n-2)}{n+10} \,, \qquad d_{\boxminus} = \operatorname{tr}\mathbf{P}_{\boxminus} = \frac{n(n+1)(n+2)}{2(n+10)} \,, \quad (19.21)$$

and the Dynkin index of the defining representation is

$$\ell = \frac{n+10}{5n-22} \,. \quad (19.22)$$

19.2 DEFINING ⊗ ADJOINT TENSORS

The $V \otimes A$ space always contains the defining rep:

$$1 = \mathbf{P}_6 + \mathbf{P}_7 . \qquad (19.23)$$

We can use d_{abc} and $(T_i)_{ab}$ to project a $V \otimes V$ subspace from $V \otimes A$:

$$\mathbf{R}_{ia,bc} = . \qquad (19.24)$$

By the invariance condition (19.12), $\mathbf{R}$ projects the symmetrized $V \otimes V$ subspace onto V

$$= -\frac{1}{2} . \qquad (19.25)$$

Hence, $\mathbf{R}$ maps the $\mathbf{P}_7$ subspace only onto the antisymmetrized $V \otimes V$:

$$\mathbf{P}_7 \mathbf{R} = \mathbf{R}\mathbf{A}$$

$$\mathbf{P}_7 = . \qquad (19.26)$$

The $V \otimes V$ space was decomposed in the preceding section. Using (19.19) and (19.20), we have

$$= + . \qquad (19.27)$$

The $\mathbf{P}_7$ space can now be decomposed as

$$\mathbf{P}_7 = \mathbf{P}_8 + \mathbf{P}_9 + \mathbf{P}_{10}$$

$$- \frac{n}{aN} = \frac{N}{} + \frac{d_{\square}}{} + \mathbf{P}_{10} . \qquad (19.28)$$

Here,

$$= \frac{1}{a} ,$$

$$= - , \qquad (19.29)$$

and the normalization factors are the usual normalizations (5.8) for 3-vertices. An interesting thing happens in evaluating the normalization for the $\mathbf{P}_8$ subspace: substituting (19.19) into $\frac{1}{\alpha}$, we obtain

$$\frac{1}{N} = \frac{1}{\alpha a^2} = \frac{26-n}{4(n+10)} ,$$

$$\frac{1}{d_{\square}} = \frac{6(n-2)}{(n+2)(n+10)} . \qquad (19.30)$$

The normalization factor is a sum of squares of real numbers:

$$= \frac{1}{\alpha a^2} \sum_{i,j,a} \left[(T_i)_{bc} d_{acd} (T_j)_{db}\right]^2 \geq 0 \,. \tag{19.31}$$

Hence, either $n = 26$ or $n < 26$. We must distinguish between the two cases: as the corresponding clebsches are identically zero,

$$n = 26 : \qquad = 0 \,, \tag{19.32}$$

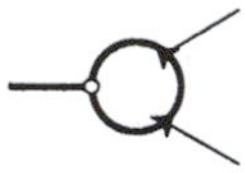

and $\mathbf{P}_7$ subspace in (19.28) does not contain the adjoint rep, (19.28) is replaced by

$$n = 26 : \qquad - \frac{n}{aN} \qquad = \frac{d_{\square}}{} \qquad + \mathbf{P}_{10} \,. \tag{19.33}$$

Another invariant matrix on $V \otimes A$ space can be formed from two $(T_i)_{ab}$ generators:

$$\mathbf{Q} = \qquad . \tag{19.34}$$

We compute $\mathbf{P}_{10}\mathbf{Q}^2$ by substituting the adjoint projection operator by (19.19), using the characteristic equation (19.15) and the invariance condition (19.12), and dropping the contributions to the subspaces already removed from $\mathbf{P}_{10}$:

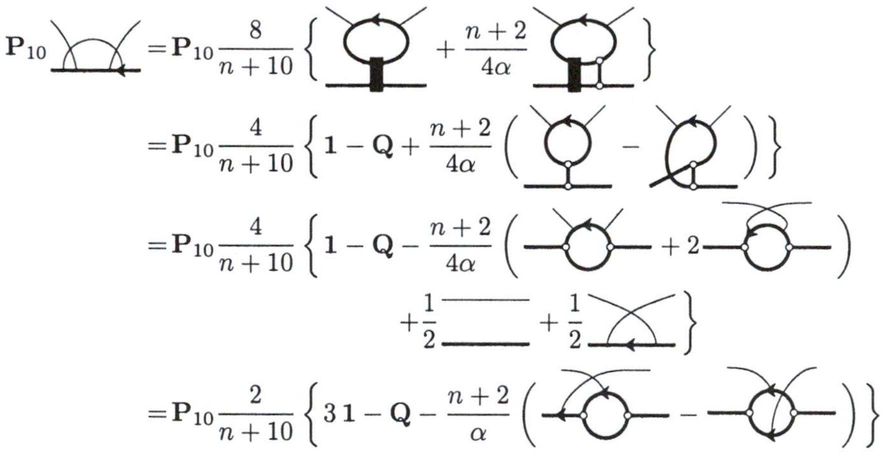

$$\mathbf{P}_{10} \qquad = \mathbf{P}_{10} \frac{8}{n+10} \left\{ \qquad + \frac{n+2}{4\alpha} \qquad \right\}$$

$$= \mathbf{P}_{10} \frac{4}{n+10} \left\{ \mathbf{1} - \mathbf{Q} + \frac{n+2}{4\alpha} \left(\qquad - \qquad \right) \right\}$$

$$= \mathbf{P}_{10} \frac{4}{n+10} \left\{ \mathbf{1} - \mathbf{Q} - \frac{n+2}{4\alpha} \left(\qquad + 2 \qquad \right) + \frac{1}{2} \qquad + \frac{1}{2} \qquad \right\}$$

$$= \mathbf{P}_{10} \frac{2}{n+10} \left\{ 3\,\mathbf{1} - \mathbf{Q} - \frac{n+2}{\alpha} \left(\qquad - \qquad \right) \right\}$$

$$= \mathbf{P}_{10} \frac{2}{n+10} \left\{ 3\,\mathbf{1} - \frac{n+4}{2} \mathbf{Q} + (\text{vanishing}) \right\} . \tag{19.35}$$

Hence $\mathbf{Q}^2$ satisfies a characteristic equation

$$0 = \mathbf{P}_{10} \left(\mathbf{Q}^2 + \frac{n+4}{n+10} \mathbf{Q} - \frac{6}{n+10} \mathbf{1} \right) , \tag{19.36}$$

with roots $\alpha_{11} = -1, \alpha_{12} = 6/(n+10)$, and the corresponding projection operators

$$\mathbf{P}_{11} = \mathbf{P}_{10} \frac{n+10}{n+16} \left(\frac{6}{n+10} \mathbf{1} - \mathbf{Q} \right) , \tag{19.37}$$

$$\mathbf{P}_{12} = \mathbf{P}_{10} \frac{n+10}{n+16} (\mathbf{1} + \mathbf{Q}) \,. \tag{19.38}$$

To use these expressions, we also need to evaluate the eigenvalues of the invariant matrix $\mathbf{Q}$ on subspaces $\mathbf{P}_6, \mathbf{P}_8$, and $\mathbf{P}_9$:

$$\mathbf{QP}_6 = \frac{n}{aN}$$  $$= \left(\frac{N}{n} - \frac{C_A}{2}\right)\mathbf{P}_6 = \frac{1}{2}\mathbf{P}_6\,. \qquad (19.39)$$

We find it somewhat surprising that this eigenvalue does not depend on the dimension n.

$$\mathbf{QP}_8 =$$ 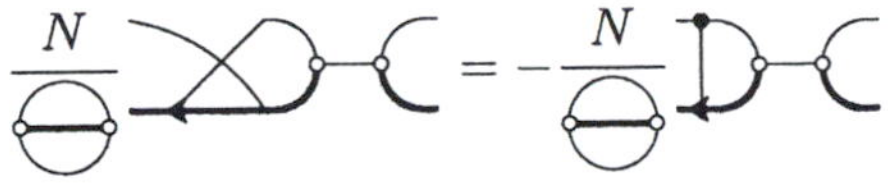

$$= -\frac{N}{2n}\mathbf{P}_8 = -\frac{3(n-2)}{2(n+10)}\mathbf{P}_8$$

$$\mathbf{QP}_9 = -\frac{n-8}{n+10}\mathbf{P}_9\,. \qquad (19.40)$$

These relations are valid for any n.

Now we can evaluate the dimensions of subspaces $\mathbf{P}_{11}, \mathbf{P}_{12}$. We obtain for $n < 26$

$$d_{11} = \mathrm{tr}\,\mathbf{P}_{11} = \frac{n(n-2)(n-5)(14-n)}{2(n+10)(n+16)}\,, \qquad (19.41)$$

$$d_{12} = \mathrm{tr}\,\mathbf{P}_{12} = \frac{3n(n+1)(n-5)}{n+16}\,. \qquad (19.42)$$

A small miracle has taken place: only $n = 26$ and $n \leq 14$ are allowed. However, $d_{12} < 0$ for $n < 5$ does not exclude the $n = 2$ solution, as in that case the dimension of the adjoint rep (19.19) is identically zero, and $V \otimes A$ decomposition is meaningless.

For $n = 26$, $\mathbf{P}_{10}$ is defined by (19.33), the adjoint rep does not contribute, and the dimensions are given by

$$n = 26: \qquad d_{11} = 0, \qquad d_{12} = 1053\,. \qquad (19.43)$$

If a dimension is zero, the corresponding projection operator vanishes identically, and we have a relation between invariants:

$$0 = \mathbf{P}_{11} = \mathbf{P}_{10}\left(\frac{1}{6}\mathbf{1} - \mathbf{Q}\right) = (\mathbf{1} - \mathbf{P}_6 - \mathbf{P}_9)\left(\frac{1}{6}\mathbf{1} - \mathbf{Q}\right).$$

Substituting the eigenvalues of $\mathbf{Q}$, we obtain a relation specific to F_4

$n = 26:$  $= \frac{1}{6}$  $+ \frac{1}{6}$  $- \frac{14}{3}$ 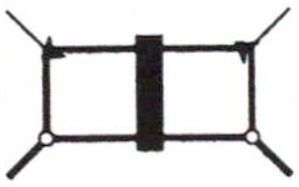 . (19.44)

Hence, for F_4 Lie algebra ($n = 26$) the two invariants, R in (19.26) and Q in (19.34), are not independent.

By now the (very gifted) reader has the hang of it, and can complete the calculation on her own: if so, the author would be grateful to see it. The 2-index adjoint tensors decomposition proceeds in what, by now, is a routine: one first notes that $A \otimes A$ always decomposes into at least four reps (17.6). Then one constructs an invariant tensor that satisfies a characteristic equation on the $A \otimes A$ space, and so on. Some of these calculations are carried out in ref. [74], sections 15, 20, and appendix, p. 97.

□ ⊗ □	=	□□	+	□	+	•	+	□ □	+	⊠				
26^2	=	324	+	26	+	1	+	273	+	52				
⊠ ⊗ □	=	⊠□	+	□	+	□ □								
$52 \cdot 26$	=	1053	+	26	+	273								
⊠ ⊗ ⊠	=	⊠⊠	+	⊠ ⊠	+	⊠	+	•	+	□ □				
52^2	=	1053	+	1274	+	52	+	1	+	324				
□□ ⊗ □	=	□□□	+	□□ □	+	□	+	□ □	+	□□	+	⊠□		
$324 \cdot 26$	=	2652	+	4096	+	26	+	273	+	324	+	1053		
□ □ ⊗ □	=	□□ □	+	⊠ ⊠	+	□□	+	□ □	+	⊠	+	□	+	⊠□
$273 \cdot 26$	=	4096	+	1274	+	324	+	273	+	52	+	26	+	1053

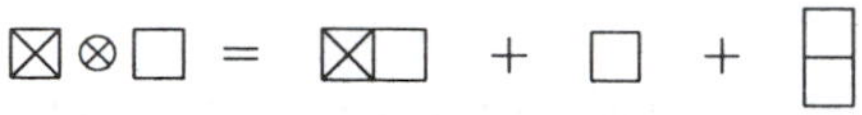

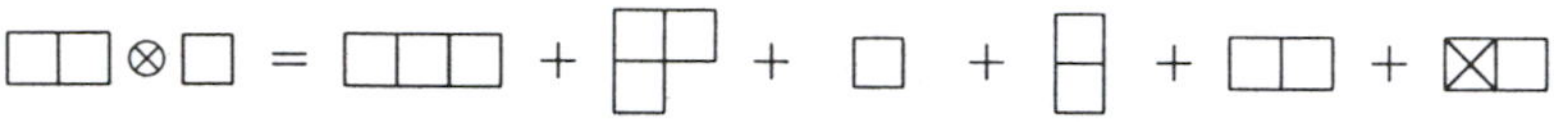

Table 19.1 Kronecker products for the five lowest-dimensional reps of F_4, where □ is the 26-dimensional defining rep, and ⊠ the 52-dimensional adjoint rep. See Patera *et al.* [236] and ref. [194] for tabulations of higher-order series.

19.3 JORDAN ALGEBRA AND $F_4(26)$

As in section 18.11, consider the exceptional simple Jordan algebra of hermitian [3×3] matrices with octonionic matrix elements. The nonassociative multiplication rule for *traceless* octonionic matrices x can be written, in a basis $x = x_a \mathbf{e}_a$, as

$$\mathbf{e}_a \mathbf{e}_b = \mathbf{e}_b \mathbf{e}_a = \frac{\delta_{ab}}{3} \mathbf{I} + d_{abc} \mathbf{e}_c \,, \qquad a, b, c \in \{1, 2, \ldots, 26\} \,, \tag{19.45}$$

where $\mathrm{tr}(\mathbf{e}_a) = 0$, and $\mathbf{I}$ is the [3×3] unit matrix. Traceless [3×3] matrices satisfy the characteristic equation

$$x^3 - \frac{1}{2}\,\mathrm{tr}(x^2)\, x - \frac{1}{3}\,\mathrm{tr}(x^3)\,\mathbf{I} = 0 \,. \tag{19.46}$$

Substituting (19.45) we obtain (19.14), with normalization $\alpha = 7/3$. It is interesting to note that the Jordan identity [304],

$$(xy)x^2 = x(yx^2) \tag{19.47}$$

(which defines Jordan algebra in the way Jacobi identity defines Lie algebra) is a trivial consequence of (19.14). Freudenthal [130] and Schafer [304] show that the group of isomorphisms that leave forms $\mathrm{tr}(xy) = \delta_{ab} x_a x_b$ and $\mathrm{tr}(xyz) = d_{abc} x_a y_b z_c$ invariant is $F_4(26)$. The "derivation" (*i.e.*, Lie algebra generators) is given by Tits:

$$Dz = (xz)y - x(zy) \qquad \text{[eq. (28) in ref. [325]]}. \tag{19.48}$$

Substituting (19.45), we recover the $n = 26$ case of the adjoint rep projection operator (19.19):

$$(Dz)_d = -x_a y_b \left(\frac{1}{3}(\delta_{ad}\delta_{bc} - \delta_{ac}\delta_{bd}) + (d_{bce}d_{ead} - d_{ace}d_{ebd}) \right) z_c \,. \qquad (19.49)$$

19.4 DYNKIN LABELS AND YOUNG TABLEAUX FOR F_4

The correspondence between the f_4 Dynkin diagram from table 7.6, the four Dynkin labels, irreducible tensor Young tableaux, and the dimensions of the lowest corresponding reps is

$$\overset{1\quad 2\quad 3\quad 4}{\circ\!-\!\circ\!=\!\bullet\!-\!\bullet} \leftrightarrow (a_1 a_2 a_3 a_4) \leftrightarrow$$

$$\left(\boxtimes, \boxtimes\boxtimes, \square\square, \square \right) \leftrightarrow (52, 1274, 273, 26)\,. \qquad (19.50)$$

Chapter Twenty

E_7 family and its negative-dimensional cousins

Parisi and Sourlas [269] have suggested that a Grassmann vector space of dimension n can be interpreted as an ordinary vector space of dimension $-n$. As we have seen in chapter 13, semisimple Lie groups abound with examples in which an $n \to -n$ substitution can be interpreted in this way. An early example was Penrose's binors [280], reps of $SU(2) = Sp(2)$ constructed as $SO(-2)$, and discussed here in chapter 14. This is a special case of a general relation between $SO(n)$ and $Sp(-n)$ established in chapter 13; if symmetrizations and antisymmetrizations are interchanged, reps of $SO(n)$ become $Sp(-n)$ reps. Here we work out in detail a 1977 example of a negative-dimensions relation [74], subsequently made even more intriguing [78] by Cremmer and Julia's discovery of a global E_7 symmetry in supergravity [68].

We extend the Minkowski space into Grassmann dimensions by requiring that the invariant length and volume that characterize the Lorentz group ($SO(3,1)$ or $SO(4)$ — compactness plays no role in this analysis) become a quadratic and a quartic supersymmetric invariant. The symmetry group of the Grassmann sector will turn out to be one of $SO(2)$, $SU(2)$, $SU(2) \times SU(2) \times SU(2)$, $Sp(6)$, $SU(6)$, $SO(12)$, or E_7, which also happens to be the list of possible global symmetries of extended supergravities.

As shown in chapter 10, $SO(4)$ is the invariance group of the Kronecker delta $g_{\mu\nu}$ and the Levi-Civita tensor $\varepsilon_{\mu\nu\sigma\rho}$; hence, we are looking for the invariance group of the supersymmetric invariants

$$\begin{aligned}(x, y) &= g_{\mu\nu} x^{\mu} y^{\nu} , \\ (x, y, z, w) &= e_{\mu\nu\sigma\rho} x^{\mu} y^{\nu} z^{\sigma} w^{\rho} ,\end{aligned} \tag{20.1}$$

where $\mu, \nu, \ldots = 4, 3, 2, 1, -1, -2, \ldots, -n$. Our motive for thinking of the Grassmann dimensions as $-n$ is that we define the dimension as a trace (3.52), $n = \delta^{\mu}_{\mu}$, and in a Grassmann (or fermionic) world each trace carries a minus sign. For the quadratic invariant $g_{\mu\nu}$ alone, the invariance group is the orthosymplectic $OSp(4, n)$. This group [177] is orthogonal in the bosonic dimensions and symplectic in the Grassmann dimensions, because if $g_{\mu\nu}$ is symmetric in the $\nu, \mu > 0$ indices, it must be antisymmetric in the $\nu, \mu < 0$ indices. In this way the supersymmetry ties in with the $SO(n) \sim Sp(-n)$ equivalence developed in chapter 13.

Following this line of reasoning, a quartic invariant tensor $e_{\mu\nu\sigma\rho}$, antisymmetric in ordinary dimensions, is symmetric in the Grassmann dimensions. Our task is then to determine all groups that admit an antisymmetric quadratic invariant, together with a symmetric quartic invariant. The resulting classification can be summarized by

$$\text{symmetric } g_{\mu\nu} + \text{antisymmetric } f_{\mu\nu\sigma\rho} : \tag{20.2}$$

$$(A_1 + A_1)(4), G_2(7), B_3(8), D_5(10)$$

antisymmetric $f_{\mu\nu}$ + symmetric $d_{\mu\nu\sigma\rho}$:

$$SO(2), A_1(4), (A_1 + A_1 + A_1)(8), C_3(14), A_5(20), D_6(32), E_7(56),$$

where the numbers in ( ) are the defining rep dimensions. The second case generates a row of the Magic Triangle (figure 1.1).

From the supergravity point of view, it is intriguing to note that the Grassmann space relatives of our $SO(4)$ world include E_7, $SO(12)$, and $SU(6)$ in the same reps as those discovered by Cremmer and Julia. Furthermore, it appears that *all* seven possible groups can be realized as global symmetries of the seven extended supergravities, if one vector multiplet is added to $N = 1, 2, 3, 4$ extended supergravities.

In sections. 20.1–20.3, we determine the groups that allow a symmetric quadratic invariant together with an antisymmetric quartic invariant. The end result of the analysis is a set of Diophantine conditions, together with the explicit projection operators for irreducible reps. In section 20.4, the analysis is repeated for an antisymmetric quadratic invariant together with a symmetric quartic invariant. We find the same Diophantine conditions, with dimension n replaced by $-n$, and the same projection operators, with symmetrizations and antisymmetrizations interchanged.

Parenthetically, you might wonder, how does one figure out such things without birdtracks? I cannot guess, and I suspect one does not. In this chapter the E_7 family is derived diagrammatically, following ref. [74], but as experts with a more algebraic mindset used to find birdtracks very foreign, in ref. [78] we hid our tracks behind the conventional algebraic notation of Okubo [255]. The reader can decide what is easier to digest, algebraic notation or birdtracks.

20.1 $SO(4)$ FAMILY

According to table 10.1, the flip σ from (6.2) together with the index contraction T from (10.8) decompose $V \otimes \overline{V}$ of $SO(n)$ into singlet (10.11), traceless symmetric (10.10), and antisymmetric adjoint (10.12) subspaces, $V \otimes \overline{V} = V_1 \oplus V_2 \oplus V_3$. Now demand, in addition to the above set of V^4 invariant tensors, the existence of a fully antisymmetric primitive *quartic invariant*,

$$f_{\mu\nu\rho\delta} = -f_{\nu\mu\rho\delta} = -f_{\mu\rho\nu\delta} = -f_{\mu\nu\delta\rho} =$$

$$f^{\mu\nu\rho\delta} = . \qquad (20.3)$$

As $f_{\mu\nu\rho\delta}$ is of even rank and thus anticyclic, $f_{\mu\nu\rho\delta} = -f_{\nu\rho\delta\mu}$, we deploy the black semicircle birdtrack notation (6.57) in order to distinguish the first leg.

The only $V \otimes \overline{V} \to V \otimes \overline{V}$ invariant matrix that can be constructed from the new invariant and the symmetric bilinear tensor (10.2) is

$$Q^{\mu\nu'}_{\nu\mu'} = g^{\mu\varepsilon} f_{\nu\varepsilon\mu'\sigma} g^{\sigma\nu'} = \qquad (20.4)$$

(we find it convenient to distinguish the upper, lower indices in what follows). Due to its antisymmetry, the **Q** invariant does not decompose the symmetric subspaces

(10.10), (10.11):

$$\mathbf{P}_1\mathbf{Q} = 0\,, \qquad \mathbf{P}_2\mathbf{Q} = \frac{1}{2}(1+\sigma)\mathbf{Q} = 0\,.$$

The $\mathbf{Q}$ invariant can, however, decompose the antisymmetric V_3 subspace (10.12) into the new adjoint subspace A and the remaining antisymmetric subspace V_7:

$$\text{adjoint:} \quad \mathbf{P}_A = \mathbf{Q} + b\mathbf{P}_3\,, \qquad b = N/d_3$$

$$\text{antisymmetric:} \quad \mathbf{P}_7 = -\mathbf{Q} + (1-b)\mathbf{P}_3 \tag{20.5}$$

where $d_3 = n(n-1)/2$ is the dimension of the $SO(n)$ adjoint representation, b is fixed by $N = \operatorname{tr}\mathbf{P}_A$, and the N is the dimension of the adjoint representation of the $f_{\mu\nu\rho\delta}$ invariance subgroup of $SO(n)$, to be determined.

By the *primitiveness assumption* (3.39) no further invariant matrices $\in \otimes V^4$ exist, linearly independent of $\mathbf{Q}$. In particular, $\mathbf{Q}^2$ is not independent and is reducible to $\mathbf{Q}$ and $\mathbf{P}_3$ by the projection operator indempotency,

$$0 = \mathbf{P}_A^2 - \mathbf{P}_A = \mathbf{Q}^2 + (2b-1)\mathbf{Q} + b(b-1)\mathbf{P}_3 \tag{20.6}$$

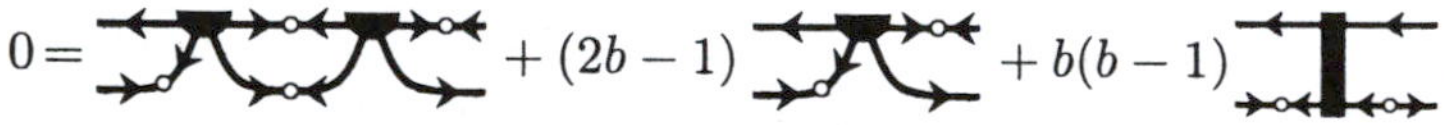

Rewriting the indempotency relation as

$$\mathbf{P}_A^2 = (\mathbf{Q} + b\mathbf{1})\mathbf{P}_A = \mathbf{P}_A$$

yields the eigenvalue $\lambda_A = 1 - b$ of the matrix $\mathbf{Q}$ on the adjoint space A:

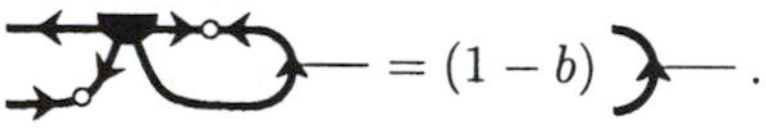

$$= (1-b) \,. \tag{20.7}$$

Condition (20.6) also insures that the $V \to V$ matrix

$$(\mathbf{Q}^2)^{\mu\nu'}_{\nu\mu} = \frac{N d_7}{n d_3}\,\delta^{\nu'}_{\nu}$$

is proportional to unity. Were this not true, distinct eigenvalues of the $\mathbf{Q}^2$ matrix would decompose the defining n-dimensional rep, contradicting the primitiveness assumption that the defining rep is irreducible.

Now antisymmetrize fully the relation (20.6). The $\mathbf{P}_3$ contribution drops out, and the antisymmetrized $\mathbf{Q}^2$ is reduced to $\mathbf{Q}$ by:

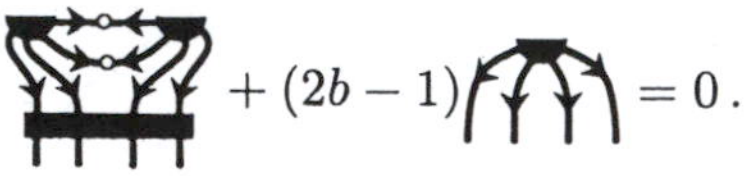

$$+ (2b-1) = 0\,. \tag{20.8}$$

The *invariance condition* (4.35)

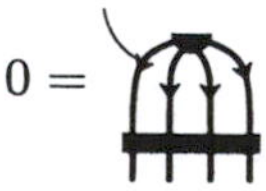

$$0 = \tag{20.9}$$

yields the second constraint on the $\mathbf{Q}^2$:

(20.10)

The quadratic casimir for the defining rep and the "4-vertex" insertion are computed by substituting the adjoint projection operator $\mathbf{P}_A$,

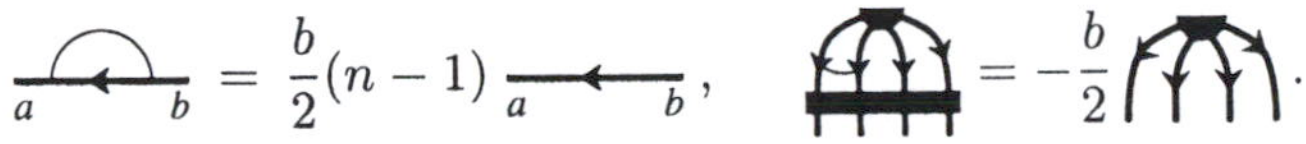

(20.11)

In this way the invariance condition (20.9)

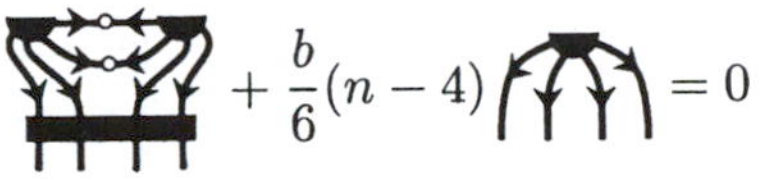

(20.12)

fixes the value of $b = 6/(16-n)$. The projection operators (20.5)

$$\text{adjoint:}\ \mathbf{P}_A = \mathbf{Q} + \frac{6}{16-n}\mathbf{P}_3 \qquad (20.13)$$

$$\text{antisymmetric:}\ \mathbf{P}_7 = -\mathbf{Q} + \frac{10-n}{16-n}\mathbf{P}_3 \qquad (20.14)$$

decompose the $n(n-1)/2$-dimensional adjoint space V_3 of $SO(n)$ into two subspaces of dimensions

$$N = \mathrm{tr}\,\mathbf{P}_A = \frac{3n(n-1)}{16-n}, \qquad d_7 = \mathrm{tr}\,\mathbf{P}_7 = \frac{n(n-1)(10-n)}{2(16-n)}. \qquad (20.15)$$

This completes the decomposition $V \otimes \overline{V} = V_1 \oplus V_5 \oplus A \oplus V_7$. From the Diophantine conditions (20.15) it follows that the subspaces V_A, V_7 have positive integer dimension only for $n = 4, 6, 7, 8, 10$. However, the reduction of $A \otimes V$ undertaken next eliminates the $n = 6$ possibility.

20.2 DEFINING ⊗ ADJOINT TENSORS

The reduction of the $V \otimes \overline{V}$ space, induced by the symmetric $g_{\mu\nu}$ and antisymmetric $f_{\mu\nu\sigma\rho}$ invariants, has led to very restrictive Diophantine conditions (20.15). Further Diophantine conditions follow from the reduction of higher product spaces $\otimes V^q$. We turn to the reduction of (adjoint) ⊗ (defining)$=A \otimes V$ Kronecker product, proceeding as in sections 9.11, 10.2, 18.5, and 19.2.

The three simplest $A \otimes V \to A \otimes V$ invariant matrices one can write down are the identity matrix, and

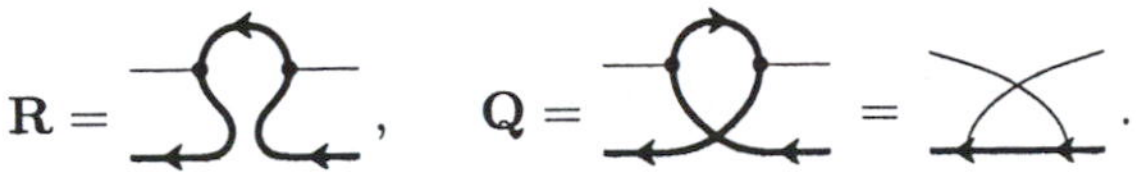

(20.16)

$\mathbf{R}$ projects onto the defining space, $A \otimes V \to V \to A \otimes V$. Its characteristic equation

and the associated projection operators (3.48)

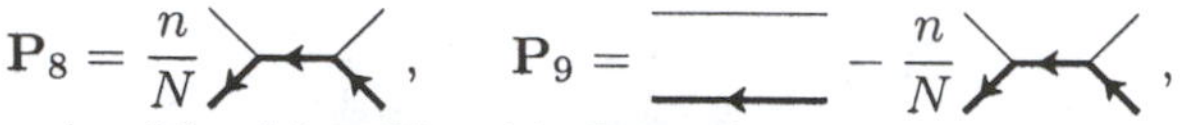

$$\mathbf{P}_8 = \frac{n}{N} \quad , \qquad \mathbf{P}_9 = \quad - \frac{n}{N} \quad , \tag{20.17}$$

decompose $A \otimes V = V_8 \oplus V_9$, with dimensions

$$d_8 = n \,, \qquad d_9 = \operatorname{tr} \mathbf{P}_9 = n(N-1) \,. \tag{20.18}$$

The characteristic equation for

$$\mathbf{Q}^2 =$$

is computed by inserting the adjoint rep projection operator (20.13) and using the invariance condition (20.9) and the $\mathbf{Q}$ eigenvalue (20.7). The result (projected onto the V_9 subspace) is a surprisingly simple quadratic equation,

$$0 = \left(\mathbf{Q}^2 - (1/2 + b)\mathbf{Q} + b/2\right) \mathbf{P}_9 = (\mathbf{Q} - 1/2)\,(\mathbf{Q} + b)\mathbf{P}_9 \,, \tag{20.19}$$

with roots

$$\lambda_{10} = -b \,, \qquad \lambda_{11} = 1/2 \,. \tag{20.20}$$

The $n(N-1)$-dimensional space V_9 is now decomposed into

$$\mathbf{P}_9 = \mathbf{P}_{10} + \mathbf{P}_{11}$$

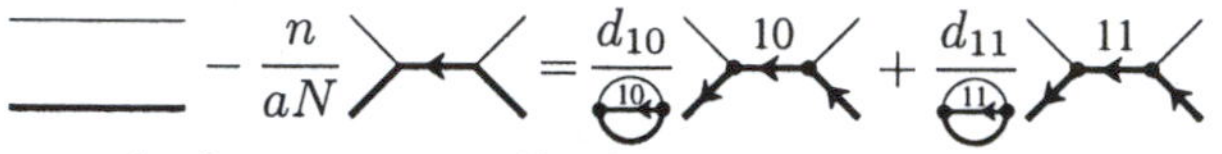

$$- \frac{n}{aN} \quad = \frac{d_{10}}{(10)} \quad + \frac{d_{11}}{(11)} \tag{20.21}$$

(the prefactors are the 3-vertex normalizations (5.8)), with the associated projection operators (3.48)

$$\mathbf{P}_{10} = \frac{2(16-n)}{28-n} \left(-\mathbf{Q} + \frac{1}{2}\mathbf{1} \right) \mathbf{P}_9 \,,$$

$$\mathbf{P}_{11} = \frac{2(16-n)}{28-n} \left(\mathbf{Q} + \frac{6}{16-n}\mathbf{1} \right) \mathbf{P}_9 \,. \tag{20.22}$$

This completes the decomposition $V \otimes V_A = V_8 \oplus V_{10} \oplus V_{11}$. To compute the dimensions of V_{10}, V_{11} subspaces, evaluate

$$\operatorname{tr} \mathbf{P}_9 \mathbf{Q} = -2n(2+n)/(16-n) \,, \tag{20.23}$$

to, finally, obtain

$$d_{10} = \operatorname{tr} \mathbf{P}_{10} = \frac{3n(n+2)(n-4)}{28-n} \,,$$

$$d_{11} = \operatorname{tr} \mathbf{P}_{11} = \frac{32n(n-1)(n+2)}{(16-n)(28-n)} \,. \tag{20.24}$$

The denominators differ from those in (20.15); of the solutions to (20.15), $d = 4, 7, 8, 10$ are also solutions to the new Diophantine conditions. All solutions are summarized in table 20.1.

20.3 LIE ALGEBRA IDENTIFICATION

As we have shown, symmetric $g_{\mu\nu}$ together with antisymmetric $f_{\mu\nu\sigma\rho}$ invariants cannot be realized in dimensions other than $d = 4, 7, 8, 10$. But can they be realized at all? To verify that, one can turn to the tables of Lie algebras of ref. [273] and identify these four solutions.

Rep	Dimension	$A_1 + A_1$	G_2	B_3	D_5
V=defining	n	4	7	8	10
A=adjoint	$N = \frac{3n(n-1)}{16-n}$	3	14	21	45
V_7=antisym.	$\frac{n(n-1)(10-n)}{2(16-n)}$	3	7	7	0
V_5=symmetric	$\frac{(n+2)(n-1)}{2}$	9	27	35	54
V_{10}	$\frac{3n(n+2)(n-4)}{28-n}$	0	27	48	120
V_{11}	$\frac{32n(n-1)(n+2)}{(16-n)(28-n)}$	8	64	112	320

Table 20.1 Rep dimensions for the $SO(4)$ family of invariance groups.

20.3.1 $SO(4)$ or $A_1 + A_1$ algebra

The first solution, $d = 4$, is not a surprise; it was $SO(4)$, Minkowski or euclidean version, that motivated the whole project. The quartic invariant is the Levi-Civita tensor $\varepsilon_{\mu\nu\rho\sigma}$. Even so, the projectors constructed are interesting. Taking

$$Q^{\mu\delta}_{\nu\rho} = g^{\mu\varepsilon} g^{\delta\rho} \varepsilon_{\varepsilon\sigma\nu\gamma}\,, \tag{20.25}$$

one can immediately calculate (20.6):

$$\mathbf{Q}^2 = 4\mathbf{P}_3\,. \tag{20.26}$$

The projectors (20.14) become

$$\mathbf{P}_A = \frac{1}{2}\mathbf{P}_3 + \frac{1}{4}\mathbf{Q}, \quad \mathbf{P}_7 = \frac{1}{2}\mathbf{P}_3 - \frac{1}{4}\mathbf{Q}\,, \tag{20.27}$$

and the dimensions are $N = d_7 = 3$. Also both $\mathbf{P}_A$ and $\mathbf{P}_7$ satisfy the invariance condition, the adjoint rep splits into two invariant subspaces. In this way, one shows that the Lie algebra of $SO(4)$ is the semisimple $SU(2) + SU(2) = A_1 + A_1$. Furthermore, the projection operators are precisely the η, $\overline{\eta}$ symbols used by 't Hooft [164] to map the self-dual and self-antidual $SO(4)$ antisymmetric tensors onto $SU(2)$ gauge group:

$$\begin{aligned}
(\mathbf{P}_A)^{\mu\delta}_{\nu\rho} &= \frac{1}{4}\left(\delta^{\mu}_{\rho}\delta^{\delta}_{\nu} - g^{\mu\delta}g_{\nu\rho} + \varepsilon^{\mu\delta}{}_{\nu\rho}\right) = -\frac{1}{4}\,\eta_{a}{}^{\mu}_{\nu}\,\eta_{a}{}^{\delta}_{\rho}\,,\\
(\mathbf{P}_7)^{\mu\delta}_{\nu\rho} &= \frac{1}{4}\left(\delta^{\mu}_{\rho}\delta^{\delta}_{\nu} - g^{\mu\delta}g_{\nu\rho} - \varepsilon^{\mu\delta}{}_{\nu\rho}\right) = -\frac{1}{4}\,\overline{\eta}_{a}{}^{\mu}_{\nu}\,\overline{\eta}_{a}{}^{\delta}_{\rho}\,.
\end{aligned} \tag{20.28}$$

The only difference is that instead of using an index pair ${}^{\mu}_{\nu}$, 't Hooft indexes the adjoint spaces by $a = 1, 2, 3$. All identities, listed in the appendix of ref. [164], now follow from the relations of section 20.1.

20.3.2 Defining rep of G_2

The 7-dimensional rep of G_2 is a subgroup of $SO(7)$, so it has invariants δ_{ij} and $\varepsilon_{\mu\nu\delta\sigma\rho\alpha\beta}$. In addition, it has an antisymmetric cubic invariant [43, 73] $f_{\mu\nu\rho}$, the invariant that we had identified in section 16.5 as the multiplication table for octonions.

Rep	Dynkin index	$A_1 + A_1$	G_2	B_3	D_5
V=defining	$\frac{16-n}{4(n+2)}$	$\frac{1}{2}$	$\frac{1}{4}$	$\frac{1}{5}$	$\frac{1}{8}$
A=adjoint	1	1	1	1	1
V_7=antisym.	$\frac{(10-n)(n-4)}{4(n+2)}$	0	$\frac{1}{4}$	$\frac{1}{5}$	0
V_5=symmetric	$\frac{1}{4}(16-n)$	3	$\frac{9}{4}$	2	$\frac{3}{2}$
V_{10}	$\frac{7(16-n)(n-4)}{4(28-n)}$	–	$\frac{9}{4}$	$\frac{14}{5}$	$\frac{7}{2}$
V_{11}	$\frac{8(2n+7)}{(28-n)}$	5	8	$\frac{46}{5}$	12

Table 20.2 Dynkin indices for the $SO(4)$ family of invariance groups.

The quartic invariant we have inadvertently rediscovered is

$$f_{\mu\nu\rho\sigma} = \varepsilon_{\mu\nu\rho\sigma\alpha\beta\gamma} f^{\alpha\beta\gamma} . \tag{20.29}$$

Furthermore, for G_2 we have the identity (16.15) by which any chain of contractions of more than two $f_{\alpha\beta\gamma}$ can be reduced. Projection operators of section 20.1 and section 20.2 yield the G_2 Clebsch-Gordan series (16.12):

$$7 \otimes 7 = 1 \oplus 27 \oplus 14 \oplus 7 , \qquad 7 \otimes 14 = 7 \oplus 27 \oplus 64 .$$

20.3.3 $SO(7)$ eight-dimensional rep

We have not attempted to identify the quartic invariant in this case. However, all the rep dimensions (table 20.1), as well as their Dynkin indices (table 20.2), match B_3 reps listed in tables of Patera and Sankoff [273].

20.3.4 $SO(10)$ ten-dimensional rep

This is a trivial solution; $\mathbf{P}_A = \mathbf{P}_3$ and $\mathbf{P}_7 = 0$, so that there is no decomposition. The quartic invariant is

$$f_{\mu\nu\sigma\rho} = \varepsilon_{\mu\nu\sigma\rho\alpha\beta\gamma\delta\omega\xi} C_{\alpha\beta,\gamma\delta,\omega\xi} \equiv 0 , \tag{20.30}$$

where $C_{\alpha\beta,\gamma\delta,\omega\xi}$ are the $SO(10)$ Lie algebra structure constants.

This completes our discussion of the "bosonic" symmetric $g_{\mu\nu}$, antisymmetric $e_{\alpha\beta\gamma\delta}$ invariant tensors. We turn next to the "fermionic" case: antisymmetric $g_{\mu\nu}$, symmetric $e_{\alpha\beta\gamma\delta}$.

20.4 E_7 FAMILY

We have established in chapter 12 that the invariance group of antisymmetric quadratic invariant $f_{\mu\nu}$ is $Sp(n)$, n even. We now add to the set of $Sp(n)$ invariants (12.8) a *fully symmetric* 4-index tensor,

$$d_{\mu\nu\rho\delta} = d_{\nu\mu\rho\delta} = d_{\mu\rho\nu\delta} = d_{\mu\nu\delta\rho} . \tag{20.31}$$

All of the algebra of invariants and Kronecker product decomposition that follow is the same as in section 20.1, and is left as an exercise for the reader. All the dimensions and Dynkin indices are the same, with $n \to -n$ replacement in all expressions:

$$\mathbf{P}_A = \mathbf{Q} + \frac{6}{16+n}\mathbf{P}_3, \qquad \mathbf{P}_7 = -\mathbf{Q} + \frac{10+n}{16+n}\mathbf{P}_3, \tag{20.32}$$

$$N = \frac{3n(n+1)}{16+n} = 3n - 45 + \frac{360}{8+\frac{1}{2}n} \tag{20.33}$$

$$d_7 = \frac{n(n+1)(n+10)}{2(16+n)}.$$

There are seventeen solutions to this Diophantine condition, but only ten will survive the next one.

20.4.1 Defining ⊗ adjoint tensors

Rewriting section 20.2 for an antisymmetric $f_{\mu\nu}$, symmetric $d_{\mu\nu\sigma\rho}$ is absolutely trivial, as these tensors never make an explicit appearance. The only subtlety is that for the reductions of Kronecker products of odd numbers of defining reps (in this case $\otimes V^3$), additional overall factors of -1 appear. For example, it is clear that the dimension of the defining subspace d_8 in (20.18) does not become negative; $n \to -n$ substitution propagates only through the expressions for λ_A, λ_7 and N. The dimension formulas (20.24) become

$$d_{10} = \frac{3n(n-2)(n+4)}{n+28}, \qquad d_{11} = \frac{32n(n-2)(n+1)}{(n+16)(n+28)}. \tag{20.34}$$

Out of the seventeen solutions to (20.33), ten also satisfy this Diophantine condition; $d = 2, 4, 8, 14, 20, 32, 44, 56, 164, 224$. $d = 44, 164$, and 224 can be eliminated [74] by considering reductions along the columns of the Magic Triangle and proving that the resulting subgroups cannot be realized; consequently the groups that contain them cannot be realized either. Only the seven solutions listed in table 20.3 have antisymmetric $f_{\mu\nu}$ and symmetric $d_{\mu\nu\rho\delta}$ invariants in the defining rep.

20.4.2 Lie algebra identification

It turns out that one does not have to work very hard to identify the series of solutions of the preceding section. $SO(2)$ is trivial, and there is extensive literature on the remaining solutions. Mathematicians study them because they form the third row of the Magic Square [130], and physicists study them because $E_7(56) \to SU(3)_c \times SU(6)$ once was one of the favored unified models [149]. The rep dimensions and the Dynkin indices listed in table 20.3 agree with the above literature, as well as with the Lie algebra tables [273]. Here we shall explain only why E_7 is one of the solutions.

The construction of E_7, closest to the spirit of our endeavor, has been carried out by Brown [34, 353]. He considers an n-dimensional complex vector space V with the following properties:

1. V possesses a nondegenerate skew-symmetric symplectic invariant $\{x, y\} = f_{\mu\nu}x^\mu y^\nu$

Rep	$SO(2)$	A_1	$A_1 + A_1 + A_1$	C_3	A_5	D_6		E_7		
V=defining	2	4	8	14	20	32	44	56	164	224
A=adjoint	1	3	9	21	35	66	99	133	451	630
V_7=symmetric	2	7	27	84	175	462	891	1463	13079	24570
V_5=antisym.	0	5	27	90	189	495	945	1539	13365	24975
V_{10}	0	6	48	216	540	1728	3696	6480	69741	134976
V_{11}	0	2	16	64	$70 + 70$	352	616	912	4059	5920
Dynkin indices:										
V=defining		$\frac{5}{2}$	1	$\frac{5}{8}$	$\frac{1}{2}$	$\frac{2}{5}$	$\frac{5}{14}$	$\frac{1}{3}$	$\frac{5}{18}$	$\frac{10}{37}$
A=adjoint		1	1	1	1	1	1	1	1	1
V_7=symmetric		14	9	9	10	$\frac{63}{5}$	$\frac{108}{7}$	$\frac{55}{3}$	$\frac{406}{9}$	$\frac{2233}{37}$
V_5=antisym.		5	6	$\frac{15}{2}$	9	12	15	18	45	60
V_{10}		$\frac{35}{4}$	14	$\frac{45}{2}$	$\frac{63}{2}$	$\frac{252}{5}$	70	90	$\frac{2205}{8}$	380
V_{11}		$\frac{1}{4}$	2	4	$\frac{11}{4} + \frac{11}{4}$	$\frac{38}{5}$	9	10	$\frac{107}{8}$	14

Table 20.3 Rep dimensions and Dynkin indices for the E_7 family of invariance groups.

2. V possesses a symmetric 4-linear form $q(x, y, z, w) = d_{\mu\nu\sigma\rho}x^{\mu}y^{\nu}z^{\sigma}w^{\rho}$

3. If the ternary product $\mathbf{T}(x, y, z)$ is defined on V by
   $\{\mathbf{T}(x, y, z), w\} = q(x, y, z, w)$, then
   $3\{\mathbf{T}(x, x, y), \mathbf{T}(y, y, y)\} = \{x, y\}q(x, y, y, y)$

The third property is nothing but the invariance condition (4.36) for $d_{\mu\nu\rho\delta}$ as can be verified by substituting $\mathbf{P}_A$ from (20.32). Hence, our quadratic, quartic invariants fulfill all three properties assumed by Brown. He then proceeds to prove that the 56-dimensional rep of E_7 has the above properties and saves us from that labor.

The E_7 family derived above is a row of the Magic Triangle (figure 1.1). This is an extension of the Magic Square, an octonionic construction of exceptional Lie algebras. The remaining rows are obtained [74] by applying the methods of this monograph to various kinds of quadratic and cubic invariants, while the columns are subgroup chains. In this context, the Diophantine condition (20.33) is one of a family of Diophantine conditions discussed in chapter 21. They all follow from formulas for the dimension of the adjoint rep of form

$$N = \frac{1}{3}(k-6)(l-6) - 72 + 360\left(\frac{1}{k} + \frac{1}{l}\right). \qquad (20.35)$$

(20.33) is recovered by taking $k = 24$, $n = 2l - 16$. Further Diophantine conditions, analogous to (20.34), reduce the solutions to $k, l = 8, 9, 10, 12, 15, 18, 24, 35$. The corresponding Lie algebras form the Magic Triangle (figure 1.1).

20.5 DYNKIN LABELS AND YOUNG TABLEAUX FOR E_7

A rep of E_7 is characterized by seven Dynkin labels $(a_1a_2a_3a_4a_5a_6a_7)$. As in section 18.7, tracing with respect to the invariant tensor $d_{\mu\nu\rho\delta}$ modifies the Young tableaux for $Sp(56)$. We leave details as an exercise for the reader. The correspondence between the E_7 Dynkin diagram from table 7.6, Dynkin labels, irreducible tensor Young tableaux, and the dimensions of the lowest corresponding reps is

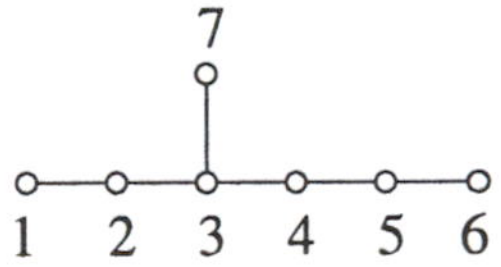

7

1 2 3 4 5 6 $\leftrightarrow (a_1a_2a_3a_4a_5a_6a_7) \leftrightarrow$ (20.36)

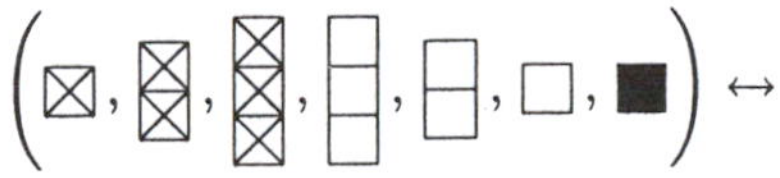

(, , , , , ,) $\leftrightarrow$

$(133, 362880, 365750, 27664, 1539, 56, 912)\,.$

The Clebsch-Gordan series for products of the five lowest-dimensional reps of E_7 are given in table 20.4.

$$56^2 = 1463 + 1539 + 1 + 133$$

$$7448 = 133 \cdot 56 = 6480 + 56 + 912$$

$$81928 = 1463 \cdot 56 = 24320 + 51072 + 56 + 6480$$

$$86184 = 1539 \cdot 56 = 51072 + 27664 + 56 + 6480 + 912$$

$$17689 = 133^2 = 7371 + 8645 + 133 + 1 + 1539$$

$$1549184 = 27664 \cdot 56 = 980343 + 365750 + 1539 + 152152 + 40755 + 8645$$

Table 20.4 The Clebsch-Gordan series for Kronecker products of the five lowest-dimensional reps of E_7.

Chapter Twenty-One

Exceptional magic

The study of invariance algebras as pursued in chapters 16–20 might appear a rather haphazard affair. Given a set of primitives, one derives a set of Diophantine equations, constructs the family of invariance algebras, and moves onto the next set of primitives. However, a closer scrutiny of the Diophantine conditions leads to a surprise: most of these equations are special cases of one and the same Diophantine equation, and they magically arrange all exceptional families into a triangular array I call the Magic Triangle.

21.1 MAGIC TRIANGLE

Our construction of invariance algebras has generated a series of Diophantine conditions that we now summarize. The adjoint rep dimensions (19.21), (18.13), (20.33), and (17.13) are

$$
\begin{aligned}
&F_4 \text{ family} \qquad N = 3n - 36 + \frac{360}{n+10} \\
&E_6 \text{ family} \qquad N = 4n - 40 + \frac{360}{n+9} \\
&E_7 \text{ family} \qquad N = 3n - 45 + \frac{360}{n/2+8} \\
&E_8 \text{ family} \qquad N = 10m - 122 + \frac{360}{m}\,.
\end{aligned}
\tag{21.1}
$$

There is a striking similarity between the Diophantine conditions for different families. If we define

$$
\begin{aligned}
&F_4 \text{ family} \qquad m = n + 10 \\
&E_6 \text{ family} \qquad m = n + 9 \\
&E_7 \text{ family} \qquad m = n/2 + 8\,,
\end{aligned}
\tag{21.2}
$$

we can parametrize all the solutions of the above Diophantine conditions with a single integer m (see table 21.1). The Clebsch-Gordan series for $A \otimes V$ Kronecker products also show a striking similarity. The characteristic equations (17.10), (18.28), (19.36), and (20.19) are one and the same equation:

$$
(\mathbf{Q} - \mathbf{1})\left(\mathbf{Q} + \frac{6}{m}\mathbf{1}\right)\mathbf{P}_r = 0\,. \tag{21.3}
$$

Here $\mathbf{P}_r$ removes the defining and $\otimes V^2$ subspaces, and we have rescaled the E_8 operator $\mathbf{Q}$ (17.10) by factor 2. The role of the $\mathbf{Q}$ operator is only to distinguish between the two subspaces; we are free to rescale it as we wish.

m	8	9	10	12	15	18	24	30	36
F_4			0	0	3	8	21	.	52
E_6		0	0	2	8	16	35	36	78
E_7	0	1	3	9	21	35	66	99	133
E_8	3	8	14	28	52	78	133	190	248

Table 21.1 All defining representation n values allowed by the Diophantine conditions (21.1) and (21.4). The $m = 30$ column of nonreductive algebras, not eliminated by the Diophantine conditions of chapters 16–20, is indicated by smaller script.

In the dimensions of the associated reps, the eigenvalue $6/m$ introduces a new Diophantine denominator $m + 6$. For example, from (17.19), table 18.4, (19.42), and (20.34), the highest-dimensional rep in $V \otimes A$ has dimension (in terms of parametrization (21.2)):

$$\begin{aligned}
&F_4 \text{ family} \quad 3(m+6)^2 - 156(m+6) + 2673 - \frac{15120}{m+6} \\
&E_6 \text{ family} \quad 4(m+6)^2 - 188(m+6) + 2928 - \frac{15120}{m+6} \\
&E_7 \text{ family} \quad 2\left\{6(m+6)^2 - 246(m+6) + 3348 - \frac{15120}{m+6}\right\} \\
&E_8 \text{ family} \quad 50m^2 - 1485m + 19350 + \frac{27 \cdot 360}{m} - \frac{11 \cdot 15120}{m+6}.
\end{aligned} \tag{21.4}$$

These Diophantine conditions eliminate most of the spurious solutions of (21.1); only the $m = 30, 60, 90$, and 120 spurious solutions survive but are in turn eliminated by further conditions. For the E_8 family, the defining rep is the adjoint rep, $V \otimes V = V \otimes A = A \otimes A$, so the Diophantine condition (21.4) includes both $1/m$ and $1/(m + 6)$ terms. Not only can the four Diophantine conditions (21.1) be parametrized by a single integer m; the list of solutions (table 21.1) turns out to be symmetric under the flip across the diagonal. F_4 solutions are the same as those in the $m = 15$ column, and so on. This suggests that the rows be parametrized by an integer ℓ, in a fashion symmetric to the column parametrization by m. Indeed, the requirement of $m \leftrightarrow \ell$ symmetry leads to a unique expression that contains the four Diophantine conditions (21.1) as special cases:

$$N = \frac{(\ell - 6)(m - 6)}{3} - 72 + \frac{360}{\ell} + \frac{360}{m}. \tag{21.5}$$

We take $m = 8, 9, 10, 12, 15, 18, 24, 30$, and 36 as all the solutions allowed in table 21.1. By symmetry, ℓ takes the same values. All the solutions fill up the *Magic Triangle* (figure 21.1). Within each entry, the number in the upper left corner is N, the dimension of the corresponding Lie algebra, and the number in the lower left corner is n, the dimension of the defining rep. The expressions for n for the top four rows are guesses. The triangle is called “magic” partly because we arrived at it by

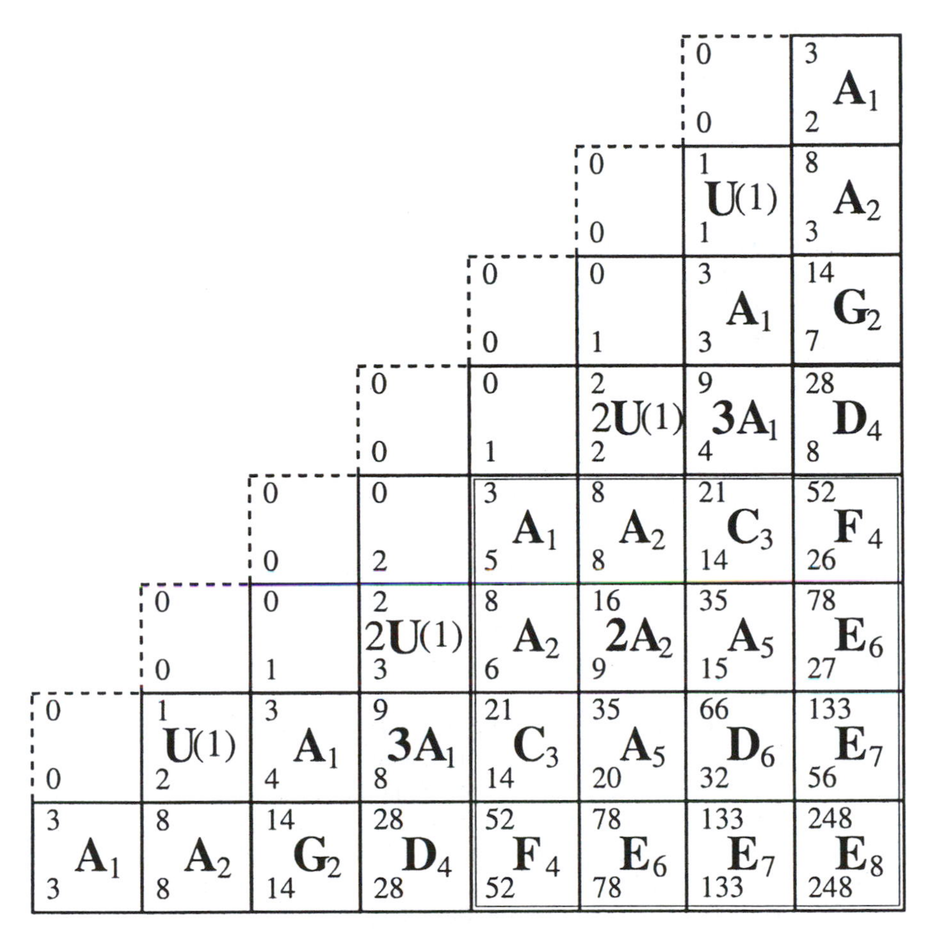

Figure 21.1 *Magic Triangle.* The admissible solutions of Diophantine conditions (21.4) and (21.5) form a triangular array that includes all of the exceptional Lie group families derived in chapters 16–20. Within each entry the number in the upper left corner is N, the dimension of the corresponding Lie algebra, and the number in the lower left corner is n, the dimension of the defining rep. The "Magic Square" is framed by the double line.

magic, and partly because it contains the Magic Square, marked by the dotted line in figure 21.1.

21.2 A BRIEF HISTORY OF EXCEPTIONAL MAGIC

To live outside the law you got to be honest.
—Bob Dylan

Literature on group theory is vast; hard work builds character and anybody who has discovered, for example, that a trace is a useful symmetry invariant writes a paper about it. The good thing about it is that there are many wonderful papers to study. The bad thing about it is that hardly anybody tracks that vast literature, and so I soldiered on with this monograph happy and undisturbed, garnering three citations to the Magic Triangle over the two decades. Theory of compact Lie groups is complete for nearly a century (Peter-Weyl theorem), and hardly anyone thinks there is a problem there, let alone a solution to it.

In 1996 Deligne changed this by rediscovering in part the construction of exceptional Lie groups described here. In quantum field theory, analytic continuation in space dimension n is a given [161]. In the classical group theory of Frobenius, Cartan, and Weyl, each group is a discrete object, with its own specific structure; Deligne's theory of GL_n tensor categories freed the representation theory of these shackles, and phrased analytic continuation in n (described here in chapter 9) in a language comfortable to mathematicians. Deligne was a student of Tits; quantum field theory has flirted with exceptional groups for at least 50 years, and so from both directions one had to explore how continuation in n fits into the theory of exceptional groups.

Deligne is a much admired prodigy (he joined IHES at age 19), and the exceptional drought was followed by new contributions that this monograph makes no attempt to incorporate. I apologize to colleagues whose important papers I have either overseen or misunderstood. Where this monograph fits into the larger picture is explained in chapter 1. A brief history of birdtracks is given in section 4.9.

There are many strands woven into the tapestry of "exceptional magic" to which this monograph is a small contribution. First noted by Rosenfeld [297], the Magic Square was rediscovered by Freudenthal, and made rigorous by Freudenthal and Tits [129, 130, 325].

The construction of the exceptional Lie algebras family described here was initiated [73, 74] in 1975–77. The "Magic Triangle" and the methods used to derive were published in the 1981 article [78] using the E_7 family (chapter 20) and its $SO(4)$-family of "negative dimensional" cousins as an example. The derivation of the E_8 family presented in chapter 17, based on the assumption of no quartic primitive invariant (see figure 16.1), was inspired by S. Okubo's observation [258] that the quartic Dynkin index (7.33) vanishes for the exceptional Lie algebras. In the intervening years several authors have independently reached similar conclusions.

In 1986 K. Meyberg [240, 241] also showed that the absence of a primitive quartic casimir leads to uniform decomposition of adjoint $\mathrm{Sym}^2 A$ and obtained the

E_8 family of chapter 17.

E. Angelopoulos is credited for obtaining (in an unpublished paper written around 1987) the Cartan classification using only methods of tensor calculus, by proving that the quadratic casimir has only two eigenvalues on the symmetric subspace Sym ^2A (the 1981 result [78] described here in section 17.1). Inspired by Angelopoulos and ref. [73], in his thesis M. El Houari applied a combination of tensorial and diagrammatic methods to the problem of classification of simple Lie algebras and superalgebras [111]. As *Algebras, Groups, and Geometries* journal does not practice proofreading (all references are of form [?,?,?]), precise intellectual antecedents to this work are not easily traced. In a subsequent publication E. Angelopoulos [12] used the spectrum of the casimir operator acting on $A \otimes A$ to classify Lie algebras, and, *inter alia*, also obtained the E_8 family of chapter 17 within the same class of Lie algebras.

In a Shimane University 1989 publication, N. Kamiya [179] constructs the F_4, E_6, E_7, and E_8 subset of the E_8 family from "balanced Freudenthal-Kantor triple systems" of dimensions $n_{FK} = 14, 20, 32, 56$. In particular, on p. 44 he states an algebra dimension formula equivalent to (17.13) under substitution $n_{FK} = 2(m - 8)$.

In a 1995 paper P. Deligne [179] attributed to P. Vogel [332] the observation that for the five exceptional groups the antisymmetric $A \wedge A$ and the symmetric Sym$^2 A$ adjoint rep tensor product decomposition, $\mathbf{P}_{\square} + \mathbf{P}_{\begin{smallmatrix}\square\\ \square\end{smallmatrix}}$ and $\mathbf{P}_{\bullet} + \mathbf{P}_{\square\square} + \mathbf{P}_{\blacksquare}$ in table 17.2, can be decomposed into irreducible reps in a "uniform way," and that their dimensions and casimirs are rational functions of the dual Coxeter number a, related to the parameter m of (17.12) by

$$a = 1/(m - 6) \,. \tag{21.6}$$

Here a is $a = \Phi(\tilde{\alpha}, \tilde{\alpha})$, where $\tilde{\alpha}$ is the largest root of the rep, and Φ the canonical bilinear form for the Lie algebra, in the notation of Bourbaki [29]. Deligne conjectured the existence of a tensor category that models the A-module structure of $\otimes A$. A consequence of the conjecture would be decomposition and dimension formulas for the irreducible modules in $\otimes A^k$, $\forall k$.

This consequence was checked on computer by Deligne, Cohen, and de Man [62, 90] for all reps up to $\otimes A^5$. They note that "*miraculously* for all these rational functions both numerator and denominator factor in $Q[a]$ as a product of linear factors." For representations computed so far, this is an immediate consequence of the methods used here to decompose symmetric subspaces (chapter 17). For $\otimes A^6$ the conjecture is open.

Cohen and de Man have also observed that D_4 should be added to the list, in agreement with our definition of the E_8 family, consisting of A_1, A_2, G_2, D_4, F_4, E_6, E_7, and E_8. Their computations go way beyond the results of chapter 17, all of which were obtained by paper and pencil birdtrack computations performed on trains while commuting between Gothenburg and Copenhagen. In all, Cohen and de Man give formulas for 25 reps, seven of which are computed here.

In the context of chapter 17, the dual Coxeter number (21.6) is the symmetric space eigenvalue of the invariant tensor $\mathbf{Q}$ defined in (17.12). The role of the tensor $\mathbf{Q}$ is to split the traceless symmetric subspace, and its overall scale is arbitrary. In chapter 17 scale was fixed in (17.4) by setting the value of the adjoint rep quadratic casimir to $C_A = 1$. Deligne [89] and Cohen and de Man [62] fix the scale by setting $\lambda_{\square\square} + \lambda_{\blacksquare} = 1$, so their dimension formulas are stated in terms of a parameter related to the $\lambda_{\square\square}$ used here by $\lambda_{CdM} = 6\lambda_{\square\square}$. They refer to the interchange of the roots $\lambda_{\square\square} \leftrightarrow \lambda_{\blacksquare}$ as "involution." Typical "translation dictionary" entries: my (17.38) is their A, (17.39) is their Y_3^*, (17.40) is their C^*, *etc.*

After a prelude on "tensor categories" that puts ruminations of this monograph into perspective, and a $GL(n)$ warm-up in which $V \otimes V \otimes \overline{V}$ irreducible reps projection operators and dimensions (here table 9.3 of section 9.11) are computed via a birdtrack-evaluated algebra of invariants multiplication table (3.42) (see section 9.11.1), in the 1999 paper [63] A. M. Cohen and R. de Man perform birdtrack computations of section 17.1, and arrive at the same projection operators and dimension formulas. While they diagonalize the full 5×5 algebra of invariants multiplication table, in this monograph the reduction proceeds in two steps, first to $SO(n)$ irreducible reps, which in turn are decomposed into E_8 family irreducible reps. This facilitates by-hand computations, but the primitiveness condition (17.10) is more elegantly stated by Cohen and de Man prior to reductions, here (17.9). They also fail to find an algorithm for reducing E_8 family vacuum bubbles whose loops are of length 6 or longer, and speculate that expansion in terms of tree diagrams will not suffice, and a new symmetric 6-index primitive invariant will have to be included in the decomposition of $\otimes A^6$. However, on the way to decomposing the $\otimes A^3$ space (section 17.2) I do eliminate the 6-loop diagram, *i.e.*, replace

by shorter loops (double line refers to $V_{\blacksquare}$ from (17.15)—details are a bit tedious for this overview). This should imply a 6-loop reduction formula analogous to (17.9), that I have not tried to extract. In the same spirit, according to table 7.1 of orders of independent casimirs [30, 288, 134, 54, 294] (the Betti numbers) for the E_8 family the next nonvanishing Dynkin index (beyond the quadratic one) corresponds to a loop of length 8.

Cohen and de Man acknowledge in passing that diagrammatic notation "is well known to physicists (*cf.* Cvitanović [83])," though I have to admit that the converse is less so: the invariant tensors basis of section 3.3.1 is "the ring $End_C(X)$, a free $\mathbb{Z}[t]$-module," birdtracks morph to "morphisms," and so on. Today no one has leisure for reading source papers in foreign tongues, so Cohen and de Man verify the E_8 family projection operators and dimension formulas of chapter 17 by the birdtrack computations identical to those already given in ref. [83].

Inspired by conjectures of Deligne, J. M. Landsberg and L. Manivel [203, 204, 205, 206, 210] utilize projective geometry and the triality model of Allison [9] to interpret the Magic Square, recover the known dimension and decomposition formulas of Deligne and Vogel, and derive an infinity of higher-dimensional rep formulas, all proved without recourse to computers. They arrive at some of the formulas derived here, including [209] the $m = 30$ column of nonreductive algebras in table 17.1. They deduce the formula (21.5) conjectured above from Vogel's [333] "universal Lie algebra" dimension formula (proposition 3.2 of ref. [205]), and interpret m, ℓ

as $m = 3(a+4)$, $\ell = 3(b+4)$, where $a, b = 0, 1, 2, 4, 6, 8$ are the dimensions of the algebras used in their construction (in case a or $b \neq 6$ these are composition algebras). For $m \geq 12$ this agrees with the Magic Square.

In 2002 Deligne and Gross [92] defined the Lie groups (*i.e.*, specified the isogeny class) whose Lie algebras were previously known to fit into the Magic Triangle of figure 21.1. B. H. Gross credits his student K. E. Rumelhart [301, 91] with introducing the Magic Triangle in the 1996 Ph.D. thesis. Also in 2002, an intriguing link between the q-state Potts models and the E_8 family was discovered by Dorey *et al.* [96]. For a related recent study of E_6 and E_7 families, see MacKay and Taylor [226].

So much for group theory from my myopic, birdtracks perspective: Are there any physical applications of exceptional magic?

21.3 EXTENDED SUPERGRAVITIES AND THE MAGIC TRIANGLE

In chapter 20 I showed that the extension of Minkowski space into negative dimensions yields the E_7 family. These $n \rightarrow -n$ relations and the Magic Triangle arose as by-products of an investigation of group-theoretic structure of gauge theories undertaken in ref. [73], written up in more detail in the 1977 Oxford preprint [74]. I obtained an exhaustive classification, but are there any realizations of it? Surprisingly, every entry in our classification appears to be realized as a global symmetry of an extended supergravity.

In 1979 Cremmer and Julia [68] discovered that in $N = 8$ (or $N = 7$) supergravity's 28 vectors, together with their 28 duals, form a $\underline{56}$ multiplet of a global E_7 symmetry. This is a global symmetry analogous to $SO(2)$ duality rotations of the doublet $(F_{\mu\nu}, F*_{\mu\nu})$ in $j^\mu = 0$ sourceless electrodynamics. The appearance of E_7 was quite unexpected; it was the first time an exceptional Lie group emerged as a physical symmetry, without having been inserted into a model by hand. While the classification I have obtained here does not explain why this happens, it suggests that there is a deep connection between the extended supergravities and the exceptional Lie algebras. Cremmer and Julia's $N = 7, 6, 5$ global symmetry groups $E_7, SO(12), SU(6)$ are included in the present classification. Furthermore, vectors plus their duals form multiplets of dimension 56, 32, 20, so they belong to the defining reps in our classification. While for $N \leq 4$ extended supergravities, the numbers of vectors do not match the dimensions of the defining reps, Paul Howe has pointed out that with one additional vector multiplet $N = 1, 2, \ldots, 7$ extended supergravities exhaust the present classification. These observations are summarized in table 5 of ref. [78].

In 1980 B. Julia introduced a different Magic Triangle [174, 175, 176, 160] unrelated to the one described here. His work was stimulated by a 1979 Gibbons and Hawking remark on gravitational instantons and Ehlers symmetry, and the vague but provocative remarks of Morel and Thierry-Mieg. The two triangles differ: Julia's "disintegration (i.e. oxidation) for E_n cosets" triangle is based on real forms that match up only with the [3×3] subsquare of the Rosenfeld-Freudenthal Magic Square. I still do not know whether there is any relation between extended supergravities and the construction presented here.

EPILOGUE

Quantum Field Theory relies heavily on the theory of Lie groups, and so I went step-by-step through the proof of the Cartan-Killing classification. Frankly, I did not like it. The proofs were beautiful, but Cartan-Weyl explicit Lie algebra matrices were inconvenient and unintuitive for Feynmann diagram computations. There must be more to symmetries observed in nature than a set of Diophantine conditions on Cartan lattices. So I junked the whole thing, and restarted in the 19th century, looking for conditions on Lie groups that would preserve invariant quantities other than length and volume. Imagine the pleasure of rediscovering all exceptional Lie algebras, arranged in a single family, in the very first step of the construction, as invariance groups that preserve an antisymmetric cubic invariant (figure 16.1)!

Monotheistic cults seek a single answer to all questions, and to a religious temperament E_8 is the great temptress. My own excursion into invariances beyond length and volume yielded no physical insights. Nature is too rich to follow a single tune; why should it care that all we know today is a bit of differential geometry? It presents us with so many questions more fundamental and pressing than whether E_8 is the mother or the graveyard of theories, so my journey into exceptional magic stops here.

Almost anybody whose research requires sustained use of group theory (and it is hard to think of a physical or mathematical problem that is wholly devoid of symmetry) writes a book about it. They, in their amazing variety of tastes, flavors, and ethnicities fill stacks in science libraries. My excuse for yet another text is that this book is like no other group-theory textbook. It's written in birdtracks. It's self-contained. Every calculation in the book is a pencil-and-paper exercise, with a rare resort to a pocket calculator. And, of course, it too is unfinished: it is up to you, dear reader, to complete it. I fear E_8 will not yield to pencil and paper.

Appendix A

Recursive decomposition

This appendix deals with the practicalities of computing projection operator eigenvalues, and is best skipped unless you need to carry out such a calculation.

Let $\mathbf{P}$ stand for a projection onto a subspace or the entire space (in which case $\mathbf{P} = \mathbf{1}$). Assume that the subspace has already been reduced into m irreducible subspaces and a remainder

$$\mathbf{P} = \sum_{\gamma=1}^{m} \mathbf{P}_\gamma + \mathbf{P}_r \,. \tag{A.1}$$

Now adjoin a new invariant matrix $\mathbf{Q}$ to the set of invariants. By assumption, $\mathbf{Q}$ does not reduce further the $\gamma = 1, 2, \ldots, m$ subspaces, *i.e.*, has eigenvalues $\lambda_1, \lambda_2, \ldots, \lambda_m$:

$$\mathbf{Q}\mathbf{P}_\gamma = \lambda_\gamma \mathbf{P}_\gamma \qquad \text{(no sum)}\,, \tag{A.2}$$

on the γth subspace. We construct an invariant, matrix $\hat{Q}$, restricted to the remaining (as yet not decomposed) subspace by

$$\hat{\mathbf{Q}} := \mathbf{P}_r \mathbf{Q} \mathbf{P}_r = \mathbf{PQP} - \sum_{\gamma=1}^{m} \lambda_\gamma \mathbf{P}_\gamma \,. \tag{A.3}$$

As $\mathbf{P}_r$ projects onto a finite-dimensional subspace, $\hat{Q}$ satisfies a *minimal* characteristic equation of order $n \geq 2$:

$$\sum_{k=0}^{n} a_k \hat{\mathbf{Q}}^k = \prod_{\alpha=m+1}^{m+n} (\hat{\mathbf{Q}} - \lambda_\alpha \mathbf{P}_r) = 0 \,, \tag{A.4}$$

with the corresponding projection operators (3.48):

$$\mathbf{P}_\alpha = \prod_{\beta \neq \alpha} \frac{\hat{\mathbf{Q}} - \lambda_\beta}{\lambda_\alpha - \lambda_\beta} \mathbf{P}_r \,, \qquad \alpha = \{m+1, \ldots, m+n\} \,. \tag{A.5}$$

"Minimal" in the above means that we drop repeated roots, so all eigenvalues are distinct. $\hat{\mathbf{Q}}$ is an awkward object in computations, so we reexpress the projection operator, in terms of $\mathbf{Q}$, as follows.

Define first the polynomial, obtained by deleting the $(\hat{\mathbf{Q}} - \lambda_\alpha \mathbf{1})$ factor from (A.4)

$$\prod_{\beta \neq \alpha} (x - \lambda_\beta) = \sum_{k=0}^{n-1} b_k x^k \,, \qquad \alpha, \beta = m+1, \ldots m+n \,, \tag{A.6}$$

where the expansion coefficient $b_k = b_k^{(\alpha)}$ depends on the choice of the subspace α. Substituting $\mathbf{P}_r = P - \sum_{\alpha=1}^m \mathbf{P}_\alpha$ and using the orthogonality of $\mathbf{P}_\alpha$, we obtain an alternative formula for the projection operators

$$\mathbf{P}_\alpha = \frac{1}{\sum b_k \lambda_\alpha^k} \sum_{k=0}^{n-1} b_k \left\{ (\mathbf{PQ})^k - \sum_{\gamma=1}^m \lambda_\alpha^k \mathbf{P}_\gamma \right\} \mathbf{P} , \tag{A.7}$$

and dimensions

$$d_\alpha = \mathrm{tr}\, \mathbf{P}_\alpha = \frac{1}{\sum b_k \lambda_\alpha^k} \sum_{k=0}^{n-1} b_k \left\{ \mathrm{tr}(\mathbf{PQ})^k - \sum_{\gamma=1}^m \lambda_\gamma^k d_\gamma \right\} . \tag{A.8}$$

The utility of this formula lies in the fact that once the polynomial (A.6) is given, the only new data it requires are the traces $\mathrm{tr}(\mathbf{PQ})^k$, and those are simpler to evaluate than $\mathrm{tr}\, \hat{\mathbf{Q}}^k$.

Appendix B

Properties of Young projections

H. Elvang and P. Cvitanović

In this appendix we prove the properties of the Young projection operators, stated in section 9.4.

B.1 UNIQUENESS OF YOUNG PROJECTION OPERATORS

We now show that the Young projection operator $\mathbf{P}_Y$ is well defined by proving the existence and uniqueness (up to an overall sign) of a nonvanishing connection between the symmetrizers and antisymmetrizers in $\mathbf{P}_Y$.

The proof is by induction over the number of columns t in the Young diagram Y — the principle is illustrated in figure B.1 For $t = 1$ the Young projection operator consists of one antisymmetrizer of length s, and s symmetrizers of length 1. Clearly the connection can only be made in one way, up to an overall sign.

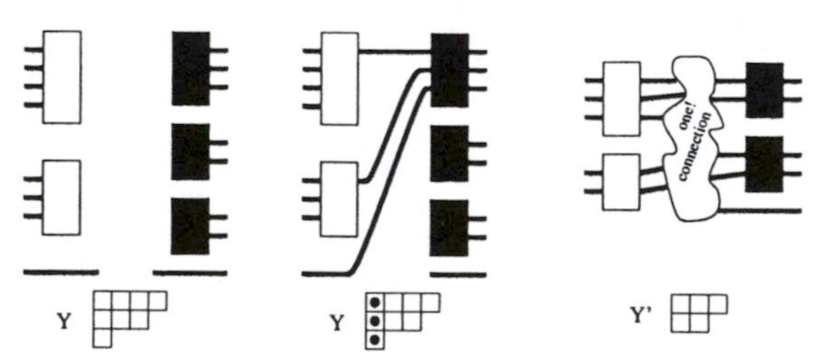

Figure B.1 There is a unique (up to an overall sign) connection between the symmetrizers and the antisymmetrizers, so the Young projection operators are well defined by the construction procedure explained in the text. The figure shows the principle of the proof. The dots on the middle Young diagram mark boxes that correspond to contracted lines.

Assume the result to be valid for Young projection operators derived from Young diagrams with $t - 1$ columns. Let Y be a Young diagram with t columns. The lines from A_1 in $\mathbf{P}_Y$ must connect to different symmetrizers for the connection to be nonzero. There are exactly $|A_1|$ symmetrizers in P_Y, so this can be done in essentially one way; which line goes to which symmetrizer is only a matter of an overall sign, and where a line enters a symmetrizer is irrelevant due to (6.8).

After having connected A_1, connecting the symmetry operators in the rest of

$\mathbf{P}_{\mathrm{Y}}$ is the problem of connecting symmetrizers to antisymmetrizers in the Young projection operator $\mathbf{P}_{\mathrm{Y}'}$, where Y' is the Young diagram obtained from Y by slicing off the first column. Thus, Y' has $k-1$ columns, so by the induction hypothesis, the rest of the symmetry operators in $\mathbf{P}_{\mathrm{Y}}$ can be connected in exactly one nonvanishing way (up to sign).

By construction, the identity is always present in the expansion. The overall sign of the Young projection operator is fixed by requiring that upon expansion of the symmetry operators, the identity has a positive coefficient.

B.2 ORTHOGONALITY

If Y_a and Y_b denote standard tableaux derived from the same Young diagram Y, then $\mathbf{P}_{\mathrm{Y}_a}\mathbf{P}_{\mathrm{Y}_b} = \mathbf{P}_{\mathrm{Y}_b}\mathbf{P}_{\mathrm{Y}_a} = \delta_{ab}\mathbf{P}^2_{\mathrm{Y}_a}$, since there is a nontrivial permutation of the lines connecting the symmetry operators of Y_a with those of Y_b, and by uniqueness of the nonzero connection the result is either $\mathbf{P}^2_{\mathrm{Y}_a}$ (if $\mathrm{Y}_a = \mathrm{Y}_b$) or 0 (if $\mathrm{Y}_a \neq \mathrm{Y}_b$).

Next, consider two different Young diagrams Y and Z with the same number of boxes. Since at least one column must be bigger in (say) Y than in Z and the p lines from the corresponding antisymmetrizer must connect to different symmetrizers, it is not possible to make a nonzero connection between the antisymmetrizers of $\mathbf{P}_{\mathrm{Y}_a}$ to the symmetrizers in $\mathbf{P}_{\mathrm{Z}_b}$, where subscripts a and b denote any standard tableaux of Y and Z. Hence $\mathbf{P}_{\mathrm{Y}_a}\mathbf{P}_{\mathrm{Z}_b} = 0$, and by a similar argument, $\mathbf{P}_{\mathrm{Z}_b}\mathbf{P}_{\mathrm{Y}_a} = 0$.

B.3 NORMALIZATION AND COMPLETENESS

We now derive the formula for the normalization factor α_{Y} such that the Young projection operators are idempotent, $\mathbf{P}^2_{\mathrm{Y}_a} = \mathbf{P}_{\mathrm{Y}_a}$. By the normalization of the symmetry operators, Young projection operators corresponding to fully symmetrical or antisymmetrical Young tableaux will be idempotent with $\alpha_{\mathrm{Y}} = 1$.

Diagrammatically, $\mathbf{P}^2_{\mathrm{Y}_a}$ is $\mathbf{P}_{\mathrm{Y}_a}$ connected to $\mathbf{P}_{\mathrm{Y}_a}$, hence it may be viewed as a set of *outer* symmetry operators connected by a set of *inner* symmetry operators. Expanding all the inner symmetry operators and using the uniqueness of the nonzero connection between the symmetrizers and antisymmetrizers of the Young projection operators, we find that each term in the expansion is either 0 or a copy of $\mathbf{P}_{\mathrm{Y}_a}$. For a Young diagram with s rows and t columns there will be a factor of $1/|\mathrm{S}_i|!$ $(1/|\mathrm{A}_j|!)$ from the expansion of each inner (anti)symmetrizer, so we find

$$
\begin{aligned}
\mathbf{P}^2_{\mathrm{Y}_a} &= \alpha^2_{\mathrm{Y}_a} \\
&= \frac{\alpha^2_{\mathrm{Y}_a}}{\prod_{i=1}^{s} |\mathrm{S}_i|! \, \prod_{j=1}^{t} |\mathrm{A}_j|!} \sum_{\sigma} \\
&= \alpha_{\mathrm{Y}_a} \frac{\kappa_{\mathrm{Y}}}{\prod_{i=1}^{s} |\mathrm{S}_i|! \, \prod_{j=1}^{t} |\mathrm{A}_j|!} \mathbf{P}_{\mathrm{Y}_a} ,
\end{aligned}
$$

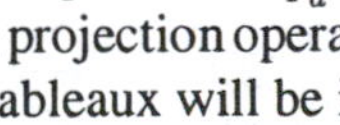

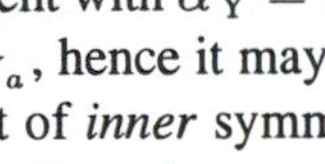

where the sum is over permutations σ from the expansion of the inner symmetry operators. Note that by the uniqueness of the connection between the symmetrizers and antisymmetrizers, the constant κ_{Y} is independent of which tableau gives rise to the projection, and consequently the normalization constant α_{Y} depends only on the Young diagram and not the tableau.

For a given k, consider the Young projection operators $\mathbf{P}_{\mathrm{Y}_a}$ corresponding to all the k-box Young tableaux. Since the operators $\mathbf{P}_{\mathrm{Y}_a}$ are orthogonal and in 1-1 correspondence with the Young tableaux, it follows from the discussion in section 9.3.2 that there are no other operators of k lines orthogonal to this set. Hence the $\mathbf{P}_{\mathrm{Y}_a}$'s form a complete set, so that

$$\mathbf{1} = \sum_{\mathrm{Y}_a} \mathbf{P}_{\mathrm{Y}_a} \,. \tag{B.1}$$

Expanding the projections the identity appears only once, so we have

$$\mathbf{P}_{\mathrm{Y}_a} = \alpha_{\mathrm{Y}} \frac{1}{\prod_{i=1}^{s} |\mathrm{S}_i|! \prod_{j=1}^{t} |\mathrm{A}_j|!} \left(\; p\,\vdots \; + \ldots \right) ,$$

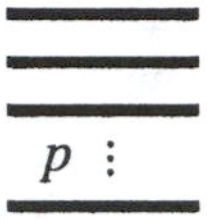

and using this, equation (B.1) states

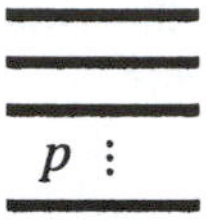

$$p\,\vdots \; = \left(k! \sum_{\mathrm{Y}} \frac{\alpha_{\mathrm{Y}}/|\mathrm{Y}|}{\prod_{i=1}^{s} |\mathrm{S}_i|! \prod_{j=1}^{t} |\mathrm{A}_j|!} \right) \; p\,\vdots \; , \tag{B.2}$$

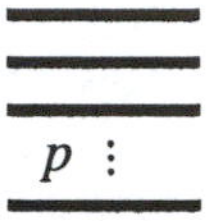

since all permutations different from the identity must cancel. When changing the sum from a sum over the tableaux to a sum over the Young diagrams, we use the fact that that α_{Y} depends only on the diagram and that there are $\Delta_Y = k!/|Y|$ k-standard tableaux for a given diagram. Choosing

$$\alpha_{\mathrm{Y}} = \frac{\prod_{i=1}^{s} |\mathrm{S}_i|! \prod_{j=1}^{t} |\mathrm{A}_j|!}{|\mathrm{Y}|} , \tag{B.3}$$

the factor on the right-hand side of (B.2) is 1 by (9.19).

Since the choice of normalization (B.3) gives the completeness relation (B.1), it follows that it also gives idempotent operators: multiplying by $\mathbf{P}_{\mathrm{Z}_b}$ on both sides of (B.1) and using orthogonality, we find $\mathbf{P}_{\mathrm{Z}_b} = \mathbf{P}^2_{\mathrm{Z}_b}$ for any Young tableau Z_b.

B.4 DIMENSION FORMULA

Here we derive the dimension formula (9.28) of the $U(n)$ irreps recursively from the Young projection operators.

Let Y be a standard tableau and Y$'$ the Young diagram obtained from Y by removal of the right-most box in the last row. Note that Y$'$ is a standard tableau. Next, draw the Young projection operator corresponding to Y and Y$'$ and note that $\mathbf{P}_{\mathrm{Y}}$ with the last line traced is proportional to $\mathbf{P}_{\mathrm{Y}'}$.

Quite generally, this contraction will look like

Rest of P_{Y} . (B.4)

Using (6.10) and (6.19), we have

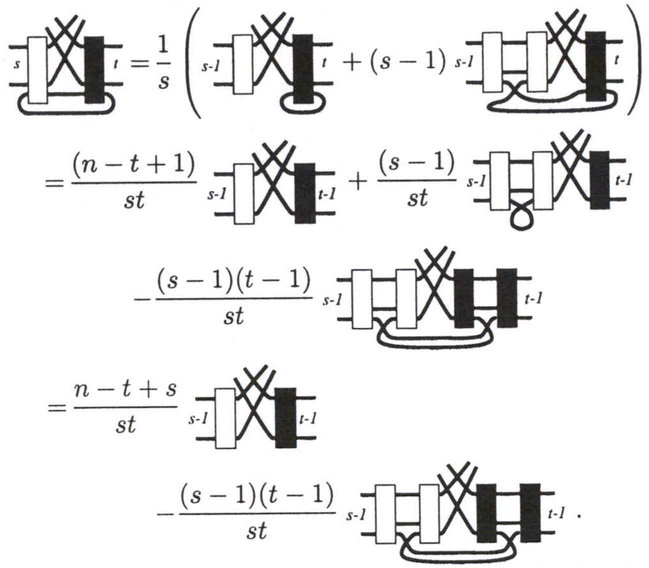

$$= \frac{1}{s}\left(\quad + (s-1) \quad \right)$$

$$= \frac{(n-t+1)}{st} \quad + \frac{(s-1)}{st}$$

$$- \frac{(s-1)(t-1)}{st}$$

$$= \frac{n-t+s}{st}$$

$$- \frac{(s-1)(t-1)}{st} \quad .$$

Inserting this into (B.4) we see that the first term is proportional to the projection operator $\mathbf{P}_{Y'}$. The second term vanishes:

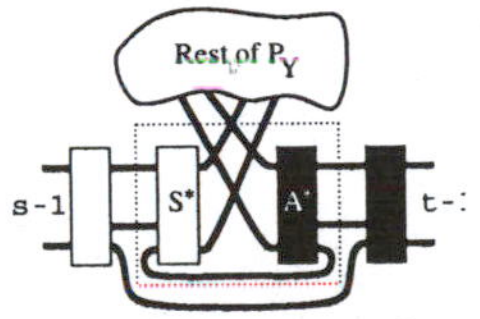

Rest of P_Y

s-1 S* A* t-1

$$= 0 \, .$$

The lines entering S* from the right come from antisymmetrizers in the rest of the $\mathbf{P}_Y$-diagram. One of these lines, from A_a, say, must pass from S* through the lower loop to A* and from A* connect to one of the symmetrizers, say S_s in the rest of the $\mathbf{P}_Y$-diagram. But due to the construction of the connection between symmetrizers and antisymmetrizers in a Young projection operator, there is already a line connecting S_s to A_a. Hence the diagram vanishes.

The dimensionality formula follows by induction on the number of boxes in the Young diagrams, with the dimension of a single box Young diagram being n. Let Y be a Young diagram with p boxes. We assume that the dimensionality formula is valid for any Young diagram with $p-1$ boxes. With $\mathbf{P}_{Y'}$ obtained from $\mathbf{P}_Y$ as above, we have (using the above calculation and writing D_Y for the diagrammatic part of $\mathbf{P}_Y$):

$$\dim \mathbf{P}_Y = \alpha_Y \operatorname{tr} D_Y = \frac{n-t+s}{st} \alpha_Y \operatorname{tr} D_{Y'} \tag{B.5}$$

$$= (n-t+s)\alpha_{Y'} \frac{|Y'|}{|Y|} \operatorname{tr} D_{Y'} \tag{B.6}$$

$$= (n-t+s) \frac{f_{Y'}}{|Y|} = \frac{f_Y}{|Y|} \, . \tag{B.7}$$

This completes the proof of the dimensionality formula (9.28).

Bibliography

[1] A. Abdesselam, "A physicist's proof of the Lagrange-Good multivariable inversion formula," *J. Phys. A* **36**, 9471 (2002); math.CO/0208174.

[2] A. Abdesselam, "The Jacobian conjecture as a problem of perturbative quantum field theory," *Ann. Inst. Henri Poincaré* **4**, 199 (2003); math.CO/0208173.

[3] A. Abdesselam, "Feynman diagrams in algebraic combinatorics," *Sém. Lothar. Combin.* **49**, article B49c (2003); math.CO/0212121.

[4] A. Abdesselam and J. Chipalkatti, "A regularity result for a locus of Brill type," math/0405236.

[5] S. L. Adler, J. Lieberman, and Y. J. Ng, "Regularization of the stress-energy tensor for vector and scalar particles propagating in a general background metric," *Ann. Phys.* **106**, 279 (1977).

[6] V. K. Agrawala and J. G. Belinfante, "Graphical formulation of recoupling for any compact group," *Ann. Phys. (N. Y.)* **49**, 130 (1968).

[7] V. K. Agrawala, "Micu type invariants of exceptional simple Lie algebras," *J. Math. Phys.* **21**, 1577 (1980).

[8] O. Aharony, S. S. Gubser, J. Maldacena, H. Ooguri, and Y. Oz, "Large N field theories, string theory and gravity," *Phys. Rept.* **323**, 183 (2000); hep-th/9905111.

[9] B. N. Allison, "A construction of Lie algebras from J-ternary algebras," *Amer. J. Math.* **98**, 285 (1976).

[10] E. M. Andreev, E. B. Vinberg, and A. G. Elashvili, *Funct. Analysis and Appl.* **1**, 257 (1967).

[11] G. E. Andrews, *The Theory of Partitions* (Addison-Wesley, Reading, MA 1976).

[12] E. Angelopoulos, "Classification of simple Lie algebras," *Panamerican Math. Jour.* **2**, 65 (2001).

[13] J. C. Baez, "The octonions," *Bull. Amer. Math. Soc.* **39**, 145 (2002).

[14] D. Bar-Natan, "Weights of Feynman diagrams and the Vassiliev knot invariants," www.math.toronto.edu/~drorbn/LOP.html (unpublished preprint, 1991); "On the Vassiliev knot invariants," *Topology* **34**, 423 (1995).

[15] K. J. Barnes and R. Delbourgo, *J. Phys. A* **5**, 1043 (1972).

[16] C. H. Barton and A. Sudbery, "Magic squares of Lie algebras" (2000), unpublished; math.RA/0001083.

[17] C. H. Barton and A. Sudbery, "Magic squares and matrix models of Lie algebras," *Adv. in Math.* **180**, 596 (2003); math.RA/0203010

[18] R. E. Behrends, J. Dreitlein, C. Fronsdal, and B. W. Lee, "Simple groups and strong interaction symmetries," *Rev. Mod. Phys.* **34**, 1 (1962).

[19] J. G. Belinfante, *Group Theory*, lecture notes (Carnegie Tech 1965), unpublished.

[20] F. A. Berezin, *Introduction to superanalysis* (Reidel, Dordrecht 1987).

[21] C. Bessenrodt, *Ann. of Comb.* **2**, 103 (1998); the hook rule (9.16) follows directly from Theorem 1.1.

[22] A. M. Bincer and K. Riesselmann, "Casimir operators of the exceptional group G_2," *J. Math. Phys.* **34**, 5935 (1993).

[23] A. M. Bincer, "Casimir operators of the exceptional group F_4: The chain B_4 contains F_4 contains D_{13}," *J. Phys. A* **27**, 3847 (1994); hep-th/9312148.

[24] H. Biritz, *Nuovo Cimento* **25B**, 449 (1973).

[25] J. F. Blinn, "Quartic discriminants and tensor invariants," *IEEE Computer Graphics and Appl.* **22**, 86 (2002).

[26] H. Boerner, *Representations of Groups* (North-Holland, Amsterdam 1970).

[27] D. Boos, "Ein tensorkategorieller Zugang zum Satz von Hurwitz (A tensor-categorical approach to Hurwitz's theorem)," Diplomarbeit ETH Zurich (March 1998), available at www.math.uni-bielefeld.de/~rost/tensors.html

[28] A. K. Bose, *Computer Phys. Comm.* **11**, 1 (1976).

[29] According to a reliable source, Bourbaki is dead. The old soldier will not be missed.

[30] R. Brauer, "Sur les invariants intégraux des variétés représentatives des groupes de Lie simples clos," *C. R. Acad. Sci. Paris* **201**, 419 (1935).

[31] R. Brauer, "On algebras which are connected with the semisimple continuous groups," *Ann. Math.* **38**, 854 (1937).

[32] J. S. Briggs, *Rev. Mod. Phys.* **43**, 189 (1971).

[33] D. M. Brink and G. R. Satchler, *Angular Momentum* (Oxford Univ. Press, Oxford 1968).

[34] R. B. Brown, *J. Reine Angew. Math.* **236**, 79 (1969).

[35] P. Butera, G. M. Cicuta, and M. Enriotti, "Group weight and vanishing graphs," *Phys. Rev. D* **21**, 972 (1980).

[36] P. H. Butler, R. W. Haase, and B. G. Wybourne, "Calculation of $3jm$ factors and the matrix elements of E_7 group generators," *Austr. J. Phys.* **32**, 137 (1979).

[37] R. F. Cahalan and D. Knight, "Construction of planar diagrams," *Phys. Rev. D* **14**, 2126 (1976).

[38] C. G. Callan, N. Coote, and D. J. Gross, *Phys. Rev. D* **13**, 1649 (1976).

[39] G. P. Canning, *Phys. Rev. D* **8**, 1151 (1973).

[40] G. P. Canning, *Phys. Rev. D* **12**, 2505 (1975).

[41] G. P. Canning, "Diagrammatic group theory in quark models," *Phys. Rev. D* **18**, 395 (1978).

[42] R. Carles, *Acad. Sci. Paris A* **276**, 451 (1973).

[43] E. Cartan (1894), *Oeuvres Complètes* (Gauthier-Villars, Paris 1952).

[44] E. Cartan, *Ann. Sci. Ecole Norm. Sup. Paris* **31**, 263 (1914).

[45] R. W. Carter, "Conjugacy classes and complex characters," *Finite groups of Lie type* (John Wiley & Sons Ltd., Chichester 1985, reprint 1993).

[46] P. Cartier, "Mathemagics (a tribute to L. Euler and R. Feynman)," *Sém. Lothar. Combin.* **44**, article B44d (2000); www.mat.univie.ac.at/~slc .

[47] R. Casalbuoni, G. Domokos, and S. Kövesi-Domokos, *Nuovo Cimento* **31A**, 423 (1976).

[48] K. M. Case, "Biquadratic spinor identities," *Phys. Rev.* **97**, 810 (1955).

[49] H. Casimir, *Proc. Roy. Acad. Amstd.* **34**, 844 (1931).

[50] A. Cayley, "On the theory of the analytical forms called trees," *Philos. Mag.* **13**, 19 (1857).

[51] S. Chadha and M. E. Peskin, "Implications of chiral dynamics in theories of technicolor. 2. The mass of the P^+," *Nucl. Phys. B* **187**, 541 (1981).

[52] V. Chari and A. Pressley, "Fundamental representations of Yangians and singularities of R-matrices," *J. Reine Angew. Math.* **417**, 87 (1991).

[53] T. P. Cheng, E. Eichten, and L. Li, "Higgs phenomena in asymptotically free gauge theories," *Phys. Rev. D* **9**, 2259 (1974).

[54] C. Chevalley, *Math. Reviews* **14**, 948 (1953).

[55] J. S. R. Chisholm, *Nuovo Cimento* **30**, 426 (1963).

[56] H. J. Chung and I. G. Koh, "Solutions to the quantum Yang-Baxter equation for the exceptional Lie algebras with a spectral parameter," *J. Math. Phys.* **32**, 2406 (1991).

[57] E. Ciapessoni and G. M. Cicuta, "Gauge sets and $1/N$ expansion," *Nucl. Phys. B* **219**, 513 (1983).

[58] G. M. Cicuta, "Topological expansion for $SO(n)$ and $Sp(2n)$ gauge theories," *Lett. Nuovo Cim.* **35**, 87 (1982).

[59] G. M. Cicuta and D. Gerundino, "High-energy limit and internal symmetries," *Phys. Rev. D* **29**, 1258 (1984).

[60] G. M. Cicuta and A. Pavone, "Diagonalization of a coloring problem (on a strip)," *J. Phys. A* **22**, 4921 (1989).

[61] W. Clifford, a letter to Mr. Sylvester, *Amer. J. Math.* **1**, 126 (1878).

[62] A. M. Cohen and R. de Man, "Computational evidence for Deligne's conjecture regarding exceptional Lie groups," *C. R. Acad. Sci. Paris, Sér. I*, **322**, 427 (1996).

[63] A. M. Cohen and R. de Man, "On a tensor category for the exceptional groups," in P. Drexler, G. O. Michler, and C. M. Ringel, eds., *Computational Methods for Representations of Groups and Algebras, Progress in Math.* **173**, chapter 6, 121 (Birkhäuser, Basel 1999)

[64] A. J. Coleman, *Advances in Quantum Chemistry* **4**, 83 (1968).

[65] E. Corrigan and D. B. Fairlie, *Phys. Lett.* **67B**, 69 (1977).

[66] E. Corrigan, D. B. Fairlie, and R. G. Yates, "SU(2) gauge potentials and their strength," Print-77-0375 (U. of Durham preprint, April 1977); unpublished.

[67] H. S. M. Coxeter, *Regular Polytopes* (Dover, New York 1973); see p. 92.

[68] E. Cremmer and B. Julia, "The $SO(8)$ supergravity," *Nucl. Phys. B* **159**, 141 (1979).

[69] M. Creutz, "On invariant integration over $SU(N)$," *J. Math. Phys.* **19**, 2043 (1978).

[70] C. W. Curtis, *Studies in modern algebra* **2**, in A. A. Albert, ed. (Prentice Hall, New York 1963).

[71] R. E. Cutkosky, *Ann. Phys. (N.Y.)* **23**, 415 (1963).

[72] P. Cvitanović and T. Kinoshita, "Sixth order magnetic moment of the electron," *Phys. Rev. D* **10**, 4007 (1974).

[73] P. Cvitanović, "Group theory for Feynman diagrams in non-Abelian gauge theories," *Phys. Rev. D* **14**, 1536 (1976).

[74] P. Cvitanović, "Classical and exceptional Lie algebras as invariance algebras," Oxford preprint 40/77 (June 1977, unpublished); available on birdtracks.eu/refs.

[75] P. Cvitanović, "Asymptotic estimates and gauge invariance," *Nucl. Phys. B* **127**, 176 (1977).

[76] P. Cvitanović, B. Lautrup, and R. B. Pearson, "The number and weights of Feynman diagrams," *Phys. Rev. D* **18**, 1939 (1978).

[77] P. Cvitanović, "Planar perturbation expansion," *Phys. Lett.* **99B**, 49 (1981).

[78] P. Cvitanović, "Negative dimensions and E_7 symmetry," *Nucl. Phys. B* **188**, 373 (1981).

[79] P. Cvitanović, P. G. Lauwers, and P. N. Scharbach, "Gauge invariance structure of Quantum Chromodynamics," *Nucl. Phys. B* **186**, 165 (1981).

[80] P. Cvitanović, J. Greensite, and B. Lautrup, "The crossover points in lattice gauge theories with continuous gauge groups," *Phys. Lett. B* **105**, 197 (1981).

[81] P. Cvitanović and A. D. Kennedy, "Spinors in negative dimensions," *Phys. Scripta* **26**, 5 (1982).

[82] P. Cvitanović, "Group theory, part I" (Nordita, Copenhagen 1984); incorportated into this monograph.

[83] P. Cvitanović, "Group theory" (Nordita and Niels Bohr Institute, Copenhagen 1984-2007). Web book version of this monograph; birdtracks.eu.

[84] P. Cvitanović, C. P. Dettmann, R. Mainieri, and G. Vattay, "Trace formulas for stochastic evolution operators: Smooth conjugation method," *Nonlinearity* **12**, 939 (1999); chao-dyn/9811003.

[85] P. Cvitanović, "Chaotic field theory: a sketch," *Physica A* **288**, 61 (2000); nlin.CD/0001034.

[86] J. A. de Azcárraga, A. J. Macfarlane, A. J. Mountain, and J. C. Pérez Bueno, "Invariant tensors for simple groups," *Nucl. Phys. B* **510**, 657 (1998); physics/9706006.

[87] J. A. de Azcárraga and A. J. Macfarlane, "Optimally defined Racah-Casimir operators for $su(n)$ and their eigenvalues for various classes of representations," *Nucl. Phys. B* **510**, 657 (2000); math-ph/0006013.

[88] J. A. de Azcárraga and A. J. Macfarlane, "Compilation of identities for the antisymmetric tensors of the higher cocyles of $su(n)$," *Internat. J. Mod. Phys.***A16** 1377 (2001); math-ph/0006026.

[89] P. Deligne, "La série exceptionelle de groupes de Lie," *C. R. Acad. Sci. Paris, Sér. I,* **322**, 321 (1996).

[90] P. Deligne and R. de Man, "La série exceptionnelle de groupes de Lie. II," *C. R. Acad. Sci. Paris Sér. I* **323**, 577 (1996).

[91] P. Deligne, letter to P. Cvitanović (December 1996); birdtracks.eu/extras/reviews.html.

[92] P. Deligne and B. H. Gross, "On the exceptional series, and its descendents," *C. R. Acad. Sci. Paris, Sér. I,* **335**, 877 (2002).

[93] G. W. Delius, M. D. Gould, and Y. Z. Zhang, "On the construction of trigonometric solutions of the Yang-Baxter equation," *Nucl. Phys. B* **432**, 377 (1994).

[94] B. de Wit and G. 't Hooft, "Nonconvergence of the $1/N$ expansion for $SU(N)$ gauge fields on a lattice," *Phys. Lett. B* **69**, 61 (1977).

[95] P. A. Dirac, "The quantum theory of electron," *Proc. Roy. Soc. Lond. A* **117**, 610 (1928).

[96] P. Dorey, A. Pocklington, and R. Tateo, "Integrable aspects of the scaling q-state Potts models I: Bound states and bootstrap closure," *Nucl. Phys. B* **661**, 425 (2003); hep-th/0208111.

[97] P. Dittner, *Commun. Math. Phys.* **22**, 238 (1971); **27**, 44 (1972).

[98] J. M. Drouffe, "Transitions and duality in gauge lattice systems," *Phys. Rev. D* **18**, 1174 (1978).

[99] V. Del Duca, L. J. Dixon, and F. Maltoni, "New color decompositions for gauge amplitudes at tree and loop level," *Nucl. Phys. B* **571**, 51 (2000); hep-ph/9910563.

[100] R. Dundarer, F. Gursey, and C. Tze, "Selfduality and octonionic analyticity of $S(7)$ valued antisymmetric fields in eight-dimensions," *Nucl. Phys. B* **266**, 440 (1986).

[101] R. Dundarer, F. Gursey, and C. Tze, "Generalized vector products, duality and octonionic identities in $D = 8$ geometry," *J. Math. Phys.* **25**, 1496 (1984).

[102] G. V. Dunne, "Negative-dimensional groups in quantum physics," *J. Phys. A* **22**, 1719 (1989).

[103] H. P. Dürr and F. Wagner, *Nuovo Cimento* **53A**, 255 (1968).

[104] E. B. Dynkin, *Transl. Amer. Math. Soc. (2)* **6**, 111 (1957).

[105] E. B. Dynkin, *Transl. Amer. Math. Soc. (1)* **9**, 328 (1962).

[106] F. J. Dyson, "The radiation theories of Tomonaga, Schwinger, and Feynman," *Phys. Rev.* **75**, 486 (1949).

[107] C. Eckart, "The application of group theory to the quantum dynamics of monatomic systems," *Rev. Mod. Phys.* **2**, 305 (1930).

[108] J. M. Ekins and J. F. Cornwell, *Rep. Math. Phys.* **7**, 167 (1975).

[109] E. El Baz and B. Castel, *Graphical Methods of Spin Algebras in Atomic, Nuclear and Particle Physics* (Dekker, New York 1972).

[110] A. Elduque, "The Magic Square and symmetric compositions II," *Rev. Mat. Iberoamericana* **23**, 57 (2007); math.RT/0507282.

[111] M. El Houari, "Tensor invariants associated with classical Lie algebras: a new classification of simple Lie algebras," *Algebras, Groups, and Geometries* **17**, 423 (1997).

[112] H. Elvang, "Birdtracks, Young projections, colours," MPhys project in Mathematical Physics (1999); www.nbi.dk/~elvang/rerep.ps.

[113] H. Elvang, P. Cvitanović, and A. D. Kennedy, "Diagrammatic Young projection operators for $U(n)$," *J. Math. Phys.* **46**, 043501 (2005); hep-th/0307186.

[114] M. Evans, F. Gursey, and V. Ogievetsky, "From 2-D conformal to 4-D selfdual theories: Quaternionic analyticity," *Phys. Rev. D* **47**, 3496 (1993); hep-th/9207089.

[115] G. R. Farrar and F. Neri, "How to calculate 35640 $O(\alpha^5)$ Feynman diagrams in less than an hour," *Phys. Lett.* **130B**, 109 (1983).

[116] J. R. Faulkner, *Trans. Amer. Math. Soc.* **155**, 397 (1971).

[117] J. R. Faulkner and J. C. Ferrar, "Exceptional Lie algebras and related algebraic and geometric structures," *Bull. London Math. Soc.* **9**, 1 (1977).

[118] E. Fermi, "An attempt of a theory of beta radiation. 1," *Z. Phys.* **88**, 161 (1934).

[119] R. P. Feynman, "Space-time approach to nonrelativistic quantum mechanics," *Rev. Mod. Phys.* **20**, 367 (1948).

[120] M. Fierz, *Z. Physik* **88**, 161 (1934).

[121] D. Finkelstein, J. M. Jauch, and D. Speiser, *J. Math. Phys.* **4**, 136 (1963).

[122] M. Fischler, "Young-tableau methods for Kronecker products of representations of the classical groups," *J. Math. Phys.* **22**, 637 (1981).

[123] J. S. Frame, D. de B. Robinson, and R. M. Thrall, *Canad. J. Math.* **6**, 316 (1954).

[124] P. H. Frampton and T. W. Kephart, "Exceptionally simple $E(6)$ theory," *Phys. Rev. D* **25**, 1459 (1982).

[125] P. H. Frampton and T. W. Kephart, "Dynkin weights and global supersymmetry in grand unification," *Phys. Rev. Lett.* **48**, 1237 (1982).

[126] L. Frappat, A. Sciarrino, and P. Sorba, *A Dictionary of Lie Algebras and Superalgebras* (Academic Press, New York 2001).

[127] G. Frege, *Begriffsschrift, eine der arithmetischen nachgebildete Formelsprache des reinen Denkens* (Halle 1879), translated as *Begriffsschrift, formula language, modeled upon that of arithmetic, for pure thought*, ref. [328]. I am grateful to P. Cartier for bringing Frege's work to my attention.

[128] G. Frege, "On Mr. Peano's conceptual notation and my own," *Berichte über die Verhandlungen der Königlich-Sächsischen Gesellschaft der Wissenschaften zu Leipzig. Mathematisch-Physische Klasse,* **XLVIII**, 361 (1897).

[129] H. Freudenthal, "Beziehungen der E_7 und E_8 zur Oktavenebene, I, II," *Indag. Math.* **16**, 218, 363 (1954); *"III, IV," Indag. Math.* **17**, 151, 277 (1955); *"V - IX," Indag. Math.* **21**, 165 (1959); *" X, XI," Indag. Math.* **25**, 457 (1963).

[130] H. Freudenthal, "Lie groups in the foundations of geometry," *Adv. Math.* **1**, 145 (1964).

[131] H. Fritzsch, M. Gell-Mann, and H. Leutwyler, *Phys. Lett.* **47B**, 365 (1973).

[132] W. Fulton and J. Harris, *Representation Theory* (Springer, Berlin 1991).

[133] W. Fulton, *Young Tableaux, with Applications to Representation Theory and Geometry* (Cambridge Univ. Press, Cambridge 1999).

[134] I. M. Gel'fand, *Math. Sbornik* **26**, 103 (1950).

[135] I. M. Gel'fand, *Lectures on Linear Algebra* (Dover, New York 1961).

[136] I. M. Gel'fand and A. V. Zelevinskiĭ, "Models of representations of classical groups and their hidden symmetries," *Funktsional. Anal. i Prilozhen.* **18**, 14 (1984).

[137] M. Gell-Mann, "The eightfold way: Theory of strong interaction symmetry," *Caltech Report No. CTSL-20* (1961), unpublished. Reprinted in M. Gell-Mann and Y. Ne'eman, *The Eightfold Way: A Review - with Collection of Reprints* (Benjamin, New York 1964).

[138] H. Georgi, *Lie Algebras in Particle Physics* (Perseus Books, Reading, MA 1999).

[139] B. Giddings and J. M. Pierre, "Some exact results in supersymmetric theories based on exceptional groups," *Phys. Rev. D* **52**, 6065 (1995); hep-th/9506196.

[140] R. Gilmore, *Lie Groups, Lie Algebras and Some of Their Applications* (Wiley, New York 1974).

[141] J. Gleick, *Genius: The Life and Science of Richard Feynman* (Pantheon, New York 1992).

[142] R. Goodman and N. R. Wallach, *Representations and Invariants of the Classical Groups* (Cambridge Univ. Press, Cambridge 1998).

[143] M. Gorn, "Problems in comparing diagrams with group theory in nonleptonic decays," *Nucl. Phys. B* **191**, 269 (1981).

[144] M. Gourdin, *Unitary Symmetries* (North-Holland, Amsterdam 1967).

[145] D. J. Gross and E. Witten, "Possible third order phase transition in the large N lattice gauge theory," *Phys. Rev. D* **21**, 446 (1980).

[146] M. Günaydin, *Nuovo Cimento* **29A**, 467 (1975).

[147] M. Günaydin and F. Gürsey, *J. Math. Phys.* **14**, 1651 (1973).

[148] F. Gürsey, in *Proceedings of the Kyoto Conference on Mathematical Problems in Theoretical Physics* (Kyoto 1975), unpublished.

[149] F. Gürsey and P. Sikivie, "E_7 as a universal gauge group," *Phys. Rev. Lett.* **36**, 775 (1976).

[150] F. Gürsey, P. Ramond, and P. Sikivie, *Phys. Lett.* **60B**, 177 (1976).

[151] F. Gürsey and M. Serdaroglu, "E_6 gauge field theory model revisited," *Nuovo Cim.* **65A**, 337 (1981).

[152] F. Gürsey and C.-H. Tze, *Division, Jordan and Related Algebras in Theoretical Physics* (World Sci., Singapore, 1996).

[153] M. Hamermesh, *Group Theory and Its Application to Physical Problems* (Dover, New York 1962).

[154] N. Habegger, "The topological IHX relation," *J. Knot Theory and Its Ramifications* **10**, 309 (2001).

[155] W. G. Harter, *J. Math. Phys.* **10**, 4 (1969).

[156] W. G. Harter and N. Dos Santos, "Double-group theory on the half-shell and the two-level system. I. Rotation and half-integral spin states," *Am. J. Phys.* **46**, 251 (1978).

[157] W. G. Harter, *Principles of Symmetry, Dynamics, and Spectroscopy* (Wiley, New York 1974).

[158] J. A. Harvey, "Patterns of symmetry breaking in the exceptional groups," *Nucl. Phys. B* **163**, 254 (1980).

[159] A. C. Hearn, "REDUCE User's Manual," *Stanford Artificial Intelligence Project, Memo AIM-133*, (1970).

[160] P. Henry-Labordére, B. L. Julia, and L. Paulot, "Borcherds symmetries in M-theory," hep-th/0203070.

[161] G. 't Hooft and M. Veltman, "Regularization and renormalization of gauge fields," *Nucl. Phys. B* **44**, 189 (1972).

[162] G. 't Hooft and M. Veltman, "DIAGRAMMAR," *CERN report 73/9*, (1973).

[163] G. 't Hooft, "A planar diagram theory for strong interactions," *Nucl. Phys. B* **72**, 461 (1974).

[164] G. 't Hooft, "Computation of the quantum effects due to a four-dimensional pseudoparticle," *Phys. Rev. Lett.* **37**, 8 (1976); *Phys. Rev. D* **14**, 3432 (1976).

[165] A. Hurwitz, "Über die Komposition der quadratischen Formen von beliebig vielen Variabeln," *Nachr. Ges. Wiss. Göttingen*, 309 (1898).

[166] A. Hurwitz, "Über die Komposition der quadratischen Formen," *Math. Ann. 88*, 1 (1923).

[167] C. G. J. Jacobi, "De functionibus alternantibus earumque divisione per productum e differentiis elementorum conflatum," in *Collected Works*, Vol. 22, 439; *J. Reine Angew. Math. (Crelle)* (1841).

[168] N. Jacobson, *Exceptional Lie Algebras* (Dekker, New York 1971).

[169] N. Jacobson, *Basic Algebra I* (Freeman, San Francisco 1974).

[170] G. James and A. Kerber, *The Representation Theory of the Symmetric Group* (Addison-Wesley, Reading, MA 1981).

[171] M. Jimbo, "Quantum R matrix for the generalized Toda system," *Comm. Math. Phys.* **102**, 537 (1986).

[172] B.-Q. Jin and Z.-Q. Ma, "R matrix for $U_q E_7$," *Comm. Theoret. Phys.* **24**, 403 (1995).

[173] P. Jordan, J. von Neumann, and E. Wigner, "On an algebraic generalization of the quantum mechanical formalism," *Ann. Math.* **35**, 29 (1934).

[174] B. L. Julia, "Group disintegrations," in S. Hawking and M. Rocek, eds., *Superspace and Supergravity* (Cambridge Univ. Press, Cambridge 1981).

[175] B. L. Julia, "Below and beyond U-duality," hep-th/0002035.

[176] B. L. Julia, "Magics of M-gravity," hep-th/0105031.

[177] V. G. Kac, "Lie superalgebras," *Adv. Math.* **26**, 8 (1977).

[178] J. Kahane, *J. Math. Phys.* **9**, 1732 (1968).

[179] N. Kamiya, "A structure theory of Freudenthal-Kantor triple systems III," *Mem. Fac. Sci. Shimane Univ.* **23**, 33 (1989).

[180] N. Kamiya, "The construction of all simple Lie algebras over $\mathbb{C}$ from balanced Freudenthal-Kantor triple systems," *Contr. to General Algebra* **7**, 205 (1991).

[181] I. L. Kantor, *Soviet Math. Dokl.* **14**, 254 (1973).

[182] L. M. Kaplan and M. Resnikoff, *J. Math. Phys.* **8**, 2194 (1967).

[183] A. B. Kempe, "On the application of Clifford's graphs to ordinary binary quantics," *Proc. London Math. Soc.* **17**, 107 (1885).

[184] A. D. Kennedy, "Clifford algebras in two Omega dimensions," *J. Math. Phys.* **22**, 1330 (1981).

[185] A. D. Kennedy, "Spinography: Diagrammatic methods for spinors in Feynman diagrams," *Phys. Rev. D* **26**, 1936 (1982).

[186] A. D. Kennedy, "Group algebras, Lie algebras, and Clifford algebras," Colloquium, Moscow State University (1997);
www.ph.ed.ac.uk/~adk/algebra-slides/all.html .

[187] T. W. Kephart and M. T. Vaughn, "Renormalization of scalar quartic and Yukawa couplings in unified gauge theories," *Z. Phys. C* **10**, 267 (1981).

[188] T. W. Kephart and M. T. Vaughn, "Tensor methods for the exceptional group $E6$," *Annals Phys.* **145**, 162 (1983).

[189] W. Killing, *Math. Ann.* **31**, 252 (1888); **33**, 1 (1889); **34**, 57 (1889); **36**, 161 (1890).

[190] J. D. Kim, I. G. Koh, and Z. Q. Ma, "Quantum $\check{R}$ matrix for E_7 and F_4 groups," *J. Math. Phys.* **32**, 845 (1991).

[191] R. C. King, *Can. J. Math.* **33**, 176 (1972).

[192] R. C. King and B. G. Wybourne, "Holomorphic discrete series and harmonic series unitary irreducible representations of non-compact Lie groups: $Sp(2n, R), U(p, q)$ and $SO^*(2n)$," *J. Math. Phys.* **18**, 3113 (1985).

[193] R. C. King and B. G. Wybourne, "Analogies between finite-dimensional irreducible representations of $SO(2n)$ and infinite-dimensional irreducible representations of $Sp(2n, R)$. I. Characters and products," *J. Math. Phys.* **41**, 5002 (2000).

[194] R. C. King and B. G. Wybourne, "Multiplicity-free tensor products of irreducible representations of the exceptional Lie groups," *J. Physics A* **35**, 3489 (2002).

[195] T. Kinoshita and W. B. Lindquist, "Eighth-Order Anomalous Magnetic Moment of the Electron," *Phys. Rev. Lett.* **47**, 1573 (1981).

[196] O. Klein and Y. Nishina, *Z. Physik* **52**, 853 (1929).

[197] I. G. Koh and Z. Q. Ma, "Exceptional quantum groups," *Phys. Lett. B* **234**, 480 (1990).

[198] M. Kontsevich, "Formal (non)commutative symplectic geometry," *The Gel'fand Mathematical Seminars 1990–1992*, 173 (Birkhäuser, Boston, MA 1993).

[199] M. Kontsevich, "Feynman diagrams and low-dimensional topology," *First European Congress of Mathematics, Vol. II (Paris 1992), Progr. Math.* **120**, 97 (Birkhäuser, Basel 1994).

[200] A. P. Kryukov and A. Y. Rodionov, "Color: Program for calculation of group weights of Feynman diagrams in nonabelian gauge theories," *Comput. Phys. Commun.* **48**, 327 (1988).

[201] P. P. Kulish, N. Y. Reshetikhin, and E. K. Sklyanin, "Yang-Baxter equations and representation theory. I," *Lett. Math. Phys.* **5**, 393 (1981).

[202] A. Kuniba, "Quantum R-matrix for G_2 and a solvable 175-vertex model," *J. Phys. A* **23**, 1349 (1990).

[203] J. M. Landsberg and L. Manivel, "Triality, exceptional Lie algebras and Deligne dimension formulas," *Adv. Math.* **171**, 59 (2002); math.AG/0107032.

[204] J. M. Landsberg and L. Manivel, "On the projective geometry of Freudenthal's magic square," *J. of Algebra* **239**, 477 (2001); math.AG/9908039.

[205] J. M. Landsberg and L. Manivel, "Series of Lie groups," *Michigan Math. J.* **52**, 453 (2004); math.AG/0203241.

[206] J. M. Landsberg and L. Manivel, "Representation theory and projective geometry," in V. Popov, ed., *Algebraic transformation groups and algebraic varieties, Encyclopaedia of Mathematical Sciences, Invariant Theory and Algebraic Transformation Groups III* (Springer-Verlag, New York 2004); math.AG/0203260.

[207] J. M. Landsberg and L. Manivel, "Construction and classification of simple Lie algebras via projective geometry," *Selecta Mathematica* **8**, 137 (2002).

[208] J. M. Landsberg, L. Manivel, and B. W. Westbury, "Series of nilpotent orbits," *Experimental Math* **13**, 13 (2004); math.AG/0212270.

[209] J. M. Landsberg and L. Manivel, "The sextonions and $E_{7\,1/2}$," *Adv. Math.* **201**, 143-179 (2006); math.RT/0402157.

[210] J. M. Landsberg and L. Manivel, "A universal dimension formula for complex simple Lie algebras," *Adv. Math.* **201**, 379 (2006); math.RT/0401296.

[211] S. Lang, *Linear Algebra* (Addison-Wesley, Reading, MA 1971).

[212] S. Lawton and E. Peterson, "Spin networks and $SL(2,C)$-Character varieties." math.QA/0511271

[213] I. B. Levinson, "Sums of Wigner coefficients and their graphical representation," *Proceed. Physical-Technical Inst. Acad. Sci. Lithuanian SSR* **2**, 17 (1956).

[214] D. B. Lichtenberg, *Unitary Symmetry and Elementary Particles* (Academic Press, New York 1970).

[215] J. Lurie, "On simply laced Lie algebras and their minuscule representations," undergraduate thesis (Harvard Univ. 2000); *Comment. Math. Helv.* **76**, 515 (2001).

[216] Z. Q. Ma, "Rational solution for the minimal representation of G_2," *J. Phys. A* **23**, 4415 (1990).

[217] Z. Q. Ma, "The spectrum-dependent solutions to the Yang-Baxter equation for quantum E_6 and E_7," *J. Phys. A* **23**, 5513 (1990).

[218] Z. Q. Ma, "The embedding e_0 and the spectrum-dependent R-matrix for q-F_4," *J. Phys. A* **24**, 433 (1991).

[219] A. J. Macfarlane, A. Sudbery, and P. M. Weisz, *Comm. Math. Phys.* **11**, 77 (1968); *Proc. Roy. Soc. Lond. A* **314**, 217 (1970).

[220] A. J. Macfarlane and H. Pfeiffer, "On characteristic equations, trace identities and Casimir operators of simple Lie algebras," *J. Math. Phys.* **41**, 3192 (2000); Erratum: **42**, 977 (2001).

[221] A. J. Macfarlane, "Lie algebra and invariant tensor technology for g_2," *Internat. J. Mod. Phys. A* **16**, 3067 (2001); math-ph/0103021.

[222] A. J. Macfarlane, H. Pfeiffer, and F. Wagner, "Symplectic and orthogonal Lie algebra technology for bosonic and fermionic oscillator models of integrable systems," *Internat. J. Mod. Phys. A* **16**, 1199 (2001); math-ph/0007040.

[223] N. J. MacKay, "Rational R-matrices in irreducible representations," *J. Phys. A* **24**, 4017 (1991).

[224] N. J. MacKay, "The full set of C_n-invariant factorized S-matrices," *J. Phys. A* **25**, L1343 (1992).

[225] N. J. MacKay, "Rational K-matrices and representations of twisted Yangians," *J. Phys. A* **35**, 7865 (2002); QA/0205155.

[226] N. J. MacKay and A. Taylor, "Rational R-matrices, centralizer algebras and tensor identities for e_6 and e_7 exceptional families of Lie algebras," (2006); math/0608248.

[227] J. B. Mandula, "Diagrammatic techniques in group theory," notes taken by S. N. Coulson and A.J.G. Hay (Univ. of Southampton, 1981), unpublished.

[228] L. Manivel, *Symmetric Functions, Schubert Polynomials and Degeneracy Loci* (American Math. Society, Providence, RI 2001), www-fourier.ujf-grenoble.fr/~manivel/cours.html.

[229] N. Maru and S. Kitakado, "Negative dimensional group extrapolation and a new chiral-nonchiral duality in $N = 1$ supersymmetric gauge theories," *Mod. Phys. Lett. A* **12**, 691 (1997); hep-th/9609230.

[230] M. L. Mehta, *J. Math. Phys.* **7**, 1824 (1966).

[231] M. L. Mehta and P. K. Srivastava, *J. Math. Phys.* **7**, 1833 (1966).

[232] A. McDonald and S. P. Rosen, *J. Math. Phys.* **14**, 1006 (1973).

[233] W. G. McKay, J. Patera, and R. T. Sharp, *J. Math. Phys.* **17**, 1371 (1977).

[234] W. G. McKay and J. Patera, "Tables of dimensions, indices and branching rules for representations of simple Lie algebras" (Dekker, New York 1981).

[235] W. G. McKay, R. V. Moody, and J. Patera, "Decomposition of tensor products of $E8$ representations," *Alg. Groups Geom.* **3**, 286 (1986).

[236] W. G. McKay, J. Patera, and D. W. Rand, *Tables of Representations of Simple Lie Algebras*, vol. I: *Exceptional Simple Lie Algebras* (Les publications CRM, Université de Montréal 1990).

[237] J. Mehra, *The Beat of a Different Drum* (Oxford Univ. Press, Oxford 1994).

[238] A. Messiah, *Quantum Mechanics* (North-Holland, Amsterdam 1966).

[239] K. Meyberg, *Ned. Akad. Wetensch. Proc. A* **71**, 162 (1960).

[240] K. Meyberg, "Spurformeln in einfachen Lie-Algebren," *Abh. Math. Sem. Univ. Hamburg* **54**, 177 (1984).

[241] K. Meyberg, "Trace formulas in various algebras and L-projections," *Nova J. of Algebra and Geometry* **2**, 107 (1993).

[242] L. Michel and L. A. Radicati, *Ann. Inst. Henri Poincaré* **18**, 13 (1973).

[243] M. Michel, V. L. Fitch, F. Gursey, A. Pais, R. U. Sexl, V. L. Telegdi, and E. P. Wigner, "Round table on the evolution of symmetries," in *Sant Feliu de Guixols 1983, Proceedings, Symmetries in Physics (1600–1980)*.

[244] R. Mirman, *Group Theory: An Intuitive Approach* (World Scientific, Singapore 1995).

[245] R. L. Mkrtchyan, "The equivalence of $Sp(2n)$ and $SO(-2n)$ gauge theories," *Phys. Lett. B* **105**, 174 (1981).

[246] S. Morita, "Casson's invariant for homology 3-spheres and characteristic classes of surface bundles, I," *Topology* **28**, 305 (1989).

[247] S. Morita, "On the structure of the Torelli group and the Casson invariant," *Topology* **30**, 603 (1991).

[248] A. J. Mountain, "Invariant tensors and Casimir operators for simple compact Lie groups," *J. Math. Phys.* **39**, 5601 (1998); physics/9802012.

[249] N. Mukunda and L. K. Pandit, *J. Math. Phys.* **6**, 746 (1975).

[250] T. Murphy, *Proc. Camb. Philos. Soc.* **71**, 211 (1972).

[251] G. Murtaza and M. A. Rashid, *J. Math. Phys.* **14**, 1196 (1973).

[252] D. E. Neville, *Phys. Rev.* **132**, 844 (1963).

[253] K. Nomizu, *Fundamentals of Linear Algebra* (Chelsea Publ., New York 1979).

[254] E. Ogievetsky and P. Wiegmann, "Factorized S-matrix and the Bethe ansatz for simple Lie groups," *Phys. Lett. B* **168**, 360 (1986).

[255] S. Okubo, "Casimir invariants and vector operators in simple and classical Lie algebras," *J. Math. Phys.* **18**, 2382 (1977).

[256] S. Okubo, "Gauge groups without triangular anomaly," *Phys. Rev. D* **16**, 3528 (1977).

[257] S. Okubo, "Constraint on color gauge group," *Phys. Rev. D* **16**, 3535 (1977).

[258] S. Okubo, "Quartic trace identity for exceptional Lie algebras," *J. Math. Phys.* **20**, 586 (1979).

[259] S. Okubo, "Modified fourth order Casimir invariants and indices for simple Lie algebras," *J. Math. Phys.* **23**, 8 (1982).

[260] S. Okubo and J. Patera, "Symmetrization of product representations and general indices and simple Lie algebras," *J. Math. Phys.* **24**, 2722 (1983).

[261] S. Okubo and J. Patera, "General indices of representations and Casimir invariants," *J. Math. Phys.* **25**, 219 (1984).

[262] S. Okubo, "Branching index sum rules for simple Lie algebras," *J. Math. Phys.* **26**, 2127 (1985).

[263] M. A. Olshanetskii and V.B.K. Rogov, "Adjoint representations of exceptional Lie algebras," *Theor. Math. Phys.* **72**, 679 (1987).

[264] P. J. Olver and C. Shakiban, "Graph theory and classical invariant theory," *Adv. Math.* **75**, 212 (1989).

[265] P. J. Olver, *Classical Invariant Theory* (Cambridge Univ. Press, Cambridge 1999).

[266] R. J. Ord-Smith, *Phys. Rev.* **94**, 1227 (1954).

[267] O. Ore, *The Four-Color Problem* (Academic Press, New York 1967).

[268] A. C. Pang and C. Ji, "A spinor technique in symbolic Feynman diagram calculation," *J. Comput. Phys.* **115**, 267 (1994).

[269] G. Parisi and N. Sourlas, "Random magnetic fields, supersymmetry and negative dimensions," *Phys. Rev. Lett.* **43**, 744 (1979).

[270] J. Patera, *J. Math. Phys.* **11**, 3027 (1970).

[271] J. Patera and A. K. Bose, *J. Math. Phys.* **11**, 2231 (1970).

[272] J. Patera, *J. Math. Phys.* **12**, 384 (1971).

[273] J. Patera and D. Sankoff, *Tables of Branching Rules for Representations of Simple Lie Algebras* (Université de Montréal 1973).

[274] J. Patera, R. T. Sharp, and P. Winternitz, *J. Math. Phys.* **17**, 1972 (1977).

[275] J. Patera and R. T. Sharp, "On the triangle anomaly number of $SU(N)$ representations," *J. Math. Phys.* **22**, 2352 (1981).

[276] B. Patureau-Mirand, "Caractères sur l'algèbre de diagrammes Λ," *Geom. Topol.* **6**, 563 (2002).

[277] B. Patureau-Mirand, "Caractères sur l'algèbre de diagrammes Λ," *C. R. Acad. Sci. Paris* **329**, 803 (1999).

[278] W. Pauli, *Ann. Inst. Henri Poincaré* **6**, 109 (1636).

[279] R. Penrose, "Structure of Spacetime," in *Battelle Rencontres*, C. M. de Witt, and J. A. Wheeler, eds., p. 121 (Benjamin, New York 1968).

[280] R. Penrose, "Applications of negative dimensional tensors," in *Combinatorial mathematics and its applications*, D.J.A. Welsh, ed. (Academic Press, New York 1971), 221.

[281] R. Penrose, "Angular momentum: An approach to combinatorial space-time," in *Quantum Theory and Beyond*, T. Bastin, ed. (Cambridge Univ. Press, Cambridge 1971).

[282] R. Penrose and M.A.H. MacCallum, *Physics Reports* **65** (1973).

[283] R. Penrose and W. Rindler, *Spinors and Space-time* (Cambridge Univ. Press, Cambridge 1984, 1986).

[284] E. Peterson, "A not-so-characteristic equation: the art of linear algebra." 0712.2058

[285] P. Pouliot, "Spectroscopy of gauge theories based on exceptional Lie groups," *J. Phys. A* **34**, 8631 (2001).

[286] G. Racah, "Theory of complex spectra. II," *Phys. Rev.* **62**, 438 (1942).

[287] G. Racah, *Phys. Rev.* **76**, 1352 (1949).

[288] G. Racah, *Rend. Lincei* **8**, 108 (1950).

[289] G. Racah, in *Ergebnisse der exakten Naturwissenschaften* **37**, p. 28 (Springer-Verlag, Berlin 1965).

[290] P. Ramond, "Is there an exceptional group in your future? $E(7)$ and the travails of the symmetry breaking," *Nucl. Phys. B* **126**, 509 (1977).

[291] P. Ramond, "Introduction to exceptional Lie groups and algebras," *CALT-68-577*, 58 (CalTech 1976).

[292] M. A. Rashid and Saifuddin, *J. Phys. A.* **5**, 1043 (1972).

[293] R. C. Read, *J. Combinatorial Theory* **4**, 52 (1968).

[294] T. van Ritbergen, A. N. Schellekens, and J. A. Vermaseren, "Group theory factors for Feynman diagrams," *Int. J. Mod. Phys. A* **14**, 41 (1999); hep-ph/9802376.

[295] D. de B. Robinson, *Representation Theory of the Symmetric Group* (Univ. Toronto Press, Toronto 1961).

[296] R. Rockmore, *Phys. Rev. D* **11**, 620 (1975). The method of this paper is applicable only to $SU(3)$.

[297] B. A. Rosenfeld, "Geometrical interpretation of the compact simple Lie groups of the class," *Dokl. Akad. Nauk USSR* **106**, 600 (1956) (in Russian).

[298] B. A. Rosenfeld, in *Algebraical and Topological Foundations of Geometry*, J. L. Tits, ed. (Pergamon, Oxford 1962).

[299] M. Rost, "On the dimension of a composition algebra," *Documenta Mathematica* **1**, 209 (1996); www.mathematik.uni-bielefeld.de/DMV-J/vol-01/10.html

[300] D. J. Rowe, B. G. Wybourne, and P. H. Butler, "Unitary representations, branching rules and matrix elements for the non-compact symplectic groups," *J. Math. Phys.* **18**, 939 (1985).

[301] K. E. Rumelhart, "Minimal representations of exceptional p-adic groups," *Representation Theory* **1**, 133 (1997).

[302] B. E. Sagan, *The Symmetric Group* (Springer-Verlag, New York 2001).

[303] S. Samuel, "$U(N)$ integrals, $1/N$, and the Dewit-'t Hooft anomalies," *J. Math. Phys.* **21**, 2695 (1980).

[304] R. D. Schafer, *Introduction to Nonassociative Algebras* (Academic Press, New York 1966).

[305] A. Schrijver, "Tensor subalgebras and first fundamental theorems in invariant theory," math/0604240math.

[306] I. Schur, "Über eine Klasse von Matrizen, die Sich einer Gegebenen Matrix zuordnen Lassen," in *Gesammelte Abhandlungen*, 1 (Springer-Verlag, Berlin 1973).

[307] S. S. Schweber, *QED and the Men Who Made It: Dyson, Feynman, Schwinger, and Tomonaga* (Princeton Univ. Press, Princeton, NJ 1994).

[308] J. Schwinger, ed., *Selected Papers on Quantum Electrodynamics* (Dover, New York 1958).

[309] G. Segal, "Unitary representations of some infinite-dimensional groups," *Comm. Math. Phys.* **80**, 301 (1981).

[310] S. M. Sergeev, "Spectral decomposition of R-matrices for exceptional Lie algebras," *Modern Phys. Lett. A* **6**, 923 (1991).

[311] A. Sirlin, "A class of useful identities involving correlated direct products of gamma matrices," *Nucl. Phys. B* **192**, 93 (1981).

[312] R. Slansky, "Group theory for unified model building," *Phys. Rep.* **79**, 1 (1981).

[313] N. J. A. Sloane, *A Handbook of Integer Sequences* (Academic Press, New York 1973).

[314] T. A. Springer, *Ned. Akad. Wetensch. Proc. A* **62**, 254 (1959).

[315] T. A. Springer, *Ned. Akad. Wetensch. Proc. A* **65**, 259 (1962).

[316] R. P. Stanley, *Enumerative Combinatorics 2* (Cambridge Univ. Press, Cambridge 1999); www-math.mit.edu/~rstan/ec/.

[317] G. E. Stedman, "A diagram technique for coupling calculations in compact groups," *J. Phys. A* **8**, 1021 (1975); A **9**, 1999 (1976).

[318] G. E. Stedman, *Diagram Techniques in Group Theory* (Cambridge Univ. Press, Cambridge 1990).

[319] J. R. Stembridge, "Multiplicity-free products and restrictions of Weyl characters," www.math.lsa.umich.edu/~jrs/papers.html (2002).

[320] D. T. Sviridov, Yu. F. Smirnov, and V. N. Tolstoy, *Rep. Math. Phys.* **7**, 349 (1975).

[321] J. J. Sylvester, "On an application of the new atomic theory to the graphical representation of the invariants and covariants of binary quantics, with three appendices," *Amer. J. Math.* **1**, 64 (1878).

[322] J. Thierry-Mieg, *C. R. Acad. Sci. Paris* **299**, 1309 (1984).

[323] H. Thörnblad, *Nuovo Cimento* **52A**, 161 (1967).

[324] M. Tinkham, *Group Theory and Quantum Mechanics* (McGraw-Hill, New York 1964).

[325] J. Tits, "Algèbres alternatives, algèbres de Jordan et algèbres de Lie exceptionnelles," *Indag. Math.* **28**, 223 (1966); *Math. Reviews* **36**, 2658 (1966).

[326] J. Tits, *Lecture Notes in Mathematics* **40** (Springer-Verlag, New York 1967).

[327] L. Tyburski, *Phys. Rev. D* **13**, 1107 (1976).

[328] J. Van Heijenoort, *Frege and Gödel: Two Fundamental Texts in Mathematical Logic* (Harvard Univ. Press, Cambridge, MA 1970).

[329] J. A. Vermaseren, "Some problems in loop calculations," hep-ph/9807221.

[330] È. B. Vinberg, *Lie groups and Lie algebras, III, Encyclopaedia of Mathematical Sciences* **41** (Springer-Verlag, Berlin 1994); "Structure of Lie groups and Lie algebras," a translation of *Current problems in mathematics. Fundamental directions* **41** (Russian), (Akad. Nauk SSSR, Vsesoyuz. Inst. Nauchn. i Tekhn. Inform., Moscow 1990); [MR 91b:22001], translation by V. Minachin [V. V. Minakhin], translation edited by A. L. Onishchik and È. B. Vinberg.

[331] È. B. Vinberg, "A construction of exceptional simple Lie groups (Russian)," *Tr. Semin. Vektorn. Tensorn. Anal.* **13**, 7 (1966).

[332] P. Vogel, "Algebraic structures on modules of diagrams," unpublished preprint (1995).

[333] P. Vogel, "The universal Lie algebra" unpublished preprint (1999).

[334] B. L. van der Waerden, *Algebra*, vol. 2, 4th ed. (Springer-Verlag, Berlin 1959).

[335] A. Weil, "Sur certains groupes d'opérateurs unitaires," *Acta Math.* **111**, 143–211 (1964).

[336] S. Weinberg, *Gravitation and Cosmology* (Wiley, New York 1972).

[337] S. Weinberg, *Phys. Rev. Lett.* **31**, 494 (1973).

[338] E. Weisstein, "Young tableaux," mathworld.wolfram.com/YoungTableau.html.

[339] H. Wenzl, "On tensor categories of Lie type E_N, $N \neq 9$," *Adv. Math.* **177**, 66 (2003); 66-104 math.ucsd.edu/ wenzl/.

[340] B. W. Westbury, "R-matrices and the magic square," *J. Phys. A* **36**, 2857 (2003).

[341] B. W. Westbury, "Sextonions and the magic square," *J. London Math. Soc.* **73**, 455 (2006).

[342] H. Weyl, *The Theory of Groups and Quantum Mechanics* (Methuen, London 1931).

[343] H. Weyl, *The Classical Groups, Their Invariants and Representations* (Princeton Univ. Press, Princeton, NJ 1946).

[344] H. Weyl and R. Brauer, "Spinors in n dimensions," *Amer. J. Math.* **57**, 425 (1935).

[345] E. P. Wigner, *Group Theory and Its Application to the Quantum Mechanics of Atomic Spectra* (Academic Press, New York 1959).

[346] F. Wilczek and A. Zee, "Families from spinors," *Phys. Rev. D* **25**, 553 (1982).

[347] K. G. Wilson, "Confinement of quarks," *Phys. Rev. D* **10**, 2445 (1974).

[348] B. G. Wybourne, *Symmetry Principles and Atomic Spectroscopy* (Wiley, New York 1970).

[349] B. G. Wybourne, *Classical Groups for Physicists* (Wiley, New York 1974).

[350] B. G. Wybourne and M. J. Bowick, "Basic properties of the exceptional Lie groups," *Austral. J. Phys.* **30**, 259 (1977).

[351] B. G. Wybourne, "Enumeration of group invariant quartic polynomials in Higgs scalar fields," *Austral. J. Phys.* **33**, 941 (1980).

[352] B. G. Wybourne, "Young tableaux for exceptional Lie groups."

[353] K. Yamaguti and H. Asano, *Proc. Japan Acad.* **51** (1975).

[354] P. S. Yeung, *Phys. Rev. D* **13**, 2306 (1976).

[355] T. Yoshimura, "Some identities satisfied by representation matrices of the exceptional Lie algebra G_2" *(King's Coll., London, Print-75-0424*, 1975).

[356] A. Young, *The Collected Papers of Alfred Young* (Univ. Toronto Press, 1977), with the eight articles on quantitative substitutional analysis, including: *Proc. London Math. Soc.* **33**, 97 (1900); **28**, 255 (1928); **31**, 253 (1930).

[357] A. P. Yutsis, I. B. Levinson, and V. V. Vanagas, *Matematicheskiy apparat teorii momenta kolichestva dvizheniya* (Gosudarstvennoe izdatel'stvo politicheskoy i nauchnoy literatury Litovskoy SSR, Vilnius 1960); English translation: *The Theory of Angular Momentum*, Israel Scientific Translation, Jerusalem 1962 (Gordon and Breach, New York 1964).

[358] D. P. Želobenko, *Compact Lie Groups and Their Representations*, trans. of Math. Monographs **40** (American Math. Society, Providence, RI 1973).

[359] R. B. Zhang, M. D. Gould, and A. J. Bracken, "From representations of the braid group to solutions of the Yang-Baxter equation," *Nucl. Phys. B* **354**, 625 (1991).

Index